Archives of Toxicology • Supplement 19

Springer

Berlin
Heidelberg
New York
Barcelona
Budapest
Hong Kong
London
Milan
Paris
Santa Clara
Singapore
Tokyo

Applied Toxicology: Approaches Through Basic Science

Proceedings of the
1996 EUROTOX Congress Meeting
Held in Alicante, Spain,
September 22-25, 1996

Edited by the Publication Commitee
J. P. Seiler and E. Vilanova

With 62 Figures and 47 Tables

Springer

Editors

DR. JÜRG P. SEILER
Interkantonale Kontrollstelle für Heilmittel
Erlachstrasse 8
CH-3000 Bern 9
Switzerland

PROF. DR. EUGENIO VILANOVA
University of Alicante
Dept. of Neurochemistry
Campus San Juan
Aptdo. 374
E-03080 Alicante
Spain

ISBN-13: 978-642-64505-1 Springer-Verlag Berlin Heidelberg New York

Cataloging-in-Publication Data applied for

Die Deutsche Bibliothek – CIP-Einheitsaufnahme
Applied toxicology: approaches through basic science; proceedings of the 1996 Eurotox Congress Meeting held in Alicante, Spain, September 22-25, 1996 – Berlin; Heidelberg; New York; Barcelona; Budapest; Hong Kong; London; Milan; Paris; Santa Clara; Singapore; Tokyo : Springer 1997
(Archives of toxicology: Supplement 19)

ISBN-13: 978-3-642-64505-1 e-ISBN-13: 978-3-642-60682-3
DOI: 10.1007/978-3-642-60682-3

NE: Seiler, Jürg P. [Hrsg.]; EUROTOX <35, 1996, Alicante>

Softcover reprint of the hardcover 1st edition 1997

Typesetting: camera ready by editors
Cover Design: design & production, Heidelberg

SPIN: 10560002 27/3136 – 5 4 3 2 1 0 – Printed on acid free paper

Contents

The Gerhard Zbinden Memorial Lecture

Detection of Chemically-Induced Changes in p53 Expression

Norman Aldridge Memorial Symposium: Toxicologists versus Toxicological Disasters: Toxic Oil Syndrome

Altered Gene Expression and Immunotoxicity

The Use of Ecotoxicology and Human Toxicology in the Regulation of Chemical Safety in the European Union

Challenges in Analytical Toxicology

Air Pollution and Allergy

Prediction of Individual Susceptibility to Toxicants

Cellular and Molecular Mechanisms of Cutaneous Toxicity

Progress in Antidotic Therapy

In Vitro Cell Models for Investigating Molecular Mechanisms of Toxicity

Protein Targets of Neurotoxicity

Carcinogenesis Mechanisms in Transgenic Mouse Models

Algal Toxins and Human Health

The Young Scientist Poster Award

The Gerhard Zbinden Memorial Lecture

(Chair: E. Dybing, Norway)

EUROTOX has instituted this lecture to honor the memory of Gerhard Zbinden (1924 - 1993), EUROTOX Honorary Member and recipient of the EUROTOX Merit Award. The Gerhard Zbinden Memorial Lecture aims at recognising scientific excellence in the area of drug and chemical safety. The lecture is held at the annual EUROTOX Congress by a scientist chosen for his outstanding research contributions to the science of toxicology. The lecture is sponsored by the Chemical Industries Basel (KGF: Ciba-Geigy AG, F. Hoffmann-La Roche AG, Lonza AG, Sandoz AG), Switzerland.

Genetic Polymorphism of Drug Metabolizing Enzymes. Implications for Toxicity of Drugs and Other Xenobiotics

Magnus Ingelman-Sundberg
Institute of Environmental Medicine, Karolinska Institutet, S-171 77 Stockholm, Sweden

Introduction

There are pronounced interindividual and interethnic differences in the capacity to metabolize drugs and other xenobiotics. This variation is due to genetic polymorphism of the corresponding genes, as well as to physiological-, environmental- and pathological factors. From a teleological point of view it is evident that genetically based interindividual variability is seen in the human lymphocyte antigen (HLA) system, the immune system and in the inherited defences towards infection by malaria or tuberculosis and that this creates survival of some part of the population during infectious epidemics. In an analogous manner the variability in the drug metabolizing system will be beneficial for survival during chemical catastrophes. In this overview I will focus on the genetic aspects of the drug metabolizing enzymes with emphasis on some P450 forms and the mechanistic basis for their variability. The methods for predicting the genotype of a certain individual and on the possible use of this information in the future will also be considered.

It is evident that polymorphic genes encoding drug metabolizing enzymes are very attractive candidates as being responsible for interindividual differences in xenobiotic toxicity. The phase I enzymes metabolically activate drugs and procarcinogens to toxic or carcinogenic electrophilic intermediates and phase II enzymes are implicated in the detoxification of such products. An unfavourable combination of alleles of those genes might reasonably be considered to be an understandable cause for enhanced or decreased susceptibility to certain environmental carcinogenic or toxic agents.

The list of drug metabolizing enzymes that exhibit documented functional polymorphism is now growing as shown in Table 1.

Table 1. Polymorphic drug metabolizing enzymes and examples of their substrates.

Enzymes	Examples of substrates
Cytochromes P450	
CYP2A6	Coumarin
CYP2C9	S-warfarin, tolbutamide, NSAID
CYP2C18	
CYP2C19	Omeprazole, proguanil
CYP2D6	Neuroleptics, antidepressants
CYP2E1	Ethanol, nitrosoamines
Other enzymes	
Xanthine oxidase	Caffeine
Alcohol dehydrogenase	Alcohols
Aldehyde dehydrogenase	Acetaldehyde
Buturylcholine esterase	Cocaine
Dihydropyrimidindehydrogenase	5-Fluorouracil
Epoxide hydrolase	Benzo(a)pyrene epoxide
Flavine monooxygenase	Nicotine
N-acetyltransferase 1	Aromatic amines
N-acetyltransferase 2	Isoniazid
Glutathione transferase M1	Arene oxides
Glutathione transferase T1	Chlorinated hydrocarbons
Glutathione transferase P1	Chlorinated hydrocarbons
UDP-glucuronosyltransferase 1.6	Phenols
UDP-glucuronosyltransferase 1.7	Phenols
Thiopurine S-methyltransferase	Mercaptopurines

Mutations in these genes can cause the formation of an enzyme that is defect, has an altered substrate specificity, or is partially defect.

Mutations of Putative Regulatory Importance

There are also examples where mutations have been implicated to associate with interindividual differences in the induction of enzyme expression. Such allelic variants have been discussed for *CYP1A1*, *CYP1A2* and *CYP2E1*. Mutant alleles with one or several mutations in the flanking regulatory regions of these genes have been described. (See Rannug et al., 1995). The "rare" alleles are generally much more common in Oriental populations. It is evident that there are phenotypes in the human population where the subjects could be divided into *efficient* and *poor inducers*. This is the case for both CYP1A1 and CYP1A2as well as for CYP2E1. However no evidence has been presented for any molecular genetic basis to the different phenotypes described or to the different levels of constitutive enzyme expression. This area needs further research.

Polymorphic Cytochrome P450 Genes and Their Evolution.

The human cytochromes P450 of greatest importance for the metabolism of drugs and other xenobiotics are CYP1A1, CYP1A2, CYP2A6, CYP2C9, CYP2C19, CYP2D6, CYP2E1 and CYP3A4. The corresponding genes lie in clusters in tandem on chromosomes 7, 10, 15, 19 and 22, where the genes on the same chromosome are quite homologous to each other. However, with the exceptions of CYP1A1 and CYP2E1, there is no impressive homology between e g the rodent orthologues and the human enzymes. This complicates toxicity testing and analysis of drug metabolism using rodent models. The known polymorphic P450s and their major mutations are listed in Table 2 (cf. Ingelman-Sundberg and Johansson, 1995).

Table 2. The polymorphic P450 enzymes, the major detrimental mutations and their consequences.

Enzyme	Major detrimental mutation	Consequence of the mutation	Frequency of major mutant allele in Caucasians	
CYP1A1	Ile 462 → Val	None	4%	
CYP2A6	Leu160 → His	Defect enzyme	5%	
CYP2C9	Arg144 → Cys	Enzymewith altered substrate specificity	20%	
	Ile359 → Leu	Higher K_m (defect?)	6%	
CYP2C18	Stop codon, exon 2	No enzyme	27%	(Japanese)
CYP2C19	Cryptic splice site in exon 5	No enzyme	10%	
CYP2D6	Splice defect in intron4/exon5 junction	No enzyme	23%	
CYP2E1	Arg76→ His	Less enzyme	<1%	

The origin of the polymorphism requires some attention. It is evident that the P450 system has evolved from one ancestor gene. The currently over 505 known P450 genes have evolved through gene duplications, gene conversions, mutations and gene deletions. We have evidence for the existence, about 400 million years ago, of at least 60 gene duplications creating a new potential of detoxifying capacity. This period coincides with the time when animals became terrestrial and it is likely that the new environment, exposing these animals to a huge amount of plant toxins, created preferential survival of those organisms which had a higher detoxifying potential in the form of a higher number of active P450 genes . Apparently each species has had the opportunity to develop a unique pattern of P450 isozymes, appropriate for the organism in question. Similarly, it is probable that in the absence of such an environmental pressure, a P450 gene – through the DNA turnover – could accumulate detrimental mutations. This has probably been the case for many human cytochrome P450 genes, which disappeared from the populations, because of lack of metabolic function. In humans, as compared to e g rodents, we now indeed see a decrease in the number of active P450 genes, and several defect P450 alleles are introduced. The reason can probably be found in changes of the dietary requirements. The lack of physiological importance is evident from the high extent of divergence of the genes between species, and dietary factors are thus most likely for the selection pressure for conservation of the genes in certain species. As a consequence of this polymorphism, we are faced with the extensive interethnic and interindividual differences in the capacity to metabolize xenobiotics.

Genes Causing Impaired Drug Metabolism.

CYP2D6 is considered to be the second most important of the human drug metabolizing P450s. Among its major substrates are most antidepressants and neuroleptics, many antiarrythmics and lipophilic β-blockers as well as opioids and serotonin uptake inhibitors (see Bertilsson et al., 1995). The enzyme is apparently non-inducible and its absence creates the inability of the subject to metabolize the substrates. At ordinary dosage this can cause too high plasma levels of the drugs and results in side effects (Bertilsson et al., 1995). About 7 % of the Caucasian population are poor metabolizers (PM) of CYP2D6 substrates and this is because of the frequent distribution of an allele with a splicing defect (Heim and Meyer, 1990). Among Orientals, this figure is much lower, about 1 %. We could show that this is because the only type of defect allele seen is the one where the whole gene has been deleted (Johansson et al., 1991, 1994).

Chinese, as opposed to Caucasians, in general have a lower capacity for metabolism of CYP2D6 drugs, although not being classified as PM (Bertilsson et al., 1995). Thus, prescribed doses of e g haloperidol and other CYP2D6 substrates are lower in China as compared to Sweden. In order to evaluate the basis for this difference we cloned two variant genes of CYP2D6 from Chinese, sequenced and characterized them and expressed the corresponding genes in COS I cells. It was found that the functionally important mutation was creating a Pro34 → Ser amino acid substitution causing a more unstable and catalytically ineffective enzyme (Johansson et al., 1994). Indeed, this allele (CYP2D6Ch, or CYP2D6*10) was found to be the most common one among Chinese and responsible for the phenotype observed. The mutation happened to occur in a highly conserved PPGP region in cytochromes P450 of the CYP2 family, which is important for the folding and stability of the protein.

In black populations a similar effect is seen as a high number of subjects have a decreased, although not abolished, capacity for hydroxylation of CYP2D6 substrates (Masimirembwa et al., 1993, 1996a,b). We cloned a variant of CYP2D6 responsible for this diminished rate of catalysis (CYP2D6Z or CYP2D6*17) and identified three functional mutations as compared to the wt allele causing Thr107→Ile, Arg286→Cys and Ser486→Thr amino acid substitutions (Masimirembwa et al., 1996). The latter two mutations are present in the common CYP2D6*2 allele (Johansson et al., 1993) and do not cause any important alterations of enzyme function. Expression of chimeric constructs in yeast carrying various combinations of the mutations revealed that the Thr107→Ile mutation by itself did not create any impairment of enzyme function; neither did so the presence of any of the other mutations by themselves. However, in combination with the Arg286→Cys mutation or together with the other two, the Thr107 →Ile mutation caused the formation of an enzyme with much lower affinity for codeine or bufuralol, as compared to the wt variant, whereas the Vmax was not affected (Oscarson et al., 1996). In addition it was found that the CYP2D6 17 enzyme formed other metabolites from codeine as compared to the wt enzyme and represents the first case where mutations in CYP2D6 causes an altered substrate specificity. The molecular reason for alteration of the active site when two of the mutations are present together warrants further investigations.

Ultrarapid metabolizers

Leif Bertilsson and colleagues found in 1985 a patient who had to be treated with doses of nortriptyline more than 3 times higher than normal in order to achieve therapeutic plasma concentrations. Characterization of the CYP2D locus from this individual revealed the presence of an additional CYP2D6 gene yielding a 12.1 kb band on EcoRI RFLP (Bertilsson et al., 1993). This gene was cloned and sequence analysis revealed the variant CYP2D6*2 allele with the with respect to functional properties unimportant Arg286→Cys and Ser486→Thr mutations

(Johansson et al., 1993). The explanation for the clinical phenomenon was thus the presence of two active CYP2D6 genes on the same allele instead of one. In another family, where the father and the two children were extremely rapid metabolizers of the CYP2D6 substrate debrisoquine, this CYP2D6*2 gene was found to be present in 13 copies and this finding represents the first case of a stably inherited gene amplification (Johansson et al., 1993). The mechanism for this gene amplification remains unknown, but it could be assumed that some selection pressure during early maturation of oocytes or spermatocytes has existed and that the amplification is of a multiduplication type.

Evaluation of the global distribution of duplicated or multiduplicated active CYP2D6 genes revealed a very high frequency of carriers in Ethiopia (29%, Aklillu et al., 1996 and in Saudi Arabia (20 %, McLellan et al., 1996). The corresponding frequency among Swedes (Dahl et al., 1995), Zimbabweans (Masimirembwa et al., 1996) and Chinese (Johansson et al., 1994) is only about 2%. By contrast, 7 % of Spaniards have been found to carry this haplotype (Agúndez et al., 1995). Phenotypically, subjects with duplicated or multiduplicated genes were the most rapid metabolizers of debrisoquine both in Sweden, Spain and in Ethiopia. We believe that a dietary selection pressure has existed in Saudi Arabia and Ethiopia to favour preservation of alleles with duplicated or multiduplicated genes and that these have been introduced in Spain during the Muslim migration. In favour of this hypothesis, we see very few defect CYP2D6 alleles among Ethiopians and Saudi Arabs and since CYP2D6 has an extremely high affinity for alkaloids, it could be feasible that plant toxins exerted this dietary pressure in the past.

We have investigated the mechanism behind this duplication event by sequence analysis of the 3′-flanking region of the additional CYP2D6*2 gene. We found a repetetive sequence (CCCACCCTTC) present in eg the *c-myc* gene, known to be amplified in tumour cells, that could explain an unequal crossover in the formation of the allele with duplicated genes (Lundqvist et al., 1996). Subsequent unequal crossover of a duplicated allele with CYP2D6*2 on another allele can explain the formation of alleles with 3, 4 and 5 gene copies. The frequency at which we find these variants in the Ethiopian population is well in line with this mechanism (see Table 3).

CYP2D6 is known to metabolize codeine to morphine (Bertilsson et al., 1995). It could be anticipated that ultrarapid metabolizers do form much more of the active drug as compared to extensive metabolizers (EM). We have presented an efficient method for PCR-mediated genotyping of individuals with CYP2D6 genes in tandem (Johansson et al., 1996). In summary the clinical importance of the different phenotypes with respect to drug metabolism could be summarized as follows:

Poor metabolizers

- Diminished first pass metabolism, increased bioavailability, exaggerated response
- Diminished drug elimination, side effects or toxic effects
- The active metabolite is not formed; loss of therapeutic efficacy

Ultrarapid metabolizers
- Lack of therapeutic effect at standard doses
- Too much of the active metabolite
- Explanation for suspected bad compliance

The question arises whether also other genes encoding drug metabolizing enzymes can be stably duplicated and recent finding sin our lab suggest that this is the case (McLellan et al., unpublished observations).

Table 3. Alleles with duplicated or multiduplicated CYP2D6 genes

Allele	Allele frequency (%)		
	Swedes[a]	Ethiopians[b]	Spaniards[c]
CYP2D6*2 x 2	1.5	13	3.5
CYP2D6*2 x 3	0.2	1.6	nd
CYP2D6*2 x 4	0	0.8	nd
CYP2D6*2 x 5	0	0.4	nd
CYP2D6*2 x 13	0.1	0	nd

nd, not determined; The data are from a/ Dahl et al., 1995, b/ Aklillu et al., 1996 and c/ Agúndez et al., 1995.

Polymorphism of ethanol-inducible CYP2E1

CYP2E1 is of high importance for the metabolic activation of paracetamol, organic solvents and many precarcinogens (Ronis et al., 1996). Furthermore it activates ethanol to hydroxyethylradicals (Albano et al., 1995), and autoantibodies toward hydroxyethyl-radical adducts to CYP2E1 are found among alcoholics (Clot et al., 1996). The enzyme release oxy radicals and is implicated to be of great importance for the development of alcohol liver disease (Ingelman-Sundberg et al., 1993, Morimoto et el., 1995). There is an extensive interindividual variation in the levels of CYP2E1 between individuals. CYP2E1 is induced by ethanol mainly at the posttranslational level (Eliasson et al., 1990, 1992) and this is one cause for the variation. Transcriptional activation of the CYP2E1 gene at high ethanol levels has been suggested (see Ronis et al., 1996) and polymorphism in the flanking regions of the gene has been implicated to be of importance for induction of the enzyme and possibly to disease (see Rannug et al., 1995). We have investigated functional polymorphism of the CYP2E1 gene and found only a few alleles carrying functional mutations. Indeed out of 200 subjects screened of Oriental or Caucasian origin, only two alleles with a mutation causing a Arg86→His mutation, and thus a less stable enzyme, was found (Hu et al., 1996). One might conclude that the human CYP2E1 gene is highly conserved and we believe that the gluconeogentic role of this enzyme by converting acetone to gluconeogenetic precursors (see Ronis et al., 1996) has created a selection for the presence of functional alleles in the human populations.

Polymorphism of Drug Metabolizing Enzymes. General Aspects

It is evident that in the past many studies have been carried out to link the presence of variant alleles of different drug metabolizing enzymes to disease and cancer. In my opinion the linkage between a single enzyme to a particular form of cancer is in many cases farfetched, and the data are in many cases controversial and inconclusive. However, with respect to more acute effects and toxicities of drugs and other xenobiotics there seems to be a tighter connection. Some of the most relevant examples are listed in Table 4 (cf Gram et al., 1995).

Table 4. Examples of polymorphic drug metabolizing enzymes and linkage to drug toxicity and disease.

Enzyme	Toxicity
CYP2D6	Side effects and toxicity during treatment with remoxipride
Epoxide hydrolase	Low activity allele associated with high levels of aflatoxin-protein adducts
N-Acetyltransferase 2	Slow acetylators are at higher risk to get neuropathy and hepatitis during treatment with isoniazide; Systemic lupus erythematosus (SLE) during treatment with hydralazine is seen entirely in PM; SLE develops much earlier in PM during treatment with procainamide Bladder cancer is more common among PM
Thiopurine-S-methyl-transferase	Toxicity during treatment of leukemia with 6-mercaptopurine is entirely seen in PM

For the future it is my opinion that genotyping of individuals for the prediction of optimal treatment with drugs having a narrow therapeutic range will constitute a most valuable tool. The development of oligonucleotide chips for genotyping assays will reduce the costs and improve the versatility of the principle. In case of the use of genes encoding drug metabolizing enzymes as risk genes for susceptibility for different types of cancer, further research has to be carried out in the field and it might be a possibility that a certain combinatory profile of genotypes can add information of preventive importance.

Acknowledgements

The research in the Author's laboratory is supported by grants from the European Commission (Biomed 2), Astra Hässle AB and from The Swedish Medical Research Council.

References.

Agúndez, J.A.G., Ledesma, M.C., Ladero, J.M. and Benítez, J. (1995) Prevalence of CYP2D6 gene duplication and its repercussion on the oxidative phenotype in a white population. Clin Pharmacol Ther 57:265-269.

Albano, E., Clot, P., Morimoto, M., Tomasi, A., Ingelman-Sundberg, M. and French, S. (1995) Role of cytochrome P4502E1-dependent formation of hydroxyethyl free radical in the development of liver damage in rats intragastrically fed with ethanol. Hepatology 23:155-163.

Aklillu, E., Persson, I., Bertilsson, L., Johansson, I., Rodrigues, F. and Ingelman-Sundberg, M. (1996) Frequent distribution of ultrarapid metabolizers of debrisoquine in an Ethiopian population carrying duplicated and multiduplicated functional CYP2D6 alleles. J Pharm Exp Ther 278:441-446

Bertilsson, L., Dahl, M-L., Sjöqvist, F., Åberg-Wistedt, A., Humble, M., Johansson I., Lundqvist, E. and Ingelman-Sundberg, M. (1993) Molecular basis for rational megaprescribing in ultrarapid hydroxylators of debrisoquine. Lancet 341: 63.

Bertilsson, L., Dahl, M.-L., Ingelman-Sundberg, M., Johansson, I. and Sjöqvist, F.(1995) Interindividual and interethnic differences in polymorphic drug oxidation - implications for drug theraphy with focus on psychoactive drugs. In: Advances in Drug Metabolism in Man, (Pacifici, G. and Fracchia, G.N., Eds), EUR 15439 EN, EC, DGXII-E-4, Luxembourg, office for publications of the European Commission, pp. 85-136.

Clot, P., Albano, E., Eliasson, E., Tabone, M., Arico, S., Israel, Y., Moncada, C. and Ingelman-Sundberg, M. (1996) Cytochrome P4502E1 (CYP2E1) hydroxyethyl radical adducts as the major antigenic determinants for autoantibody formation among alcoholics. Gastroenterology 111: 206-216.

Gram, L.F., Brõsen, K, Sindrup, S and Skjelbo, E (1995) Genetic and interethnic variability in drug metabolism: What are the clinical consequences. In: European Conference on Specificity and Variability in Drug Metabolism, (Alván, G., Balant, L.P., Bechtel, P.R., Boobis, A., Gram, L.F., Paintaud, G and Pithan, K., Eds) COST B1, Luxembourg: Office for Official Publications of the European Communities EU, pp. 235-245.

Dahl, M.-L., Johansson, I., Bertilsson, L., Ingelman-Sundberg, M. and Sjõqvist F. (1995) Ultrarapid hydroxylation of debrisoquine in a Swedish population. Analysis of the molecular genetic basis. J Pharmacol Exp Ther 274:516-520.

Eliasson, E., Johansson, I. and Ingelman-Sundberg, M. (1990) Substrate, hormone and cAMP- regulated cytochrome P450 degradation, Proc. Natl. Acad. Sci (USA) 87: 3225-3229.

Eliasson, E., Mkrtchian S, and Ingelman-Sundberg, M. (1992) Hormone- and substrate regulated intracellular degradation of cytochrome P450 (2E1) involving MgATP-activated rapid proteolysis in the endoplasmic reticulum membranes J Biol Chem 267: 15765-15769.

Heim, M.H. and Meyer, U.A. (1990) Genotyping of poor metabolisers of debrisoquine by allele-specific PCR amplification. Lancet 336:529-532.

Hu, Y., Johansson, I., Yun, Q-Y and Ingelman-Sundberg, M.(1996) Genetic polymorphism of the human CYP2E1 gene. Characterization of two variant alleles present in Caucasian and Chinese populations. Mol Pharmacol, in press.

Ingelman-Sundberg, M. and Johansson, I.(1995) The molecular genetics of the human drug metabolizing cytochrome P450s. In: Advances in Drug Metabolism in Man, (Pacifici, G. and Fracchia, G.N., Eds), EUR 15439 EN, EC, DGXII-E-4, Luxembourg, Office for Publications of the European Comission, pp. 543-586.

Ingelman-Sundberg, M., Johansson, I, Hu, Y., Terelius, Y., Eliasson, E., Clot, P., and Albano, E. (1993) Ethanol-inducible cytochrome P4502E1: Genetic polymorphism, regulation and possible role in the etiology of alcohol-induced liver disease. Alcohol 10:447-452

Johansson, I., Bertilsson, L., Dahl, M-L, Sjõqvist, F and Ingelman-Sundberg, M. (1993) Inherited amplification of an active gene in the cytochrome P450 CYP2D locus as a cause of ultrarapid metabolism of debrisoquine. Proc Natl Acad Sci (USA) 90:11825-11829.

Johansson, I., Yue, Q.-Y., Dahl, M.-L., Heim, M., Säwe, J., Bertilsson, L., Meyer, U.A,, Sjõqvist. F, and Ingelman-Sundberg, M. (1991) Genetic analysis of the interethnic difference between Chinese and Caucasians in the polymorphic metabolism of debrisoquine and codeine. Eur J Clin Pharmacol 40: 553-556.

Johansson, I., Oscarson, M., Yue, Q.-Y., Bertilsson, L., Sjõqvist, F. and Ingelman-Sundberg, M. (1994) Genetic analysis of the Chinese cytochrome P4502D locus: characterization of variant CYP2D6 genes present in subjects with diminished capacity for debrisoquine hydroxylation. Mol Pharmacol 46:452-459.

Johansson, I., Lundqvist, E., Dahl, M.-L. and Ingelman-Sundberg, M. (1996) PCR-based genotyping for duplicated and deleted CYP2D6 genes Pharmacogenetics 6:351-355.

Masimirembwa, C.M., Johansson, I., Hasler, J.A. and Ingelman-Sundberg, M. (1993) Genetic polymorphism of cytochrome P450 CYP2D6 in a Zimbabwean population. Pharmacogenetics 3:275-280.

Masimirembwa, C.M., Persson, I., Bertilsson, L., Hasler, J.A. and Ingelman-Sundberg, M.(1996a) Diminished enzyme activity and lack of metabolic coregulation of CYP2D6 probe drugs in a black Zimbabwean population. Eur J Clin Pharmacol, in press.

Masimirembwa, C.M., Persson, I., Bertilsson, L., Hasler, J.A. and Ingelman-Sundberg, M.(1996b) Molecular basis of diminished debrisoquine hydroxylase (CYP2D6)activity in a black African population: A common mutation in exon 2 of the CYP2D6 gene. Br J Clin Pharmacol, in press.

McLellan, R.A., Oscarson, M., Seidegård, J., Price Evans, D.A. and Ingelman-Sundberg, M.(1996) Frequent occurrence of CYP2D6 gene duplication in Saudi Arabia, Pharmacogenetics, in press.

Morimoto, M., Hagbjörk, A.-L., Yvonne- Y-J., Fu, P.C. Albano, E. Ingelman-Sundberg, M. and French, S.W. (1995) Modulation of alcoholic liver disease by cytochrome P450 2E1 inhibitors, Hepatology 21:1610-1617.

Oscarson M, Hidestrand M, Johansson I, Masimirembwa CM, Hasler JA, Ingelman-Sundberg M.(1996) Detrimental effect of combined mutations in the CYP2D6Z (2D6*17) allele. XIth International Symposium on Microsomes and Drug Oxidations, Los Angeles, p. 214.

Rannug, A., Alexandrie, A-K., Persson, I and Ingelman-Sundberg, M. (1995) Genetic polymorphism of cytochrome P450 1A1, 2D6 and 2E1. Regulation and toxicological significance. J Occ Env Med 37:25-36.

Ronis, M.J.J., Lindros, K.O. and Ingelman-Sundberg, M.(1996) The cytochrome P4502E subfamily. In: Cytochromes P450, Pharmacological and Toxicological Aspects (Ioannides, I., Ed.) CRC Press, pp. 211-239.

Detection of Chemically-Induced Changes in p53 Expression

(Special Workshop, organized by the Molecular Toxicology and Mechanisms Speciality Section of EUROTOX)

Looking at p53: Theoretical Implications and Methodological Aspects

Bertram Opalka, Sabine Kasimir-Bauer
Klinikum der Universität (GHS) Essen, Innere Klinik und Poliklinik (Tumorforschung)
Hufelandstr. 55, D-45122 Essen

Introduction

Since its discovery as a protein with a mass of 53 kD binding to simian virus' 40 large T antigen, the p53 gene and its protein product have gained an enormous interest in the scientific community. The literature on this topic has grown exponentially (Marx, 1993) and more than 5500 papers accumulated in the MedLine within the last 5 years. First described as one more oncogene whose products cause immortalization and malignant transformation, it was later found to be just the opposite, i.e. a tumor suppressor gene, whose functional inactivation contributes to carcinogenesis. Moreover, it was found to be among the genes most frequently mutated in human cancers. Intensive research revealed that the p53 protein is a DNA binding transcription factor which is important for maintenance of the integrity of the genome, some kind of "guardian of the genome" (Lane, 1992; for reviews see Levine et al., 1991; Harris, 1993; Levine, 1995; Ko and Prives, 1996; Jacks, 1996; Arrowsmith and Morin, 1996). For epidemiologists and toxicologists, an interesting feature of p53 is that the specific mutation patterns correlate with the inducing agent, i.e. tobacco smoke will result in other types of lesions than UV light (Greenblatt et al., 1994). In this respect, analysis of p53 mutational patterns may become a tool to evaluate the consequences of noxic exposure for the genome and for cancer risk.

The p53 gene and its protein product

The p53 gene is highly conserved in evolution. The human homologue maps to chromosomal band 17q13.1. It is about 23 kbp in size and composed of eleven exons, the first exon not being translated. A messenger RNA of 2.2 - 2.5 kb encodes a protein product of 393 amino acids (aa), which can be divided into different structural and functional domains (Fig.1). The acidic amino terminal domain contains the transcriptional stimulating activity within aa 1 - 42. The central core region is responsible for the recognition of the p53-specific consensus DNA motif RRRCA/TT/AGYYY, where R stands for a purine base and Y for a pyrimidine, respectively. Nearly all missense mutations of p53 are found in this core domain with the mostly affected hot spots at aa positions Arg175, Gly245, Arg248, Arg249, Arg 273 and Arg282 (Greenblatt et al., 1994; Ko and Prives, 1996). The basic carboxy terminus of p53 is responsible for unspecific interactions with DNA and for oligomerization of p53, which acts normally as a homotetramer. Moreover, the carboxy terminus contains three nuclear localization signals to direct the protein to the proper cellular compartment for its activity.

Besides oligomerization to a homotetramer, p53 is able to bind a variety of cellular and viral proteins. These can be classified as transcription factors (e.g. TFIID, TBP, dTAFs, HTAF31, and others), transcriptional activators (WT-1, Sp1, CBF), replication and repair proteins (e.g. RP-A, CSB, TFIIH components), kinases (c-abl, casein kinase II), calcium-binding proteins, ubiquitin-system proteins, oncogene products (mdm-2), viral oncogene products (adenovirus E1b 55kD, SV40 TAg, HPV 16 E6, HBV HBx, EBV BZLF1, EBV EBNA-5), and the uncharacterized proteins 53BP1 and 53BP2 (Ko and Prives, 1996). For some of these interactions the relevant domains of the p53 are known, these are indicated in Fig. 1.

Functional aspects of p53

Meanwhile, a number of genes has been identified which contain p53-specific binding motifs in their promoters and which are transcriptionally activated by p53. Among these are the genes encoding p21/WAF1/Cip1, a negative regulator of the cell cycle, the oncogene mdm-2, the gadd4 gene (growth arrest DNA damage inducible), cyclins D and G, the retinoblastoma tumor suppressor protein Rb, PCNA (proliferating cell nuclear antigen), EGF-R, and others. Some genes without a p53 consensus motif are negatively regulated by interactions of p53 with the basic transcription machinery, especially the TBP. These genes include fos, jun, IL-6, Rb, and bcl-2.

P53 itself is expressed ubiquitously but to such low concentrations that the protein is normally not detected by immunohistochemistry. However, upon exposure of a cell to DNA damaging agents like irradiation by UV light or gamma rays, p53 is up-regulated by post translational mechanisms and transcrptional activation. The main effect of p53 up-regulation is the arrest of the cell in the G1 phase of the cell cycle. This effect is mediated at least in part by stimulation of the p21/WAF1/Cip1 gene, which encodes an inhibitor for cyclin dependent kinases, which normally drive the cell cycle ahead. The G1 arrest

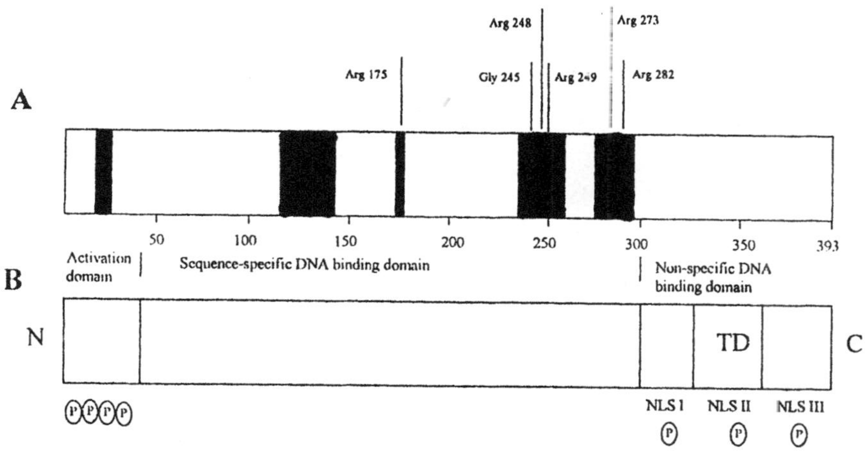

Figure 1. Schematic view of the p53 protein (after Ko and Prives, 1996, with alterations)
A: The p53 protein is drawn schematically as open bar; the five conserved domains are marked by black boxes. The amino acid positions are givenas well as the mutational hot spots (not drawn to scale).
B: The p53 protein is drawn schematically as open bar. The functional domains are given as well as positions of phosphorylation (circled Ps), nuclear localization signals (NLSI through NLSIII), and the tetramerization domain (TD).
N = amino terminus, C = carboxy terminus of the protein. For further information see text.

allows the cell to repair the damaged DNA. If, however, the damage is to an extent which cannot be repaired, the cell is stimulated to undergo apoptosis. Thus, either by repair or cell death, the accumulation of damaged DNA which may result in tumorigenesis is hindered through the action of p53. The exact mechanisms by which p53 induces apoptosis still have to be elucidated. There are hints that both the positive as well as negative transcriptional regulatory activity are required. Moreover, it is known that the decision whether repair or apoptosis occurs is dependent on cell type, growth factor availability, and a number of other factors (Ko and Prives, 1996).

A mechanism of fine tuning of the p53 concentration within the cell is the negative feedback by the mdm-2 oncoprotein. The mdm-2 gene is induced by p53 and its product negatively regulates the expression of p53, thus maintaining a homeostasis between arresting and arrest-releasing signals. Overexpression of mdm-2 has been observed in soft tissue sarcomas and is considered to be involved in tumorigenesis (Harris and Hollstein, 1993).

Apart from its transcriptional regulation, a direct role for p53 in replication and DNA repair can be deduced from its ability to bind factors required for both processes. These activities, however, await further elucidation.

Mutations in p53 and human cancer

P53 mutations are among the most abundant genomic alterations found in human cancer (for review see Harris and Hollstein, 1993; Greenblatt et al., 1994). The all-over frequency in human cancers extends to about 50 %, in some types of tumors e.g. colorectal carcinomas and lung cancer mutations are detected in 55 to 74 % of cases (Soussi et al., 1994). Most of these alterations are missense mutations with amino acid exchanges, the majority occurring at the above mentioned hot spots. Besides these missense mutations, deletions and insertions were observed, preferably in regions of repetitive sequences. In contrast to the products of other oncogenes like Rb, p53 mutations not only result in a loss of function but rather in a gain of function of the altered proteins. P53 protein variants are much more stable than p53wt and have an extended half life. Therefore, they can in most cases easily be detected by immunohistochemistry. Moreover, the mutated proteins behave mainly "dominant negative" in competition with their wild type counterpart, i.e. the mutated forms inhibit DNA binding by the p53wt molecules and thus contribute to an enhanced proliferation potential of a cell. In collaboration with mutated ras genes, p53 mutant forms are able to fully transform cells. Many human tumors have lost one p53 wt allele resulting in homozygous expression of the mutant forms (Harris and Hollstein, 1993; Greenblatt et al., 1994).

As mentioned above, most p53 mutations occur in the core domain between aa 120 and 290 and mainly at the hot spots (Figure 1; Levine, 1995). There is, in addition, a striking concordance between tumor type and type of mutation as well as with the type of mutagen and type of mutation (reviewed in Greenblatt et al., 1994; Harris and Hollstein, 1993). For example, in UV-induced skin cancers CC to TT transitions are detected most frequently. In some areas of China and in Mozambique, after exposure to high doses of aflatoxin, specific G:C to T:A transversion at codon 249 were detected. In tobacco-induced lung cancer A:T to G:C transversions, G:C to T:A transversions as well as deletions and insertions were found in small cell and non-small cell lung cancer, and in adenocarcinomas of the lung. Eighty one percent of tobacco induced oral cancers carry p53 mutations, wheras only 34 % of laryngeal or pharyngeal cancers carry mutated p53 alleles. In lung cancers of ^{222}Radon-exposed uranium miners mainly missense mutation at codon 249 have been found. In HPV-induced hepatocellular carcinomas (HCC) p53 is inactivated by binding to the HPV E6 tumor protein which induces ubiquitin-mediated degradation of p53. In the rare cases of HPV-negative HCCs missense mutations of p53 have been found (Greenblatt et al., 1994). Apart from these most probably carcinogen-induced mutations of p53, there is also evidence for spontanous mutations of this gene. A well established mechanism of endogenous mutations is the deamination at CpG dinucleotides generating G:C to A:T transitions. As p53 contains 39 CpG dinucleotides, it is reasonable to conclude that mutations at these sites most probably are due to sporadic mutagenesis. Moreover, small deletions and insertions can be explained by inaccuracies of the DNA polymarase during DNA replication, thus creating another source of spontaneous mutations. These types of mutations are frequently found in tumors of the brain, colon, stomach, thyroid, and hematological cancers, suggesting a minor role for exogenous noxes in these neoplasms (Greenblatt et al., 1994).

The question may arise whether p53 mutations are an early or late event in the multistep process of tumorigenesis. There is no simple answer to this question, as several investigations have shown that in some tumor types p53 is mutated early whereas in others it is a marker of tumor progression. Lung cancer, breast, head and neck, cervical, and stomach cancers were found to have early p53 mutations, late occurring mutations were associated with brain, thyroid, liver, and ovary cancers. In the latter group, p53 mutations often correlate with a poorer prognosis. In lung cancer it was shown that p53 alterations were almost absent in normal mucosa, but their frequency increased up to 76 % with progression from squamous or mild dysplasia to fully invasive lung carcinoma (Greenblatt et al., 1994).

The role of p53 as a tumor suppressor gene has been proven both *in vitro* and *in vivo*. Numerous experiments have shown an anti-oncogenic effect of p53 in tissue culture. In 1992, a report on p53 "knock out" mice which have no functional allele of the gene demonstrated normal development of these animals, but a high susceptibility to spontaneous tumors (Donehower et al., 1992). A hereditary form of p53 mutations in humans has been found in families with the Li-Fraumeni syndrome. These persons develop osteosarcomas, adrenocortical carcinomas, breast and brain tumors often at an early age. Moerover, these people have an increased risk of developing secondary cancers in their life.

The role of p53 for cancer treatment

The ability of p53 to induce apopstosis seems to be an important prerequisite for the treament of cancer with irradiation and cytostatic drugs. Experimentally, it was found that fibroblasts expressing the adenoviral E1a oncoprotein were resistant to gamma irradiation, 5-fluorouracil, doxorubicin, and etoposid when p53 was functionally inactive. In the presence of p53wt, these cells underwent apoptosis upon treatment with irradiation or the drugs. These findings were confirmed with explanted human tumor cells in nude mice (reviewed in Jacks, 1996). Moreover, in an *in vitro* system it was shown that introduction of a functional p53wt gene was able to induce sensitivity to cisplatin. Mutated p53 not only has lost these functions, but induces the expression of the Mdr-1 gene encoding an effective transporter protein which is responsible for the efflux of cytotoxic drug from tumor cells (Harris and Hollstein, 1993).

Summary and conclusions

The p53 gene and its protein product are important regulators of the cell cycle machinery. The p53 protein controls an essential restriction point in G1 prior to DNA replication ensuring authenticity of the genome to be multiplied, to allow DNA damage repair, and inducing apoptosis as a mechanism for the elimination of cells bearing DNA lesions too large to be repaired. Thereby, cells with a cancerogenic potential are removed from the organism. Mutations of p53 result in proteins with dominant negative effects which play a major role in the pathogeneisis of many human cancers. Their mutational spectra often allow to deduce whether a tumor was the result of spontaneous mutagenesis or exposures to external carcinogens. Moreover, in some cases the types of mutations are specific for a given carcinogen, e.g. UV irradiation, aflatoxins, or components of tobacco smoke. As p53 plays a key role in at least one important pathway for induction of apoptosis, it has a major relevance for therapeutic interventions in cancer patients intended to be treated with apoptosis-inducing drugs or irradiation. In this context, p53 may in future be a good candidate transgene for gene therapy of human malignancies. For the prevention of cancers, scoring for p53 mutation may prove furthermore a useful toool for screening of large cohorts of probands exposed to different carcinogens.

Methods for evaluation of mutant p53 proteins and p53 gene mutations

In contrast to the wild type protein, mutant p53 variants are detectable by their prolonged half life within the cells resulting from point mutations which lead to amino acid substitutions and conformational changes in the protein. The following chapter will show a broad spectrum of methods which have been established to detect the mutant p53 protein in peripheral blood cells, serum, and tissue from patients suffering from solid tumors and hematological malignancies.

Immunohistochemistry

Immunohistochemistry has still been the method of choice for evaluating the mutant p53 protein because immunohistochemical staining can be done using slides with paraffin-embedded tissue from a variety of tumor entities. This technique includes an incubation of the slides at 60øC for 1 h, deparaffinization with xylene, rehydration through series of graded alcohol and treatment of the sections with 1% hydrogen peroxide to quench endogenous peroxidase activity. Normal goat serum is used to block non-specific immunostaining. The most often chosen p53 antibodies are Pab 1801, DO-1 and DO-7, respectively. The detection of p53 is achieved by incubation of the slides with biotinylated goat antibody and subsequent incubation with the strept ABComplex/horseradish peroxidase. Immunostaining is developed by immersion in 0.06% 3.3`diaminobenzidine tetrahydrochloride solution dissolved in 0.5% Triton-x/PBS for 5 minutes. Sections are counterstained with modified Harris-hematoxylin and 0.3% ammonia water and passed through graded alcohols and xylene to dehydrate.

In case of positive-staining, the following variables are used to categorize the staining pattern: (1) percentage of positive-staining cells (<25%; 25% to 50%; 50 to 75%; >75%), and (2) intensity of staining (weak [+]; moderate [++]; strong [+++]. (For detailed information see Auvinen et al., 1994, Zeng et al., 1994, Borg et al., 1995, Horne et al., 1996.)

Although being the method of choice, immunohistochemistry has been critisized because a) it often gives false positive results, and b) it shows a lack of reproducible interpretation of staining intensity and cellularity due to different methods of fixation, including both the fixative solution and temperature which affect the stability of the p53 protein (Fisher et al., 1994, Dix et al., 1994).

Western Blotting can be applied for analyzing fresh or frozen peripheral blood cells , cell lines and tissue. Briefly, 50-75 µg of total protein are fractionated on a 12.5% polyacrylamide gel containing SDS. Following electrophoresis, the proteins are transferred to nitrocellulose and p53 is detected by Western-Blot analysis using anti-p53 antibody DO-1 (Oncogene Science) which detects both mutant and wild-type forms of p53 on immunoblots (Fan et al., 1994).

Besides Western-Blot analysis and immunohistochemistry, *flow cytometry* (Kurose et al., 1995), an *ELISA* for detecting serum p53 antibodies (Wild et al., 1995) and a *color assay* (Flaman et al., 1995) have been established for the evaluation of p53 protein but none of the methods have been shown to be quantitative and useful for routine diagnosis.

Luminometric immunoassay (LIA)

A luminometric immunoassay (LIA) has recently emerged as a rapid and reliable method for the quantitative assessment of p53 in tumor tissue from patients with breast cancer (Borg et al., 1995) and we demonstrated that the LIA represents a useful tool in clinical laboratories with regard to the detection of p53 mutations and the prediction of cellular response to chemotherapy in various malignant entities including cancer of the stomach and the colon, NHL, leukemic blasts, as well as in mononuclear cell fractions from normal donors (Kasimir-Bauer et al., 1996).

The sandwich-type luminometric immunoassay (LIA) procedure was developed by AB Sangtec Medical for quantitative measuring of mutant and wild-type p53 protein in tumor tissue. The LIA uses two monoclonal antibodies for denaturation-resistant epitopes (different sites) at the N-terminus of the p53 protein: polystyrene tubes coated with Pab 1801 monoclonal anti-p53 antibody (Banks et al., 1986) and a tracer solution consisting of an aminobutylethylisoluminol (ABEI)-conjugated DO1 monoclonal antibody (Vojtesek et al., 1992). The preparation of antibody coated tubes and the LIA tracer is described by Borg et al., 1995.

Test procedure: The LIA is conducted in a single incubation step by adding 100 µl of tracer and 100 µl (100 µg) of cytosol or p53 standard to the antibody-coated tubes. After incubation overnight at room temperature, the tubes are washed four times with 2 ml of 0.15 M sodium chloride. The luminescence is determined using the LIA-mat starter service kit (Byk-Sangtec) and immedeately measured as intègrals over a period of 5 s in a Berthold luminometer. The assay is standardised using pure soluble recombinant wild-type human p53 protein isolated from bacteria (Midgley et al., 1992). The standard curve is calculated with a curve-fitting programme (spline smooth Multicalc, Wallace OY, Turku, Finland), the p53 content (ng/ml) being determined per RLU (relative light unit). The detection limit (zero standard + 3 standard deviations) is approximated to be 0.01 ng/ml. The concentration of p53 protein is expressed in ng/ml cytosol protein.

Methods for detection of p53 gene mutations

Single-Strand Conformation Polymorphism (SSCP)

SSCP has become one of the most rapid methods to scan genes for single base differences (Orita et al., 1989). This method has been demonstrated to be effective in detecting polymorphisms (Orita et al., 1990) and in identifying disease-causing mutations (Soto et al., 1992; Kanjiial et al., 1993). The method typically involves the amplification by PCR of a discreet segment of genomic DNA in the presence of radiolabelled or non-labelled nucleotides, melting of the PCR products, and analysis of the single strands on a non-denaturating polyacrylamide gel. Polymorphic differences in strand mobility result from the effects of primary sequence changes on the folded structure of a single DNA strand. The primary sequence differences alter the intramolecular interactions that generate a three-dimensional folded structure. The molecules will thus move at different rates through a non-denaturating gel. AT Biochem has developed a unique matrix, MDE gel (Mutation Detection Enhancement) which significantly improves the resolution of conformationally different DNA molecules and which is often superior to conventional polyacrylamide in resolving subtle sequence and conformation differences among DNA strands.

While the first applications of SSCP analysis used radiolabelled [^{32}P]dATP or [^{32}P]dCTP, recently non-radioactive SSCP using silver staining for detection has been successfully established by many groups.

PCR

DNA is prepared by e.g. phenol/chloroform extraction and oligonucleotide primers were synthesized according to the sequences published by Hensel et al., 1991. 100-200 ng of genomic DNA are routinely amplified in 50 µl reactions containing 20 mM Tris HCl (pH 8.3), 100 mM KCl, 3 mM $MgCl_2$, 0.002% gelatin, deoxy-nucleotide triphosphates 0.2 mM each dATP, dGTP, dCTP, and dTTP (e.g. Pharmacia Biotech Europe GmbH, Freiburg, Germany),12 pmoles of each primer and 1.25 units of Taq DNA polymerase (Perkin-Elmer Cetus). The reactions are overlayed with one drop of mineral oil and amplified according to the following PCR protocol: 94°C for 5 min followed by 35 cycles of 94°C 1 min, 55°C (exon 5-6 primers, exon 8-9 primers) or 66°C (exon 4 primers, exon 7 primers) for 2 min, 65°C for 1 min, and 72øC for 2 min. The last cycle is followed by an additional 5 min at 72°C to insure that all products are complete.

SSCP

8 µl of the PCR products are mixed with 3 µl sample buffer containing 20 mM EDTA, 0.05% bromophenol blue, 0.05% xylene xyanol, 3% ficoll and formamide; this mixture is heated at 95°C for 2 min, chilled on ice and loaded on a 6% neutral polyacrylamide gel (MDE gel solution), containing 90 mM Tris-borate and 2 mM EDTA. After electrophoresis at 550 V for 16 h mutations are detected by silver staining.

Temperature-gradient gel electrophoresis (TGGE)

TGGE is one further and often applied method to detect single-base alterations. This technique includes DNA amplification with exon-specific primers binding to adjacent intron regions. To detect single-base alterations more efficiently, one of each group of primers contains a GC-rich "clamp" and each exon usually is amplified seperately. Subsequently, the PCR products are purified by phenol/chloroform/isoamylalcohol (25:24:1) extraction and precipitation with ethanol. After redissolving the pellets in a special running buffer consisting of 20 mM MOPS and 1 mM EDTA, pH 8.0 TGGE gel electrophoresis of the purified amplification products is performed using horizontal 8% polyacrylamide gels in 8 M urea, 20 mM MOPS, 1 mM EDTA and 2% glycerol. p53 mutations are detected by mobility shifts of the corresponding DNA strands and visualized by silver staining. (For further details see ref. Scholz et al., 1993; Kappes et al., 1995.)

Non-Isotopic RNase Cleavage Assay (NIRCA™)

The new NIRCA™ method was developed by Ambion (Austin, Texas), has a comparable sensitivity to SSCP and seems to be very promising. It allows larger fragments to be analyzed, results are available immedeately because no autoradiography or secondary detection steps are needed, and the electrophoresis step is complete in less than one hour with the convenience of agarose gels. Also, the size of the cleavage products provides an indication of the location of the mutation.

Briefly, this method uses RNA/RNA duplexes as substrates for RNase digestion made by hybridizing complementary wild type and experimental transcripts. The templates are produced by *in vitro* transcription of PCR products containing opposable T7 and SP6 phage promotors. Wild type and test transcripts are hybridized to form double-stranded RNA targets which are treated with Rnase. Rnase will not attack double stranded RNA but cleave the single strands formed at the site of a single base mismatch. The cleavage products are then separated on agarose gels containing ethidium bromide and assessed under UV light. Experimental samples are scored as positive for mutations if they contain cleavage products not seen in the wild type control. Further studies will show whether NIRCA™ is a reliable method for detecting mutations and whether it will be superior to SSCP.

Sequencing
The only absolut reliable method to detect p53 mutations on the DNA or RNA level is sequencing. For this pupose, the dideoxy technique as described by Sanger et al., (1977; Sambrook et al., 1989) is the most widely used today. The DNA to be sequenced can be PCR amplified and/or cloned into plasmid vectors. Alternatively, amplification and sequencing can be done in one step ("cycle sequencing", e.g. with a kit from Perkin Elmer Cetus). The DNA on the sequencing gel can either be visualized autoradiographically or by labelling with fluorescent dyes. The most comfortable way of sequencing is using an automated sequencer (e.g.from manufacturers like Perkin Elmer Cetus or Pharmacia).

References

(for reason of brevity, in the non-methodological part wherever possible only review articles have been cited)

Arrowsmith CH, Morin P (1996) New insights into p53 function from structural studies. Oncogene 12: 1379-1385

Auvinen A, Isola J, Visakorpi T, Koivula T, Virtanen S. and Hakama M (1994) Overexpression of p53 and long-term survival in colon carcinoma. Br J Cancer 70: 293-296

Banks L, Matlashewski G, Crawford L (1986) Isolation of human p53 specific monoclonal antibodies and their use in the studies of human p53 expression. Eur J Biochem, 159:529-534

Borg A, Lennerstrand J, Stenmark-Askmalm M, Fernö M, Brisfors A, Öhrvik A, Stal O, Killander D, Lane D, Brundell J (1995) Prognostic significance of p53 overexpression in primary breast cancer; a novel luminometric immunoassay applicable on steroid receptor cytosols. Br J Cancer 71: 1013-1017

Dix B, Robbins P, Carello S, House A, Iacopetta B (1994) Comparison of p53 gene mutation and protein overexpression in colorectal carcinomas. Br J Cancer 70: 585-590

Donehower LA, Harvey M, Slagle BL, McArthur MJ, Montgomery jr CA, Butel JS, Bradley A (1992) Mice deficient for p53 are developmentally normal but susceptible to spontaneous tumors. Nature 356: 215-221

Fan S, El-Deiry WS, Bae I, Freeman J, Jondle D, Bhatia K, Fornace Jr. AJ, Magrath I, Kohn KW, O'Connor PM (1994) p53 gene mutations are associated with decreased sensitivity of human lymphoma cells to DNA damaging agents. Cancer Res 54: 5824-5830

Fisher CJ, Gillett CE, Vojtesek B, Barnes DM, Millis RR (1994) Problems with p53 immunohistochemical staining: the effect of fixation and variation in the methods evaluation. Br J Cancer 69: 26-31

Flaman JM, Frebourg T, Moreau V, Charbonnier F, Martin C, Chapuist P, Sappino AP, Limacher JM, Brons L, Benhattar J, Tada M, Van Meir EG, Estreicher A, Iggo RD (1995) A simple p53 functional assay for screening cell lines, blood and tumors. Proc Natl Acad Sci USA, 92: 3963-3967

Fujiwara T, Grimm EA, Mukhopadhyay T, Zhang W-W, Owen-Schaub LB, Roth JA (1994) Induction of chemosensitivity in human lung cancer cells in vivo by adenovirus-mediated transfer of the wild-type p53 gene. Cancer Res. 54: 2287-2291

Harris CC (1993) p53: At the crossroad of molecular carcinogenesis and risk assessment. Science 262: 1980-1981

Harris CC, Hollstein M (1993) Clinical implications of the p53 tumor-suppressor gene. New Engl J Med 329: 1318-1327

Hensel CH, Xiang RH, Sakaguchi AY, Naylor SL (1991) Use of single strand conformation polymorphism technique and PCR to detect p53 gene mutations in small cell lung cancer. Oncogene 6: 1067-1071

Jacks T (1996) Lessons from the p53 mutant mouse. J Cancer Res Clin Oncol 122: 319-327

Kanjiial S, Pierceall WE, Cummings KK, Kripke ML, Ananthaswamy HN (1993) High frequency of p53 mutations in ultraviolet radiation-induced murine skin tumors: Evidence for strand bias and tumor heterogeneity. Cancer Res 53: 1-4

Kappes S, Milde-Langosch K, Kressin P, Passlack B, Dockhorn-Dworniczak B, Rõhlke P, Lõning T (1995) p53 mutations in ovarian tumors, detected by temperature-gradient gel electrophoresis, direct sequencing and immunohistochemistry. Int J Cancer 64: 52-59.

Kasimir-Bauer S, Ottinger H, Belka C, Stellberg W, Halfmeyer K, Gewaltig I, Seeber S, Grosse-Wilde H, Scheulen ME (1996) Assessment of p53 in cancer cells by SSCP and LIA may fail to predict resistance to, or rate of apoptosis induced by chemotherapy. (submitted)

Ko LJ, Prives C (1996) p53: puzzle and paradigm. Genes Development 10: 1054-1072

Kurose A, Sasaki K, Ishida Y, Shibata Y, Yanagisawa S, Kanno C, Uesugi N, Wada T, Miura Y (1995) Flow cytometric analysis of p53 expression during the cell cycle. Oncology 52: 123-127

Lane DP (1992) p53, guardian of the genome. Nature 358: 15

Levine AJ (1990) The p53 protein and its interactions with the oncogene products of the small DNA tumor viruses. Virology 177:419-426

Levine AJ, Momand J, Finlay, CA (1991) The p53 tumor suppressor gene. Nature 351: 453-456

Levine AJ. (1995) Tumor suppressor genes. In: Mendelsohn J, Howley PM, Israel MA, Liotta LA (eds) The molecular basis of cancer. Saunders, Philadelphia, London, Toronto, Montreal, Sydney, Tokyo, pp 86-104

Marx J (1992) Learning how to suppress cancer. Science 261: 1385-1387

Midgley CA, Fisher CJ, Bartek J, Vojtesek B, Lane DP, Barnes DM (1992) Analysis of p53 expression in human tumors; an antibody raised against human p53 expressed in Escherichia coli. J Cell Sci 101: 183-189

Orita M, Sekiya T, Hayashi K (1990) DNA sequence polymorphisms in ALU repeats. Genomics: 8: 271-278

Orita M, Suzuki Y, Sekiya T, Hayashi K (1989) Rapid and sensitive detection of point mutations and DNA polymorphisms using the polymerase chain reaction. Genomics 5: 874-879.

Sambrook J, Fritsch EF, Maniatis T (1989) Molecular cloning: a laboratory handbook. Cold Spring Harbor Press, pp13.3-13.104

Sanger F, Nicklen S, Coulson AR (1977) DNA sequencing with chain-terminating inhibitors. Proc Natl Acad Sci USA 74: 5463-5467

Scholz RB, Milde-Langosch K, Jung R, Schlechte H, Kabisch H, Wagener C, LöningT (1993) Rapid screening to Tp53 mutations by temperature-gradient gel electrophoresis: a comparison with SSCP analysis. Hum Molec Genetics 2: 2155-2158.

Slingerland JM, Minden MD, Benchimol S (1991) Mutation of the p53 gene in human acute myelogenous leukemia. Blood 77: 1500-1507

Soto D, Sukumar S (1992) Improved detection of mutations in the p53 gene in human tumors as single-stranded conformation polymorphs and double stranded heteroduplex DNA. PCR Methods and Applications 2: 96-98

Soussi T, Legros Y, Lubin R, Ory K, Schlichtholz B (1994) Multifactorial analysis of p53 alteration in human cancer: a review. Int J Cancer 57: 1-9

Vojtesek B, Bartek J, Midgley CA, Lane DP (1992) An immunochemical analysis of the human nuclear phosphoprotein p53. New monoclonal antibodies and epitope mapping using recombinant p53. J Immunol Methods 151: 237-244

Wild CP, Ridanpää M, Antila S, Lubin R, Soussi T, Husgafvel-Pursiainen K, Vainio H (1995) p53 antibodies in the sera of lung cancer patients: Comparison with p53 mutations in the tumor tissue. Int J Cancer 64: 176-181

This paper represents a combination of the two presentations given at the Workshop on „Detection of Chemically-Induced Changes in p53 Expression: Methodological Aspects“, organised by the Molecular Toxicology and Mechanism Specialty Section of EUROTOX.

Norman Aldridge Memorial Symposium

Toxicologists versus Toxicological Disasters: Toxic Oil Syndrome

(Chair: E. Gelpí and M. Posada, Spain)

This Symposium has been dedicated to the memory of
Norman Aldridge († 30.6.1996)
EUROTOX Merit Awardee 1994

Toxicologists versus Toxicological Disasters: Toxic Oil Syndrome, Clinical Aspects

A. Gómez de la Cámara, I. Abaitua Borda and M. Posada de la Paz
Toxic Oil Syndrome Research Center. Instituto de Salud Carlos III, Ministerio de Sanidad y Consumo. Sinesio Delgado 6. 28029 Madrid. Spain

In spring 1981, the sudden and massive epidemic of the Toxic Oil Syndrome (TOS) struck Spain, mainly in the central and northwestern areas of the country. The majority of cases were diagnosed during May, June and July of that year, with only a few documented at later dates, including one as long as a year after the epidemic outbreak (Posada, 1989). Neither clinically nor pathologically had any such disease been previously recognized, nor had its characteristics been seen in animal models.

An infective etiology was first suspected, but exhaustive laboratory tests proved negative (Centers for Disease Control, Atlanta, USA, unpublished observations, 1981). Then, epidemiological studies pointed to its cause being rapeseed oil denatured with 2% aniline and further processed before being illicitly sold in 5- litre plastic containers by itinerant traders (Tabuenca, 1981).

The incidence reached a peak about one month after the first case was reported, then rapidly declined. This decline could be attributed to two factors: first, the official warning about this particular oil and its subsequent withdrawal from the market; and second, the dilution of this oil by the purveyors a few days before the warning was issued.

The conclusion was that the etiology was toxicologically induced, presumably by some agent in the oil. Isolated cases of unknown etiology but with an analogous clinical presentation in some respects have been described elsewhere (Bascom, 1988; Allen, 1989) but these have not so far helped to pinpoint the cause of TOS. On the other hand, the relationship between TOS and the eosinophilia-myalgia syndrome (Hertzman, 1990) makes even more interesting the search for the causal agent of TOS.

Previous clinical accounts of TOS have been derived largely from reports on hospital patients. This paper summarizes the descriptive epidemiological data, gives an account of a systematic 31-month follow-up of a group of patients that includes other relevant clinical data published since 1983, and discusses the current status of TOS patients.

Incidence and Mortality

A total of 20 643 people were afflicted with the disorder, with a female:male ratio of 1.5:1. This sex difference was more pronounced in those around 40 years of age. When the epidemic began, 21.8% of the cases were in those under 15 years of age (Fernadez Rubio, 1983).

Accurate figures on all deaths from TOS have been recently assembled. After 1983, mortality rates seem to be lower than those for the Spanish population at large (after exhaustive researching, this fact is confirmed but not explained; I. Abaitua, Fondo de Investigación Sanitaria, unpublished observation). The total number of deaths registered among censused TOS patients was 1518 during the period 1981-1993.

Patients died from several causes. During the acute phase respiratory failure from noncardiogenic oedema was most common, and in the intermediate phase thromboembolic incidents were frequently fatal. In the chronic phase, death has usually been due to respiratory failure, with secondary infectious and/or haemorrhagic complications, as a consequence of gross respiratory insufficiency from nerve and muscle dysfunction. A form of pulmonary hypertension with many clinical and pathological similarities to primary pulmonary hypertension has figured prominently among recent deaths (Gómez-Sánchez, 1989).

Evolution of the Disease

Systematic data of the entire affected population are available at present. Two random samples (one of adults, the other of children: Meeting of the Liason Group of the WHO Scientific Steering Committee for the Toxic Oil Syndrome, Madrid, 27-28 January 1987. Copenhagen, WHO Regional Office for Europe, 1987, unpublished document SPA/CEH 502) were selected among those patients diagnosed within the first four months of the epidemic (Villamor, 1982). From these samples, 914 patients completed the follow-up programme till month 31 after onset of the disease. Clinical findings are summarized in Tables 1-4. Percentages reported in these tables are estimated from both samples together. They represent averages weighted to the real proportion of affected children and adults in the population.

TOS is a multisystem disease with three successive phases in its evolution but with a common physiopathological link (Table 1) (Kilbourne, 1983; Toxic Epidemic Syndrome Study Group, 1982; Tabuenca, 1983 a, Serrano-Rios, 1984).

Table 1. Major clinical signs of TOS as it develops

Acute phase (months 1-2)	Intermediate phase (months 2-4)	Chronic phase (month 4 onwards)
Fever Rash	Skin oedema Alopecia Sicca syndrome Pulmonary hyper-tension	Scleroderma Dysphagia Sicca syndrome Pulmonary hyper-tension
Myalgia	Myalgia Sensory neuropathy Weight loss Hepatopathy Thromboembolism	Peripheral neuropathy Weight loss Hepatopathy
Eosinophilia	Eosinophilia	Eosinophilia (gradually disappears)

Acute phase

The notable early symptoms (1-2 months after onset) were nonproductive cough, pain, tightness of the chest and dyspnoea. Radiologically, the lungs showed a widespread, alveolar-interstitial infiltration, unilaterally or bilaterally distributed, with conspicuous Kerley A and B lines and often a pleural effusion, but with a normal cardiac silhouette.

Nonrespiratory complaints were general malaise, asthenia, anorexia, headache, tissue oedema, pruritis, a polymorphous rash, arthralgia, myalgia, muscle cramps, and sometimes hepatomegaly and lymphadenopathy. Quite early on, many of these patients suffered from abdominal pains with and without diarrhoea. The most prevalent clinical signs are given in Table 2.

Only about 1% of patients exhibited central nervous system dysfunction, seen as confusion and stupor and measured by a diffuse slowing of the electroencephalogram. The cerebrospinal fluid remained normal. Resolution of this central nervous system dysfunction was complete.

Haematologically, the eosinophilia in the peripheral blood exceeded 500 cells/mm^3, and in a large number of patients it rose to over 3000 cells/mm^3. Other laboratory findings were high levels of immunoglobulin E, hypertriglyceridaemia, thrombocytopenia and elevated hepatic transaminase levels.

Table 2. Most prevalent clinical signs during the acute phase (N = 914)

Clinical sign	Percentage
Eosinophilia	86.3
Pulmonary oedema	72.4
Myalgia	71.5
Fever	43.9
Rash	41.5

Intermediate phase

Approximately 59% of these patients developed some of the clinical features of the intermediate phase (2 - 4 months after onset of TOS). This clinical stage was characterized by pulmonary hypertension (Gómez-Recio, 1983), which was more common among children and young adults, by thromboembolic phenomena (Castro, 1986), and by oedema of the skin progressing to thickening and loss of flexibility, with or without whitish papular lesions interspersed with hyperpigmented areas, to give rise to what has been termed a guttate skin (Fonseca, 1983). Alopecia and severe myalgia were experienced, together with muscular weakness, hyporeflexia, alterations in superficial and deep sensibility, dysphagia, sicca syndrome, severe weight loss without apparent cause, arterial hypertension in young patients without other contributory factors, Raynaud's phenomenon, and endocrinological disturbances such as hyperglycaemia, hypertrichosis and amenorrhoea (Table 3).

Pancreatic enzyme changes were often detected, along with glucose intolerance. Autopsies carried out at this stage showed histological disorganization of the gland.

A number of patients developed slowly progressing dysphagia. Manometry revealed uncoordinated motility in the middle and lower regions of the oesophagus, apparently due to atrophy by denervation and/or fibrosis of the oesophageal wall (Diaz-Ruslo, 1986; Abaitua, 1986).

At this stage, deaths occurred with manifestations of mesenteric thrombosis and ischaemic colitis. Laboratory tests pointed frequently to hyperglycaemia, hypertriglyceridaemia, hypercholesterolaemia, hypoalbuminaemia, thrombocytopenia (or thrombocytosis), unabated eosinophilia, and hepatic enzyme abnormalities with raised transaminases and cholestasis.

Table 3. Most prevalent clinical signs during the intermediate phase (N = 914)

Clinical sign	Percentage
Myalgia	48.8
Weight loss	36.3
Hepatopathy	20.7
Sicca syndrome	19.6
Skin oedema	13.6
Sensory neuropathy	9.6
Alopecia	9.5
Pulmonary hypertension	3.1
Thromboembolism	0.9

Chronic phase

The intermediate phase merged almost imperceptibly into the chronic phase between months 4 and 6 after the initial onset of the disease. Percentages of the most prevalent signs among the 914 patients selected for follow-up are given in Table 4. This phase seems to persist indefinitely in many patients, with increasing disability and sometimes death from the incapacitating complications. In others, however, the symptomatology has improved and physical abilitv has been more or less restored.

Table 4. Most prevalent clinical signs during the chronic phase (N = 914)

Clinical sign	Percentage
Peripheral neuropathy	37
Hepatopathy	32
Scleroderma	22
Pulmonary hypertension	10

Moreover, the significant features of the chronic phase are pulmonary arterial hypertension (Martín, 1986), atrophy and induration of the skin of a sclerodermic nature (Conde, 1983; Iglesia, 1983) and muscular wasting and weakness, with hyporeflexia or areflexia and some sensory neuropathy (Cruz,

1984). Also seen is respiratory distress arising from neuromuscular deficit, together with limb deformity and muscle retraction (Robles, 1984). Muscle tremor, cramp and myotonia have also been described (Leiva, 1985).

Some of the most severely afflicted patients have had a 90% reduction in lung function and required intubation, which has frequently led to complications from infection. At the same time, any eosinophilia that persisted through the intermediate phase gradually disappears in the chronic phase. Some cases have shown marked liver derangement, with chronic hepatitis, cholestasis or nodular regenerative hyperplasia (Solis, 1987). With involvement of the salivary glands the sicca syndrome appears (Oliva, 1988), with ensuing dental caries (Barberie, 1985). Other secondary features include osteoporosis (Mateo, 1984), osteonecrosis (Leal, 1984), carpal tunnel syndrome (Olmedo, 1983), acro-osteolysis (Alonso, 1986) and defective memory (Ser, 1986 a).

Treatment

None of the treatments tried (e.g. corticosteroids, azathioprine, D-penicillamine, plasmapheresis, vitamin E, superoxide dismutase, vasodilators, analgesics, antiinflammatories) produced any convincing effect in the control of the disease. Steroids administered during the acute and intermediate phases of the disease did not appear to prevent the chronic phase, although the general thought was that they reduced eosinophil levels, as well as producing an improvement in patients with pulmonary oedema. On the other hand, one well designed study (A. Luengo Dos Santos, unpublished observations, 1985) indicated control of the involuntary muscular movements by diphenylhydantoin (but not valproic acid) as long as the serum levels of the drug were maintained.

The only really effective relief to patients with severe neuromuscular involvement is physical rehabilitation (Pastrana, 1984).

Current Status

Alimentary system

While the biochemical parameters of hepatic function have, in general, slowly improved, cholestasis and portal hypertension can persist. Cirrhosis and nodular regenerative hyperplasia are present in some patients (Solis, 1986, Velicia, 1986). Complaints of, for example, pyrosis, dysphagia and intestinal rhythm disorders are probably due to the uncoordinated functioning of the alimentary tract musculature (Vidal, 1987). Autopsy findings point to occasional lesions within the pancreas, while hyperglycaemia in some patients may be the result of islet cell damage.

Cardiorespiratory system

The interstitial fibrosis seen in most organs has not been found in the lungs. Nevertheless, pulmonary hypertension, with clinical and pathological characteristics similar to those of primary pulmonary hypertension, has been seen in the chronic phase. Some of these patients have exhibited a low carbon monoxide diffusion factor for a few years. Throughout the chronic phase, the incidence of this hypertension has been about 10%, but at present only about 1.2% of the patients are so affected (Martín, 1986). Complaints include palpitations, cardiac arrhythmias and dyspnoea on exertion, though their extent and importance are difficult to determine.

Nervous system

In the peripheral nervous system, the neuropathy that was secondary to vascular deprivation proceeded to complete denervation, with cramps and myalgias giving way to muscular atrophy and paresis. Correspondingly, involvement of the sensory nerves is evident as hyperaesthesia, hypoaesthesia and neuritic pain. During the chronic phase there had been some evidence of slow re-innervation in the form of cramps and myalgia (Leiva, 1985). Severe disability in the form of headaches, cramps, myalgia, myoclonia, polyneuropathy and strokes nevertheless continues to affect some patients.

Skeletal system

In contrast to the acute phase, when arthritis of the large joints was occasionally observed, the chronic phase is characterized dramatically by musculoskeletal deformities, contractures, arthralgia and spinal pain, upon which demineralization of the bones has become superimposed owing to immobilization (Molina, 1984; Martinez, 1984).

Dermal system

Symptoms attributable to the skin changes are mostly in remission, though atrophic areas are still found in the dermis.

Mental aspects

Not surprisingly, victims of this devastating disease are commonly beset with mental anxiety and depression, difficulties in adapting to their predicament and, understandably in some instances, compensation neurosis (López-Ibor, 1987 a); López-Ibor, 1987 b). Other psychological symptoms are insomnia, somnolence and memory disorders. These could simply be an expression of the incapacitating physical disease, bearing in mind that organic changes in the central nervous system during the acute phase might be responsible (Ser, 1986 b).

Most of the affected persons recovered from the acute and intermediate phases but some conditions, - appearing either alone or in various combinations - as cramps, paresthesias, contractures, sclerodermiform changes of the skin, neuropathy, and muscle wasting persisted in a sizeable number of patients, many of whom remained permanently deteriorated.

Data from a very recent large study involving 4015 patients (A. Gómez de la Cámara, unpublished observation) shows that myalgias and paresthesias are still very common symptoms affecting over sixty percent of the affected people. There exist longterm sequelae mainly focused on sclerodermiform changes of the skin, neuropathy and contractures. We can say at this point that around twenty percent of the affected population have still, in different degrees, clinical manifestations of the disease. It is possible to stress this argument from another point of view. Patients had claimed severe psychological, physical and social sequelae wich affected their normal lives. A well-being questionnaire, the Nottingham Health Profile was applied to a representative sample of 840 patients (A. Gómez de la Cámara, unpublished observation). The results showed that the scores were much higher that those of a sample of the general Spanish population. They express a clear pattern of very afflicted self-perceived state of health which increases according the degree of clinical affectation.

Conclusions

After 15 years the survivors are, for the most part, showing progressive clinical improvement. So far, no indications of excess stillbirths (Guerra, 1985) or infantile deformities (Tabuenca, 1983 b; Martinez, 1982) have been reported. What is more uncertain is the possibility of long-term vascular, neoplastic and psychiatric effects. Moreover, within the limitations of the inadequate statistics and the relatively short period of observation, no obviously elevated rate of cancer has yet been noted among these people.

Doubts of this kind necessitate an assiduous follow-up of as large a number of this patient cohort as possible to ascertain the nature of their fate, in addition to offering them the utmost care and assistance, medically and socially, to ameliorate their lot.

References

Abaitua Borda I et al. (1986) Esophagus and toxic oil syndrome. Digestive diseases and sciences, 31: 443.

Allen JN et al. (1989) Acute eosinophilic pneumonia as a reversible cause of noninfectious respiratory failure. New England Journal of Medicine, 321: 569-574.

Alonso Rulz A et al. 1986) Toxic oil syndrome: a syndrome with features overlapping those of various forms of scleroderma. Seminars in arthritis and rheumatism, 15: 20~212 .

Barberie Leache E et al. (1985) Estudio de las consecuencias bucodentarias del síndrome tóxico epidémico [Study of dental damage in the toxic oil syndrome] Revista de Sanidad e Higiene Pública, 59 815-833.
Bascom R et al. (1988) Eosinophilia, respiratory symptoms and pulmonary infiltrates in ubber workers. Chest, 93: 154-158.
Castro García M et al. (1986) Hypercoagulable states and the toxic oil syndrome. Annals of Internal Medicine. 104: 730.
Conde Zurita JM (1983) Síndrome tóxico. Manifestaciones cutaneas de la fase tardía: histología [Toxic oil syndrome. Cutaneous manifestations in the late phase: histology]. Actas dermosifilográficas, 74: 357-359.
Cruz Martínez A et al. (1984) Neuromuscular disorders in a new toxic syndrome: electrophysiological study - a preliminary report. Muscle and Nerve, 7: 12-22.
Diaz-Ruslo M et al. (1986) Trastorno motor esofágico en pacientes afectos de síndrome del aceite tóxico [Esophageal motor disorders found in toxic oil syndrome patients]. Anales de medicina interna, 3: 53-57.
Fernandez Rubio F et al. (1983) Epidemiología descriptiva del síndrome tóxico [Descriptive epidemiology of the toxic oil syndrome]. Anales del Instituto de Medicina Benéfica, 18: 49-71.
Fonseca Capdevila E (1983) Manifestaciones cutáneas del Síndrome Tóxico por aceite de colza adulterado [Cutaneous lesions in the toxic oil syndrome caused by adulterated rapeseed oil]. Actas dermosifiliográficas,74: 297-316.
Gomez de la Camara A (1995) Toxic Oil Syndrome Research Center. Instituto de Salud Carlos III. Ministerio de Sanidad y Consumo. Unpublished observation..
Gomez-Sanchez MA et al. (1989) Pulmonary hypertension due to toxic oil syndrome. A clinicopathologic study. Chest, 95: 325-331.
Gomez Recio M et al. (1983) Hipertensión pulmonar en el síndrome tóxico [Pulmonary hypertension in the toxic oil syndrome]. Archivo Facultad de Medicina de Madrid, 41: 237-244.
Guerra Flecha JM and Aguaron de la Cruz A (1985) Repercusiones del síndrome tóxico sobre la gestación [Repercussions of the toxic oil syndrome on gestation]. Revista colombiana de obstetricia y ginecología, 36: 48-51.
Hertzman PA et al. (1990) Association of the eosinophilia-myalgia syndrome with the ingestion of tryptophan. New England Journal of Medicine, 322: 869-873.
Iglesias JL and De Moragas JM (1983) The cutaneous lesions of the Spanish toxic oil syndrome. Journal of the American Academy of Dermatology, 9: 159-160.
Kilbourne EM et al. (1983) Clinical epidemiology of toxic-oil syndrome. New England Journal of Medicine, 309: 1408-1414.
Leal Gonzalez L and Sampedro Santos S (1984) Hallazgos radiológicos en el sistema osteoarticular de enfermos del síndrome del aceite tóxico [Radiological findings in the osteoarticular system in toxic oil syndrome patients]. Rehabilitación, IX: 551-560.
Leiva C et al. (1985) Abnormal movements in the Spanish toxic oil syndrome. Journal of Neurology, 232: 242.
Lopez-Ibor Aliño JJ (1987 a) Social reinsertation after catastrophes. The toxic oil syndrome experience. European Journal of Psychiatry, 1: 12-19.
Lopez-Ibor Aliño JJ and Jimenez Arriero MA (1987 b) Psychosocial rehabilitation in disasters: experience of the Spanish toxic oil syndrome. International disability studies, 9: 81-83 .
Martín Escribano P and Díaz de Atauri MJ (1989) Síndrome tóxico. Años después [Toxic oil syndrome. Years later]. Archivo bronconeumológico, 25: 143-146 .
Martínez Florez A et al. (1984) Secuelas actuales del síndrome tóxico en niños [Current sequelae of the toxic oil syndrome in children]. Rehabilitación, 18: 519-525.

Martínez Frias ML et al. (1982) Spanish toxic oil and congenital malformations The Lancet, 2: 1349.
Mateo IM et al. (1984) Toxic epidemic syndrome: musculoskeletal manifestations. Joumal of Rheumatology, 11: 333-338.
Molina Ariño A et al. (1984) Secuelas del miembro inferior en el síndrome tóxico [Sequelae in the lower limbs in the toxic oil syndrome]. Rehabilitación, 18: 469-471.
Oliva Aldamiz H et al. (1988) Patología de la glándula salival en el síndrome por aceite tóxico adulterado en España: una forma de síndrome de Sjögren secundario [Pathology of the labial salivary gland in the Spanish toxic oil syndrome: a form of secondary Sjögren's syndrome]. Revista clínica espanola, 182: 71-78.
Olmedo Garzon FJ et al. (1983) The toxic-oil syndrome: a new cause of the carpal-tunnel syndrome. New England Journal of Medicine, 309: 1455.
Pastrana R (1984) Técnicas y programas de rehabilitación utilizadas en el tratamiento del síndrome tóxico. [Rehabilitation programme and techniques used in the treatment of the toxic oil syndrome]. Rehabilitación, 18: 491-496.
Posada de la Paz M et al. (1989) Late cases of toxic-oil syndrome: evidence that the aetiological agent persisted in oil stored for up to one year. Food and Chemical Toxicology, 8: 517-521.
Robles Gomez E (1984) Síndrome tóxico. Secuelas en miembro superior [Toxic oil syndrome. Sequelae in upper limbs]. Rehabilitación, 18: 473-482.
Ser T del et al. (1986 a) Trastornos de memoria en el síndrome del aceite tóxico (SAT) [Memory disorders in the toxic oil syndrome]. Archivo neurobiológia, 49: 19-39.
Ser Quijano T del et al. (1986 b) Evolución de la afección neuromuscular en el síndrome del aceite tóxico [Evolution of neuromuscular effects in the toxic oil syndrome]. Medicina clínica, 87: 231-236.
Serrano-Ríos M and Faro V (1984) Clinical and pathological features of TOS. In: Grandjean P and Tarkowski S (eds.) Toxic oil syndrome. Mass food poisoning in Spain. Copenhagen, WHO Regional Office for Europe, pp. 53-74.
Solis Herruzo JA et al. (1987) Clinico-biochemical evolution and late hepatic lesions in the toxic oil syndrome. Gastroenterology, 93: 558-568.
Solis Herruzo JA et al. (1986) Nodular regenerative hyperplasia of the liver associated with the toxic oil syndrome: report of five cases. Hepatology, 6: 687-693.
Tabuenca Oliver JM et al. (1983 a) Algunos aspectos del estado actual de los conocimientos del síndrome tóxico [Current aspects of research on the toxic oil syndrome]. Archivo Facultad de Medicina de Madrid, 41: 225-228.
Tabuenca Oliver JM et al. (1983 b) Spanish toxic oil and congenital malformations. Lancet, 1: 181 .
Tabuenca JM (1981) Toxic allergic syndrome caused by ingestion of rapeseed oil denatured with aniline. Lancet, 2: 567-568
Toxic Epidemic Syndrome Study Group. (1982) Toxic epidemic syndrome: Spain 1981. Lancet, 2: 697-702.
Velicia R et al. (1986) Hepatic disease in the Spanish toxic oil syndrome. A thirty months follow-up. Journal of Hepatology, 3: 59-65.
Vidal Ruiz JV et al. (1987) Prevalencia de clínica digestiva en la fase crónica del síndrome del aceite toxico [Prevalence of gastrointestinal manifestations in the chronic phase of the toxic oil syndrome]. Gastroenterología y hepatología, 10: 11-15.
Villamor Leon J et al. (1982) Descripción clínica de la enfermedad en adultos [Clinical description of the the illness in adults]. In: Simposium Nacional "Síndrome Tóxico" [National Symposium on the Toxic Oil Syndrome], Madrid, 1 l-12 June 1982. Madrid, Ministry of Health and Consumer Affairs, pp. 47-73.

Epidemiology of the Toxic Oil Syndrome

Rossanne M. Philen[1], Manuel Posada de la Paz[2], Robert H. Hill[1], Helen H. Schurz[1], Ignacio Abaitua Borda[2], Agustín Gómez de la Cámara[2] and Edwin M. Kilbourne[1]

[1] Centers for Disease Control and Prevention, Public Health Service, U.S. Department of Health and Human Services, Atlanta, Georgia, U.S.A. 30341

[2] Instituto de Salud Carlos III, Subdirección General de Salud, Centro de Investigación para el Síndrome del Aceite Tóxico, Ministerio de Sanidad y Consumo, Calle Sinesio Delgado 6, E-28029 Madrid, SPAIN

Introduction

During the early 1980's importation of rapeseed oil into Spain for human consumption was illegal, except for the Canary Islands, although denatured rapeseed oil could be legally imported for industrial use. Denaturation of imported food oils with castor oil, methylene blue, or aniline was required by Spanish law and indicated that the oil was intended for industrial use due to undesirable properties of the denaturants.

In the early 1980's some fraudulent trade in food oil existed in Spain. Many people unknowingly purchased either olive oil mixed with other seed oils and sold as pure olive oil, or other vegetable oils which were marketed as pure olive oil from itinerant salesmen or through *mercadillos* (local outdoor markets). In late 1980 and early 1981 importation of aniline denatured rapeseed oil, presumably for industry, increased. Much of this oil was later discovered to have been diverted for human consumption through the area of Catalonia.

The Epidemic

The toxic oil syndrome (TOS) epidemic began on May 1, 1981, when an 8 year old boy was pronounced dead on arrival at the Hospital del Rey in Madrid (Dirección General de Planificación Sanitaria 1981). Hours later, five other members of his family were admitted to the same hospital with symptoms suggestive of an atypical pneumonia. Within a week hundreds of people a day were being admitted to the hospital with the syndrome. Although *Mycoplasma pneumoniae* or an unknown infectious agent was initially suspected to be the cause of the epidemic (Dirección General de Planificación Sanitaria 1981), patients failed to respond to antimicrobial therapy and no microbial agent was identified.

A pediatrician at Madrid's Niño Jesús children's hospital, Dr. Manuel Tabuenca, soon noted that children diagnosed with TOS developed a clinical syndrome different from adults (Tabuenca 1984). Pulmonary symptoms were less common in children, but they often developed a rash such as might be seen with a toxic exposure. Many children developed TOS, although it was rare in babies under six months old. Thus, when an infant less than six months old was diagnosed with TOS, investigators focused on the exposures the baby might have had in common with older children and adults. They found that the infant's grandmother had supplemented the baby's formula with food oil. This led to a study in which the mothers of children hospitalized with TOS and mothers of children admitted to the same hospital for trauma or surgery were compared. This case-control study showed that consumption of a specific type of cooking oil sold by itinerant street vendors was a risk factor for developing TOS (Casado Flores 1982).

Following this discovery, oils from households and *mercadillos* were analyzed in the Central Customs laboratory; these oils were determined to be rapeseed oil, denatured with 2% aniline, as required by Spanish law at that time, to prevent human consumption. On June 10, 1981, 40 days after the epidemic started, and in an attempt to prevent further cases of TOS, the Ministry of Health and Consumer Affairs alerted the population through the mass media about the relationship of the unlabeled oil with the epidemic.

Descriptive epidemiology

Over 20,000 people of both sexes and all ages were affected, although TOS was rare in children under 6 months old. Women were affected more often than men in a 1.5 to 1 ratio, and people in their twenties and thirties, in particular women between 30 and 40 years old, were more often affected than men of that age.

The implicated oil was marketed as pure olive oil, but at a much cheaper price, which likely explains why middle to lower socioeconomic groups were the most affected. Cases clustered in families or households, suggesting to investigators that the exposure factor was probably closely related to household life. Supporting this view was the finding that school students, military personnel, and others who are typically at high risk of infectious diseases were not disproportionately affected.

Geographic distribution: The first cases were reported in an area 20 kilometers northwest of Madrid (Grandjean 1984). Early descriptive studies showed that TOS occurred mainly in residents of towns around the capital city of Madrid. Many cases were also found to be associated with areas where weekly *mercadillos* were located. Later, the geographic distribution of the epidemic was closely linked to the secondary road network in Spain s central and northwestern areas where itinerant salesmen in local *mercadillos* sold unlabeled oil in 5 liter plastic containers (Canas 1987).

Fourteen of Spain's 52 provinces reported cases of TOS as shown in figure 2. The province of Madrid reported 14,294 (71.1%) of 20,096 cases (Grandjean 1984). Segovia and Palencia had the highest incidence rates. The few cases reported from the rest of the country were in relatives of persons who lived in the provinces where the epidemic occurred and who had either consumed oil in the epidemic area during a visit there or had brought oil to their residence from an affected area (Grandjean 1984). Two workers at the ITH oil refinery in Seville, the refinery most connected with the epidemic, took oil from an implicated lot home for personal use and later became ill as well (Posada 1989).

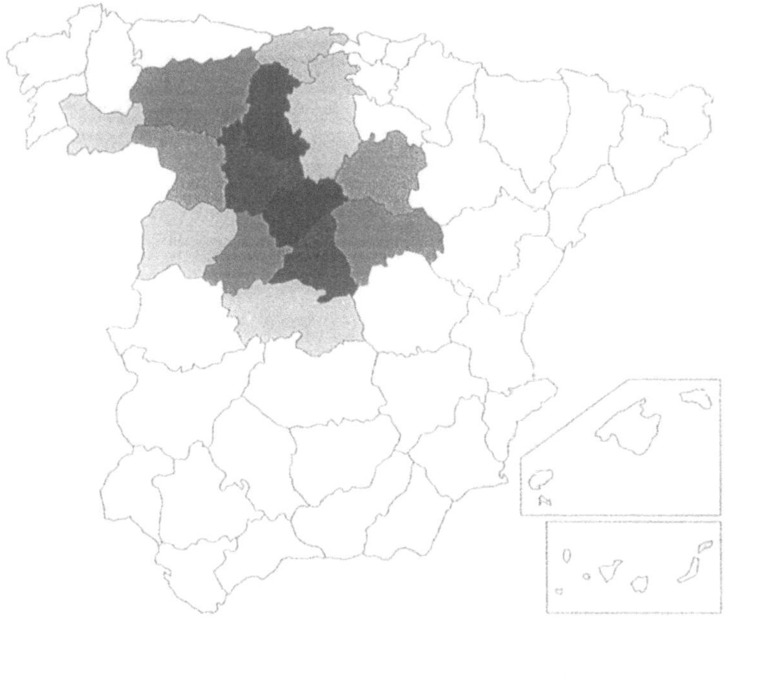

Figure 1. Rates of toxic oil syndrome by province, Spain, 1981

Epidemic curve: Two French oil companies, which sold rapeseed and other oil for human consumption in France, also sold 2 percent aniline denatured rapeseed oil to Spanish intermediaries who brought the oil overland through the north of Spain (Posada 1991). Other oils which were not denatured were imported from Germany and Switzerland through the port of Barcelona. In Catalonia the oils went to various refineries where denatured oils were first diluted with other oils, including rapeseed oil which had not been denatured, and then all oils were refined for resale for human consumption and sold through a network of distributors and traveling salesmen. This became known as the Catalonian circuit, however, oil from this circuit was never associated with TOS.

In late 1980, RAELCA, an oil distributor in Madrid, entered the illicit oil sales market and dealt with the Catalonians and other oil importers. RAELCA soon began to import oil directly from France and to use its own networks to refine the oil (Posada 1991). In March 1981, RAELCA bought five lots of aniline denatured rapeseed oil from two French food oil companies. Three of these lots were shipped in three trucks to the ITH oil refinery in Seville, and two were sent to the Danesa Bau refinery in Madrid. The first of the three lots was received by ITH on March 23rd, and the last on April 10th. The three lots were refined together at ITH and sent to RAELCA in two truck shipments on the 14th and 23rd of April, and could have been sold for consumption as early as April 15th (figure 2). In fact, a study of the few early cases who had disease onset before May 1 found that the first person became ill on April 23rd. The two lots of oil sent to Danesa Bau arrived there on April 24th and May 12th, were refined, and sent to RAELCA for bottling and distribution on May 19th and 20th. Thus, the Danesa Bau oil could have been sold no earlier than May 21st, three weeks after the epidemic began, and after the epidemic curve had already peaked, suggesting that the ITH refined oil led to the epidemic.

The epidemic curve shown in figure 2 is suggestive of a point source epidemic (Grandjean 1984). The curve begins to fall just before the official announcement of the oil-disease relationship on June 10, 1981. Most likely the number of new cases began to decline as the etiologic oil began to disappear from households and the market, and this may have coincidentally happened shortly before the announcement of the oil-disease relationship. The number of new cases fell rapidly once the announcement was made, however, many patients were still being admitted to hospitals and readmissions of those previously discharged in good condition occurred. These readmissions suggested to physicians that after discharge patients went home and were reexposed to the toxic oil. Investigation of readmitted cases showed that many people continued consuming the same oil after they went home.

Oil recall: When the Ministry of Health and Consumer Affairs realized that toxic oil was likely still in use by the public, they launched a program to remove all potentially toxic oil from public consumption. People who were concerned about their oil could exchange it for pure olive oil at government expense. Thus, the government launched an official recall of all potentially toxic oils.

The oil exchange program involved the collection and storage of millions of liters of suspect oil. This oil was often poorly kept after collection; much of the oil that was collected was later found to lack careful labeling. Failure to keep accurate records of the origin of oil samples hampered later epidemiologic studies performed with these oils (Posada 1996). However, the oil exchange program led to a drop in the number of new TOS cases as demonstrated by the fall of the epidemic curve.

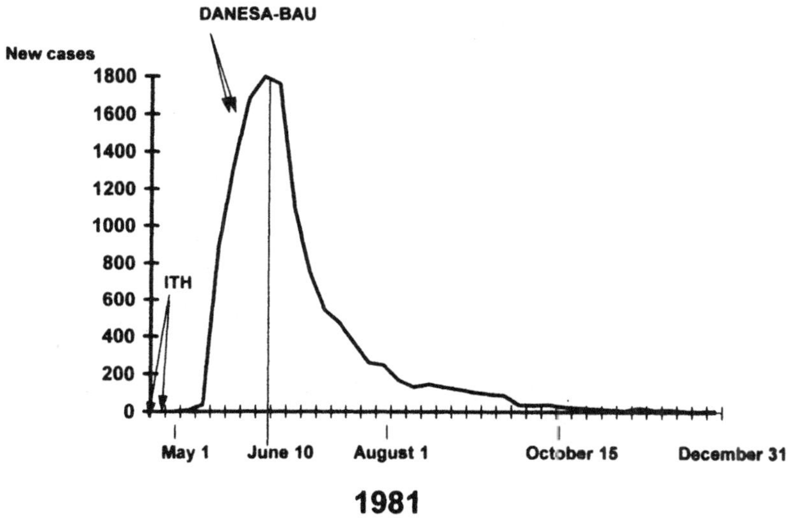

Figure 2. Epidemic curve, toxic oil syndrome, Spain, 1981

Mortality: Over 1,500 people in the TOS cohort have died from all causes since the epidemic began, giving a crude mortality rate of 7.5% (I. Abaitua, Centro de Investigación para el Síndrome del Aceite Tóxico, unpublished manuscript). Mortality was highest during 1981, with a standardized mortality ratio of 4.92 (95 percent confidence interval 4.39-5.50) compared with the mortality ratio of the Spanish population as a whole. Over the 10 year follow-up period of this study, overall mortality of the cohort was less than expected when compared with mortality among the general Spanish population.

Latency period: A presumed latency period has been estimated based on studies ofcertain groups such as convents of nuns (Diaz 1987), and on reports of family members who moved from a non-epidemic geographic zone to another location in the epidemic area. These cases allowed the dates of first oil consumption to be readily verified. Few cases were diagnosed 6 months to 1 year after the epidemic began. These late cases had bought oil at the time of the epidemic, but had stored the oil at home for a prolonged period and consumed it only a few days before becoming ill (Grandjean 1984). The latency period was thus determined to be less than 10 days, and likelier between 4 and 7 days (Grandjean 1984).

Analytic Epidemiology

Bottle shape: The case definition in table 1 includes the use of oil sold by itinerant street vendors as one criterion, leading to a large number of false positive case reports, as many people who reported having ingested an unlabeled oil did not become ill. However, a study of the shape of the plastic oil containers has recently shown that including the consumption of oil from a 5 liter plastic container without a sanitary registry number as part of the case definition was not enough to establish exposure to a toxic oil (Posada 1996).

The container study involved the chemical analysis of over 1000 oils from bottles of 10 different, but very similar shapes. Not all 5 liter bottles studied were found to have the same oil composition nor the same contaminants. In addition, the bottle shape most associated with illness was found to be one characteristic of the implicated RAELCA company. This study also found that not all bottles of oil from RAELCA contained contaminants. The finding of oil from RAELCA which was not contaminated would have tended, if anything, to bias the case control studies to the null, since many people who did report consumption of the implicated oil did not become ill (Posada 1996). Of 849 bottles from the implicated distributor, 319 (37.6 percent) contained oleyl anilides, whereas extremely few of the other bottle shapes studies contained oleyl anilides. In 1981, however, this detailed knowledge about the specific contents of similar oil bottles was not available and although it probably affected all the case control studies, it also serves to further strengthen the oil-illness association that was found. A further relationship between RAELCA and the implicated containers was established when a plastic injection mold, capable of making the exact bottles which were found to contain the greatest amounts of contaminants, was found on the company's premises (Posada 1996) as mentioned above.

Case-control studies: Prior to the official announcement about the origin of the epidemic, cases were identified by their pulmonary symptoms and the presence of eosinophilia. Two important case-control studies were carried out during the first 40 days of the epidemic. The key study (number 1 in table 2), was performed in the Niño Jesús pediatric hospital and examined the hypothesis of a possible toxic ingestion via food (Casado Flores 1982). This was the first study to suggest a relationship between the unlabeled oil sold in *mercadillos* and TOS.

Fifty days after the first case was reported, and subsequent to the Niño Jesús study and the public announcement of the oil-disease relationship, 9 other case-control studies (numbers 2-10 in table 2) were done using similar methods but performed in different geographic areas by different research teams. These studies used a case definition of the diagnosis of an atypical pneumonia. All of the studies showed a strong association between development of TOS and the ingestion of oils sold by itinerant street vendors in 5 liter plastic containers not subject to sanitary control. Study 2 in table 2 was done by an investigator who did four studies with the same case-control group (Posada 1987). This investigator first examined various hypotheses including respiratory transmission, but not including oil consumption, and obtained no significant findings. However, upon learning of the oil-disease relationship that was found at the Niño Jesús, a second study of the same case-control group was done, confirming those results.

Since oil was typically purchased by one family member, but was used in family meals, individuals were usually unaware of the quantity of oil they consumed, thus in case-control studies 2 through 10 the family was used as the case unit instead of an individual. Consumption patterns of food oils hampered later attempts to study the dose-response relationship of the oil to TOS (Doll 1987).

Table 1. Case definition of toxic oil syndrome proposed by the Spanish clinical commission, August 3, 1981.*

Major criteria

1. Consumption of oil presumed toxic before onset of illness or occurrence of the illness in the nuclear family
2. Pulmonary pathology with radiologic findings of diffuse interstitial or alveolar interstitial infiltrates, with or without pleural effusion
3. Incapacitating myalgias with functional impairment
4. Eosinophil count greater than 500 eosinophils per mm^3

Minor Criteria

1. Epidemic outbreak in the community
2. Severe skin itching
3. Rash or localized edema of skin
4. Severe and persistent mouth dryness
5. Minimal or moderate myalgias
6. Neurologic pathology
7. Abdominal pain
8. Clinical or analytical signs of hepatic involvement
9. Recent onset of exertional dyspnea
10. Recent onset of hypoxemia
11. Pulmonary hypertension
12. Cardiomyopathy
13. Vascular thrombosis

* two major or one major and four minor criteria were required for clinical case definition.

The studies in table 2 were all done independently, making the possibility of a systematic bias throughout the studies unlikely. Although recall bias is possible, the overall consistency of the studies and the highly significant p values reported in each study (all had $p < 0.0252$ or smaller) support the conclusion that neither recall bias nor any other bias or confounder could have accounted for the findings (Rigau 1984a).

Table 2. Case-control studies of toxic oil syndrome, Spain.

Study (reference)	First Author	Location ofStudy	Study Unit	Frequency Cases	Frequency Controls
1 (Casado Flores 1982)	Casado-Flores	Madrid	Individual	62/62 (100%)	4/62 (6%)
2 (Rigau 1984a)	Rigau-Pérez	Navas del Marqués (Avila)	Family	27/27 (100%)	16/38 (42%)
3 (Doll 1987)	Andrés-Martinez	Pozuelo de Alarcón (Madrid)	Family	42/48 (88%)	32/96 (33%)
4 (Doll 1987)	Sanchez Mozo	Madrid	Family	52/58 (90%)	615/1,725 (36%)
5 (Doll 1987)	Catalá Villanueva	Chozas de Abajo (León)	Family	19/19 (100%)	15/15 (100%)
6 (Doll 1987)	------	Cerezo de Arriba (Segovia)	Family	13/13 (100%)	25/44 (57%)
7 (Doll 1987)	Garcia	San Cristobal de Entrevías (Zamora)	Family	10/10 (100%)	3/10 (30%)
8 (Doll 1987)	Diez	Bocigas de Perales (Soria)	Family	11/11 (100%)	22/33 (67%)
9 (Doll 1987)	Diez	Arconada (Palencia)	Family	18/18 (100%)	12/21 (57%)
10Lopéz-Abente (Doll 1987)		Colmenar Viejo(Madrid)	Family	16/20 (80%)	6/20 (30%)
11 Diaz de (Diaz 1987)	Madrid Rojas	Individual	25/35	0/56 (66%)	(0%)
12 Diaz de (Diaz 1987)	Madrid Rojas	Individual	42/43	0/70 (98%)	(0%)
13 Cañas (Cañas 1987)	Madrid	Family	8/8	72/204 (100%)	(35%)

Later case-control studies done in two convents (numbers 11 and 12 in table 2) (Diaz 1987) also confirmed that aniline denatured rapeseed oil was the vehicle of the etiologic agent. Isolated cases, which were outliers in terms of temporal and geographic distribution, also supported this finding (Grandjean 1984, Posada 1989).

A study done in Orcasur (number 13 in table 2) showed that of all the ways in which unlabeled 5 liter bottles of oil were sold, only oil sold by traveling salesmen was associated with illness (Cañas 1987). This is consistent with information obtained through an investigation of the distribution network by the Ministry of Health (which regulates oil sales) that the system used by RAELCA, the most implicated oil distributor, was distribution through traveling salesmen rather than through the usual outlets for food products, such as grocery stores. (Cañas 1987, Posada 1987)

Toxico-epidemiologic studies: Two case-control studies (Kilbourne 1988, Posada 1994) have shown a relationship between the concentration of the fatty acid anilide oleyl anilide, and the risk of developing TOS. These two studies identified possible toxic oils more precisely, and thus gave more accurate toxicologic results. The first of these two case-control studies conducted four years after the outbreak, used stored oil samples from case and non-case families. This study showed a dose-response relationship between the amount of an aniline-derived fatty acid, oleyl anilide, and the risk of becoming ill (Kilbourne 1988). This study, carried out in two Madrid locations, was replicated two years later by a study that included all the areas affected by the epidemic (Posada 1994). These two studies support both the conclusion that consumption of contaminated oil was important and that a definite dose-response relationship existed between the chemical compounds found in the oils and the risk of developing illness. These results are shown in table 3.

Table 3. Dose response relationship of oleyl anilide to illness, toxic oil syndrome, Spain, 1981.

	Oleyl anilide concentration, µg/g (parts per million)						
	0	1-100	101-600	601-1,200	1,201+	Total	Chi-Square Linear Trend
Case	11	2	3	6	7	29	
Control	48	6	7	3	0	64	
Odds ratio	1	1.45	1.89	8.73	Infinite	93	21.83

Recently, 3-(N-phenylamino)-1,2-propanediol (DEPAP), and other related chemical compounds have been reported in the same oils (Hill 1995). Some of these compounds were described earlier (Guitart 1991), however, recent analyses, using techniques not available in 1981, have established that DEPAP is a better marker of case-oils than oleyl anilide. The results of this work for the same set of case and control oils is shown in table 4. In addition, DEPAP has been found to be specific only to the ITH refinery in Seville, some of whose workers became ill from consuming oil they took home from lots that were implicated in the epidemic (Grandjean 1984).

Table 4. Dose response relationship of DEPAP to illness, toxic oil syndrome, Spain, 1981.

DEPAP concentration in μg/gram (parts per million)

	0	0.1-8.9	9.0-18.9	19.0+	Total	Chi Square Linear Trend
Case	12	4	6	7	29	
Control	58	4	2	0	64	
Odds ratio	1.0	4.83	14.5	Infinite	93	29.94

Other research: The TOS epidemic led to the establishment of a patient registry, also known as the official census, which has become an important tool for follow up studies. Without this documentation the epidemiologic studies would have been close to impossible. The official census has allowed the clinical follow up of a substantial portion of the cohort, as well as the thorough study of mortality in the cohort. The recently established Centro de Investigación para el Síndrome del Aceite Tóxico (Center for the Study of TOS) will be greatly aided in its progress by the ability to access the census for patient information.

TOS research continues in several areas. A thorough review and revision of the registry of patients has been conducted, which led to as complete as possible a listing of all TOS patients. This revised registry also contains a wealth of information about each individual TOS patient, and plans are in place to further analyze this data. The revised registry has also been a key factor in the success of the long term mortality study to examine the rates and causes of death of all members of the TOS cohort.

Long term follow up in TOS is also very important. The potential outcomes of TOS can be studied and documented, and follow up will assist researchers in the United States who are studying the evolution of eosinophilia-myalgia syndrome (EMS) (Philen 1993). The clinical resemblance of TOS and EMS is remarkable, and investigators have proposed the hypothesis that they may have related etiologies. Features of TOS and EMS are similar to those of several immunologic diseases. Studies of drug induced lupus (Jiang 1994) have suggested that environmental exposures may lead to connective tissue diseases and that these diseases may be related etiologically. The continued investigation of chemical compounds in TOS oils or in EMS associated L-tryptophan may one day reveal the etiologic agents in these diseases and enhance research of other connective tissue diseases whose causes are as yet unknown, including scleroderma, systemic lupus erythematosus, and rheumatoid arthritis.

Finally, we hope that the continued study of TOS may one day disclose the identity and source of the etiologic agent, and in so doing will allow prevent similar episodes in the future.

References

Bernert JT Jr, Kilbourne EM, Akins JR, et al (1987) Compositional analysis of oil samples implicated in the Spanish toxic oil syndrome. J Food Sci 52:1562-9

Cañas R, Kilbourne EM (1987) Oil ingestion and the toxic-oil syndrome: results of a survey of residents of the Orcasur neighbourhood in Madrid, Spain. Int J Epidemiol 16:3-6.

Casado Flores J, Casquero J, Colomar P, et al (1982) Síndrome Tóxico por consumo de aceite adulterado: Una encuesta alimentaria esclarecedora. (Spanish) Pediátrika 2:17-20

Díaz de Rojas F, Castro García M, Abaitua Borda I, et al (1987) The association of oil ingestion with toxic-oil syndrome in two convents. Am J Epidemiol 125:907-11

Direccion General de Planificación Sanitaria (1981) Propuesta de la Comisión Clínica. Síndrome Tóxico - Investigación. (Spanish) Ministerio de Trabajo, Sanidad y Seguridad Social; Madrid; May-August.

Doll R (1987) The aetiology of the Spanish toxic syndrome: interpretation of the epidemiological evidence (including addendum June 1987). Copenhagen, WHO Regional Office for Europe (unpublished document SPA/CEH 502)

Grandjean P, Tarkowski S, eds (1984) Toxic oil syndrome: mass food poisoning in Spain. In: Report of a WHO meeting: Madrid 21-25 March 1983. Copenhagen: World Health Organization Regional Office for Europe 22:3-16

Guitart R, Gelpi E (1991) Chemical composition of TOS related oils. In: World Health Organization, editor. Toxic oil syndrome: current knowledge and future perspectives. Copenhagen: World Health Organization Regional Publications 42:95-138

Hill RH, Schurz H, Posada M, et al (1995) Possible Etiologic Agents for Toxic Oil Syndrome: Fatty Acid Esters of 3-(N-Phenylamino)-1,2-Propanediol. Arch Environmental Contam Toxicol 28:259-64

Jiang X, Khursigara G, Rubin RL (1994) Transformation of lupus inducing drugs to cytotoxic products by activated neutrophils. Science 266:810-3

Kilbourne EM, Bernert JT Jr, Posada de la Paz M, et al (1988) Chemical correlates of pathogenicity of oils related to the toxic-oil syndrome in Spain. Am J Epidemiol 127: 1210-27

Lahoz C, Rose NR, Goter Robinson CJ (1991) Immunology. In World Health Organization, editor: Toxic oil syndrome: current knowledge and future perspectives. Copenhagen: World Health Organization Regional Publications, European Series 42:139-48

Philen RM, Posada M (1993)Toxic Oil Syndrome and Eosinophilia-Myalgia Syndrome: May 8-10, 1991, World Health Organization Meeting Report. Seminars in Arthritis and Rheumatism 23:104-24

Posada de la Paz M, Philen RM, Abaitua Borda I, et al (1991) Manufacturing processes at two French rapeseed oil companies: possible relationships to toxic oil syndrome in Spain. Food Chem Toxicol 29:797-803

Posada M, Philen RM, Abaitua I, et al (1994) A Further Study of Factors Associated with Pathogenicity of Oils Related to the Toxic Oil Syndrome Epidemic in Spain. Epidemiology 5:404-9

Posada M, Philen RM, Abaitua I, et al (1996) Toxic Oil Syndrome: A Study of Oil Composition by Container Type - Evidence for a Point Source Epidemic. Food and Chemical Toxicology 34:251-257

Posada M, Castro M, Kilbourne EM, et al (1987) Toxic-oil syndrome case reports associated with the ITH Oil Refinery in Sevilla. Food Chem Toxicol 25:87-90

Posada de la Paz M, Abaitua Borda I, Kilbourne EM, et al (1989) Late cases of toxic oil syndrome: evidence that the aetiological agent persisted in oil stored for up to one year. Food Chem Toxicol 27:517-21

Rigau Pérez JG (1984) Summary of case-control studies and case or cluster investigations. In: Grandjean P, Tarkowski S, eds. Toxic oil syndrome: mass food poisoning in Spain.In: Report of a WHO meeting: Madrid 21-25 March 1983. Copenhagen: World Health Organization Regional Office for Europe 22:47-51

Rigau-Pérez JG, Pérez-Alvarez L, Dueñas-Castro S, et al (1984) Epidemiologic investigation of an oil-associated pneumonic paralytic eosinophilic syndrome in Spain. Am J Epidemiol 119:250-60

Tabuenca JM (1984) Discovery of toxic oil as the cause of the epidemic. In: Grandjean P, Tarkowski S, eds. Toxic oil syndrome: mass food poisoning in Spain. Report of a WHO meeting: Madrid 21-25 March 1983. Copenhagen: World Health Organization Regional Office for Europe 22:39-46

Analytical Measurements of Products of Aniline and Triglycerides in Oil Samples Associated with the Toxic Oil Syndrome

Helen H.Schurz[1], Robert H. Hill[1], Rossanne M. Philen[1], Manuel Posada de la Paz[2], Ignacio Abaitua Borda[2], Edwin M. Kilbourne[1], Tom Bernert[1], and Larry L. Needham[1]

[1] Centers for Disease Control and Prevention (CDC), 4770 Buford Hwy, NE, Mailstop F17, Atlanta, GA 30341-3724

[2] Dirección General de Ordenación de la Investigación y Formación, Fondo de Investigación Sanitaria, Ministerio de Sanidad y Consumo, Sinesio Delgado 6, 28029, Madrid, Spain

Introduction

A shocking and previously unknown epidemic of food-related illness gained recognition in May of 1981. The toxic oil syndrome (TOS), as it was to become known, was caused by the consumption of an oil that was purported to be edible olive oil, but was actually aniline-denatured rapeseed oil. Since the epidemic, researchers worldwide have pursued the etiologic agent of this devastating disease (Guitart and Gelpi, 1991;Kilbourne et al., 1988; Bernert, et al., 1987). At the height of the epidemic, more than 20,000 people in the central and northwestern regions of Spain had been afflicted and, of these, 11,000 required hospitalization. Since the epidemic, more than 800 of the afflicted have died.

When TOS was first presented, it exhibited itself as an atypical pneumonia suggesting a bacterial or viral origin. As time passed, however, TOS developed into a severe chronic immunologic disease. Early exhaustive laboratory investigations discounted infectious agents as potential TOS agents. Complicating the diagnosis of TOS was a high eosinophilia count occurring among the victims that suggested a parasitic disease, yet laboratory testing for parasites was negative. Tabuenca (1981) was first to recognize an association between TOS and the consumption of a food oil that was being sold in *mercadillos* (outdoor markets). The Spanish government instituted a massive oil-exchange program to remove the tainted oil from the population. This recall removed most of the tainted oil from the market, and the number of new TOS cases decreased.

Soon thereafter, the oil associated with TOS was found to be technical-grade, aniline-denatured rapeseed oil that had been illegally sold as pure olive oil (Guitart and Gelpi, 1991). Further investigations showed that aniline and fatty-acid anilides were present in TOS oils, and these compounds became the first markers of toxic oils (Tabuenca, 1981;Ventura Diaz, 1982). As the investigation into the production of the toxic oils progressed, it became obvious that refined aniline-denatured oils had been sold for a period of time before the epidemic, but these oils did not produce disease. This meant that the markers that were being used to determine toxic oils, aniline and fatty acid anilides, did not always implicate actual toxic oils, and another agent or agents needed to be sought.

Compounds other than the aniline and fatty acid anilides were reported to be in the disease-producing oils, however, their association to TOS was not clear (Tabuenca, 1981). Recently, those compounds, the fatty-acid esters of 3-N-phenylamino-1,2-propanediol (PAP) – the products of aniline and triglycerides – have been shown to have a strong association with TOS. Even though the fatty-acid anilides and the PAP esters have been associated with TOS, the etiological agent(s) responsible for this epidemic have not been identified. The main barrier to identification of the agent(s) has been that no animal model exists that fully duplicates TOS.

In an effort to search for possible causative agents in TOS-associated oils, we used liquid chromatography with atmospheric pressure chemical ionization tandem mass spectrometry (LC-MS/MS) to measure fatty acid anilides and identify more than 20 PAP esters in these oils. For the first time, we also identified a class of compounds known as the PAP ester amides in these oils as well. Studies contained in this paper will describe the analytical work done in our laboratories at the CDC.

Methods and Materials

Samples and Standards: Oil samples, known as the Toxico-Epi-I oils that were collected during the 1981 epidemic (Bernert et al., 1987; Kilbourne et al., 1988) were used in our studies. Oil samples collected from the ITH refinery (the refinery most clearly associated with toxic oils) in Seville, Spain, were also used. These include samples of two aniline-denatured rapeseed oils collected before refining and one aniline-denatured rapeseed oil collected after the refining process (Hill et al., 1995). This latter refined oil from ITH is called the source-contaminated oil, and it is believed to be the primary source of all TOS oils. Before the epidemic, the refined oil was shipped to RAELCA, a packaging facility, where it was diluted with other oils and bottled for sale to the public. We also analyzed several oils from the Catalan region of Spain that were not associated with illness.

Using several tandem mass spectral scanning techniques (multiple reaction monitoring, daughter ion mass spectrometry, parent ion mass spectrometry) we tentatively identified probable structures for several of the compounds described in this paper. Through an agreement with the Fondo de Investigacion Sanitaria, we requested and were supplied with standards from the laboratory of Dr. Angel Messeguer of the Centro de Investigación y Desarrollo in Barcelona, Spain (Tabuenca, 1981). General structures of these compounds are shown in Figure 1, and Tables 1 and 2 provide the specific structures of all compounds. Each synthesized standard was prepared as a 10 µg/ml solution in toluene from stock solutions. Oil samples were diluted with n-propanol to a concentration of 50 mg oil/ ml n-propanol. The toxicity of these substances in humans is unknown; however they may be etiologic agents of TOS and should be treated with caution.

Analytical Methods: Chromatography was performed using a Whatman Partisil 5 ODS-3 25 cm liquid chromatography column (Whatman Inc., Fairfield, N.J.) and a flow rate of 1 ml/min. Analytical measurements and structural determinations were performed using a SCIEX API III+ tandem quadrupole mass spectrometer system with atmospheric pressure chemical ionization (APCI) and ionspray. The parent-ion scanning technique was used to tentatively identify new PAP esters as parents of selected daughter ions. Aniline was the denaturant in TOS-associated oils, and daughter-ions containing the anilino moiety were selected to find all aniline-derived compounds in these oils. The specific daughter-ions that we monitored were *m/z* 94 ($C_6H_5NH_3^+$) for the anilino moiety and *m/z* 132 ($C_6H_5NHCH_2CHCH^+$). Specific daughter-ions containing fatty-acid groups (examples: *m/z* 432 (oleyl), *m/z* 430 (linoleyl) and *m/z* 428 (linolenyl)) were used in parent-ion scanning to further identify types of fatty-acid esters present in the parent structure. Daughter-ion mass spectrometry of selected parents was then used to derive and confirm the proposed structures. Several of these compounds were compared with synthesized standards; standards were not available for all of the identified compounds. Further details of the experimental portion of this paper can be found elsewhere (Schurz et al., 1996).

Results

First Analyses of the Toxico-Epi-I Oils: The Toxico-Epi-I oils were initially analyzed by Bernert et al. (1987) for their fatty acid, sterol and fatty acid anilide content. The fatty acid profiles of the oils were measured, and it was found that oleic and linoleic acids predominated in the oils, indicating their adulteration with rapeseed oil. The relative concentrations were found to be oleic acid > linoleic acid > palmitic acid > stearic acid > linolenic acid. Most of the differences in concentration that were observed for the case and control samples were statistically significant.

The measured sterol content of the oil samples indicated that the oils were not only adulterated with rapeseed oil (presence of brassicasterol), but were also adulterated with animal fats (presence of cholesterol). The differences in concentration between the case and control samples for brassicasterol, campesterol and stigmasterol were also statistically significant. Three fatty acid anilides were quantified using HPLC analysis (oleyl anilide, palmityl anilide and linoleyl anilide), and later confirmed by GC/MS. All differences between case and control samples were statistically significant, as well, with the highest concentrations of the fatty acid anilides present in the case samples.

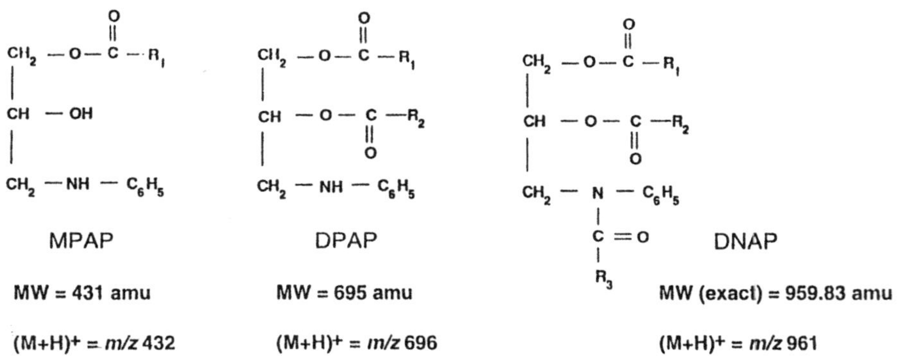

Figure 1. General Structures of Compounds Related to the Toxic Oil Syndrome. The structures are named after their base skeleton 3-N-phenylamino-1,2-propane diol (PAP). MPAP is the mono-ester of PAP, DPAP is the di-ester of PAP and DNAP is the di-ester amide of PAP. The R_1, R_2 and R_3 groups are alkyl groups associated with a specific fatty acid. Molecular weight and molecular ion information is given for each structure based upon the R_1, R_2 and R_3 groups being $C_{17}H_{33}$ (oleyl group).

Later Analysis of the Toxico-Epi-I Oils and Oils from the ITH Oil Refinery: The initial LC-MS/MS analyses that were performed on the Toxico-Epi-I oils measured eight fatty acid anilides in the oils, as well as the 1,2-di-oleyl DPAP, 2-(oleyloxy)-3-(N-phenylamino)propyl oleate (structure **8**, Table 2) and the mono-ester of PAP (MPAP)[6], 1-oleyl MPAP, 2-hydroxy-3-(N-phenylamino)propyl oleate, (structure 16, Table 2) using multiple reaction monitoring. The fatty acid anilides were found to be statistically higher in many of the case oils, however, the 1,2-di-oleyl DPAP and 1-oleyl MPAP were found to be more strongly associated with oils that caused TOS than fatty acid anilides (Hill et al., 1995). The presence of 1,2-di-oleyl DPAP and 1-oleyl MPAP in case-associated oils was significant, with an odds ratio (OR) of 13.7 (95% CI 5.0-38) for DPAP and an OR of 21.9 (95% CI 6.1-78) for MPAP.

Table 1. Fatty Acid Reference Table

Fatty acid chain		Molecular structure	Alkyl Chain
palmityl	C16:0	$(CO)(CH_2)_{14}CH_3$	$C_{15}H_{31}$
oleyl	C18:1	$(CO)(CH_2)_7CH{=}CH(CH_2)_7CH_3$	$C_{17}H_{33}$
linoleyl	C18:2	$(CO)(CH_2)_7CH{=}CHCH_2CH{=}CH(CH_2)_4CH_3$	$C_{17}H_{31}$
linolenyl	C18:3	$(CO)(CH_2)_7(CH{=}CH\ CH_2)_3\ CH_3$	$C_{17}H_{29}$
stearyl	C18:0	$(CO)(CH_2)_{16}CH_3$	$C_{17}H_{35}$
eicosenyl	C20:1	$(CO)(CH_2)_6CH{=}CH(CH_2)_{10}CH_3$	$C_{19}H_{37}$
erucyl	C22:1	$(CO)(CH_2)_{11}CH{=}CH(CH_2)_7CH_3$	$C_{21}H_{41}$
behenyl	C22:0	$(CO)(CH_2)_{20}CH_3$	$C_{21}H_{43}$

The importance of this finding was further corroborated with the analysis of oils from the ITH refinery. We received three sealed oil samples from the ITH refinery in Seville (the refinery most closely associated with oils that produced TOS). Two of the oil samples were of three shipments that were to be refined at ITH (reference numbers 4753-2 and 4753-3). The third shipment was not available for analysis. At ITH, the three shipments of oil were combined and refined, and this refined oil was our third sample (4753-5). When we analyzed these oils, we found high concentrations of fatty acid anilides in all, however, the third sample, which was the refined oil sample, was the only one that contained the 1,2-di-oleyl DPAP and 1-oleyl MPAP. This indicated that prior to the refining process, these compounds did not exist in the oils, but rather were formed by the refining process. The concentrations of the 1,2-di-oleyl DPAP and 1-oleyl MPAP were 150 and 44 μg/g, respectively, in the ITH-refined oil. The mean concentration of these compounds in the Toxico-Epi-I case oils was much lower, 10 μg/g for 1,2-di-oleyl DPAP and 0.22 μg/g for 1-oleyl MPAP. Overall, it was found that the oils were not a uniform mixture of components, i.e the distribution of aniline-related compounds varied from sample to sample. This was not entirely unexpected, as the ITH refined oil, as well as other oils were shipped to RAELCA and stored in large containers that delivered the oils by hose, gravity-fed, to be mixed together to prepare the toxic oils that were eventually sold to the public.

Aniline-denatured rapeseed oils from refineries in the Catalonian region of Spain were also analyzed. Analysis of these oils indicated that although they contained high concentrations of fatty acid anilides, they did not contain the 1,2-di-oleyl DPAP or 1-oleyl MPAP. This is significant, because consumption of these oils was not associated with TOS cases.

Table 2. Newly Identified Components of Toxic Oils - DPAPs, MPAPs and DNAPs

DPAP - (Di-esters)		
R_1(CO)	R_2(CO)	$(M+H)^+$
1. palmityl	linoleyl	668
2. palmityl	oleyl	670
3.[a] linolenyl	linolenyl	688
4.[a] linolenyl	linoleyl	690
5.[a] linolenyl	oleyl	692
6.[a] linoleyl	linoleyl	692
7.[a] linoleyl	oleyl	694
8.[a] oleyl	oleyl	696
9. stearyl	oleyl	698
10. stearyl	stearyl	700
11. linoleyl	eicosenyl	722
12. oleyl	eicosenyl	724
13. oleyl	erucyl	752
14. eicosenyl	eicosenyl	752

MPAP - (Mono-esters)	
R_1(CO)	$(M+H)^+$
15. linoleyl	430
16.[a] oleyl	432

DNAP - (Di-ester amides)			
R_1(CO)	R_2(CO)	R_3(CO)	$(M+H)^+$
17. linoleyl	oleyl	linolenyl	955
18.[a] linoleyl	linoleyl	linoleyl	955
19. linoleyl	linolenyl	oleyl	955
20. linolenyl	oleyl	stearyl	957
21. linoleyl	linoleyl	stearyl	957
22. oleyl	stearyl	linolenyl	959
23. oleyl	linoleyl	oleyl	959
24. linoleyl	oleyl	oleyl	959
25.[a] oleyl	oleyl	oleyl	961

[a] Denotes a standard has been synthesized for this structure.

The ITH-refined oil and the Toxico-Epi-I oils all indicated the presence of other compounds that were structurally related to the 1,2-di-oleyl DPAP and 1-oleyl MPAP. Because it is believed that these compounds may be significant in the etiology of TOS, we have made an effort to identify all of the various esters of PAP.

Mass Fragmentation of Esters of PAP: In an effort to identify all of the various esters of PAP, we used a combination of parent-ion scanning techniques and daughter-ion scanning techniques to further study TOS associated oils. On the basis of the varied fatty-acid composition of the oils, we suspected that the oils contained other compounds similar in structure to compounds **8** and **16**. The daughter-ion spectrum of **8** and **16** had an ion at *m/z* 132 characteristic of those structures. Monitoring for parent ions that produced an ion at *m/z* 132 allowed us to identify the protonated molecular ion and characteristic retention time of each PAP ester. We then examined the daughter spectrum of each parent-ion to further characterize and identify each PAP ester. Based upon this data and, with comparison to authentic standards for several compounds, we found 14 di-esters of PAP that are listed in Table 2 (structures 1-14), 2 mono esters of PAP (structures 15-16), and 8 di-ester amides of PAP (structures 17-25) in a refined oil from the ITH refinery, the probable source oil from which all TOS oils were derived. The fatty-acid groups, that are referred to in Table 2 are defined in Table 1. In addition to the compounds shown in Table 2, many of their isomers are also thought to exist in the oils, and some of these have been synthesized.

Parent-Ion Mass Spectrometry – DPAPs: To identify new DPAPs and MPAPs that are structurally similar to structures **8** and **16**, we analyzed the case oil containing the greatest amount of these compounds using parent-ion mass spectrometry as described in the experimental section. Parent-ion mass spectrometry sets the analyzing quadrupole (Q3) of a triple stage quadrupole mass spectrometer to a single mass (a known daughter), selects ions one mass at a time with the first quadrupole (Q1), and sends those ions into the collision cell (Q2). In Q2, the selected ions collide with argon which causes fragmentation to daughter-ions; Q3 mass analyzes the fragments produced in Q2. If the measured daughter-ion mass agrees with the fixed Q3 mass (pre-determined), the operating system will record the mass of the parent that was first transmitted to the collision cell from Q1. For the identification of compounds structurally similar to DPAP, the daughter-ion *m/z* 132 was selected. This daughter-ion does not contain any fatty-acid ester groups, and would identify the largest number of DPAP compounds independent of the type of fatty-acid ester group present.

The initial parent-ion scan, using *m/z* 132 as the fragment-ion, identified several parent-ions (*m/z* 688, 690, 692, 694, 696, 722, 724, 752) that could be tentatively identified as di-ester DPAP structures. Because it is possible to have more than one species being selected (i.e., isomers) in a parent-ion scan, further identification of the molecular ion must be done using a combination of techniques. The first of these uses parent-ion mass spectrometry to obtain more specific information on the identities of these compounds on the basis of the fragmentation of a standard. The second uses daughter-ion mass spectrometry of the molecular ion.

The daughter-ion mass spectrum of 1,2-di-oleyl DPAP contained a fragment ion at *m/z* 432 that was produced by the loss of the $(CO)C_{17}H_{33}$ group. This type of loss is reflected in the daughter-ion mass spectra of all DPAPs. Figure 2 contains the parent-ion scans of three daughter-ions formed by a similar loss: *m/z* 428 (linolenyl containing fragment), 430 (linoleyl containing fragment), and 432 (oleyl containing fragment). The parent-ion scan of *m/z* 432 revealed three possible parent-ions with masses *m/z* 696, *m/z* 694, and *m/z* 692. The parent-ion scan of *m/z* 430 produced parent-ions at *m/z* 694, *m/z* 692, and *m/z* 690. The weakest spectrum indicated only two parent-ions for the *m/z* 428 daughter-ion at *m/z* 692 and *m/z* 690. Knowing the parent-ion and the daughter-ion masses, it is possible to determine the missing part of the molecule. For example, the parent-ion scan of *m/z* 432 (an oleyl containing fragment of DPAP) has a parent-ion mass of *m/z* 696. This is a difference of 264 amu, which corresponds to a loss of the $C_{17}H_{33}(CO)$ group and addition of a proton. This originates from the oleyl group, indicating that the parent-ion *m/z* 696 is probably composed of two oleyl groups. The daughter-ion mass spectra of *m/z* 696 in the actual oil sample and the standard were identical, indicating that *m/z* 696 is likely to be composed of only the di-oleyl DPAP. Parent-ion MS/MS spectra were measured for fragments containing other fatty-acid groups (stearyl, *m/z* 434; palmityl, *m/z* 406; erucyl, *m/z* 488; behenyl, *m/z* 490; and eicosenyl, *m/z* 460) in order to identify other di-esters of PAP. The DPAPs (Structures 1-14) identified using the above described experiments are compiled in Table 2. Structural isomers were identified in the oils for compounds 1, 2, 4, 5, 6, 7, 9, 11, 12, and 13, which increases the number of identified DPAPs to 23 (Schurz et al., 1996).

Parent-ion Mass Spectrometry - Mono-esters of PAP: The mono-oleyl MPAP (structure 16, Table 2) was also detected in the refined oil from ITH. The daughter-ion spectrum of 1-oleyl MPAP ($(M+H)^+$ = *m/z* 432) has been shown to have similarities to the 1,2-di-oleyl DPAP MS/MS spectrum. A second MPAP (structure 15, Table 2) that corresponds to the 1-linoleyl MPAP ($(M+H)^+$= 430) was also detected using parent-ion scans of *m/z* 132. We have assumed that the position of the fatty-acid group on these molecules is R_1(CO), though we recognize that a combination of both may exist in the oils as structural isomers.

Parent-ion Mass Spectrometry - Di-ester Amides of PAP: While we were determining the structures of the DPAPs and MPAPs, we discovered another group of related compounds. These structures are referred to as di-ester amides of PAP (DNAP, Figure 1), owing to formation of an amide at the nitrogen of the DPAP. Tentative identifications of several structures were made, and the 1,2,N-tri-oleyl DNAP (structure 25) and 1,2,N-tri-linoleyl DNAP (structure 18) were confirmed in toxic oils. Similar to the DPAPs, the parent-ions and daughter fragment-ions suggest that combinations of different fatty-acid groups make up the variety of DNAPs that are observed (Table 2, structures 17-25), and it seems likely that there are structural isomers present. Though exact mass measurements of these compounds were not performed, their exact masses were calculated. This was done because the nominal mass of the 1,2,N-tri-oleyl DNAP, MW = 959 amu, is in disagreement with the molecular ion (M+H)+ = *m/z* 961. This apparent mass discrepancy is due to the ^{13}C isotope content and the relatively large size of the molecule. The exact mass of 1,2,N-tri-oleyl DNAP is 959.83 amu, almost 1 amu greater than its nominal mass. Therefore, in a unit mass resolving instrument (i.e., a quadrupole mass spectrometer) the $(M+H)^+$ = *m/z* 961. This is true for all of the DNAP species.

The DNAPs, indicated by parent-ions at *m/z* 961 and *m/z* 959, were initially detected in parent-ion scans of the molecular ion for 1,2-di-oleyl DPAP, 8 (*m/z* 696). The daughter-ion mass spectra of these two ions produced not only a DPAP-like fragment but similar fragmentation patterns as well. Further parent-ion scans were performed that indicated the presence of several other possible DNAP structures, in particular *m/z* 955, *m/z* 957, and *m/z* 959, in addition to *m/z* 961. Chromatography confirmed that the DPAPs and the DNAPs were discrete compounds and that DPAP was not simply a daughter of DNAP. The procedure for identifying the various fatty-acid groups was similar to that used for identifying the DPAPs.

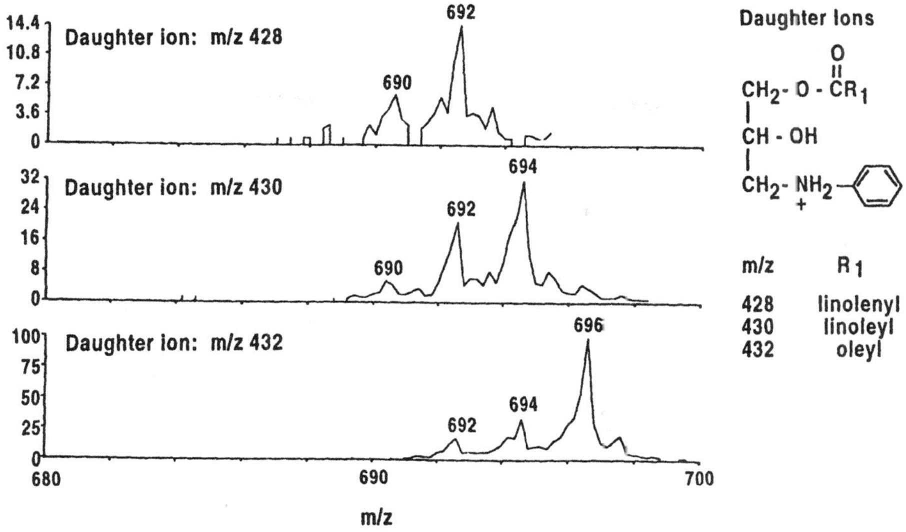

Figure 2. Parent-ion Mass Spectra of *m/z* 428, *m/z* 430 and *m/z* 432.

Discussion

The analytical studies of TOS have provided a plethora of information related to the study of the oils that caused it. Because of the appearance of the esters of PAP in the sample of oil from the ITH refinery only after refining, we believe that the presence of those compounds in TOS-associated oils is an important finding. These compounds are very similar in structure to principal components of cell membranes - diacylglycerols. If these compounds were incorporated into cell membranes their presence could evoke an immunologic response, and to that end, these compounds seem to be good candidates for etiologic agents of TOS (Hill et al., 1995). Until we have a clear understanding of the cause of TOS, we cannot prevent outbreaks of similar diseases in the future. This is evidenced by the 1989 epidemic in the United States that became known as the eosiniphilia myalgia syndrome (EMS, Ferrer et al., 1993). That disease was associated with the consumption of L-tryptophan food supplements, and like TOS, its etiologic agent(s) have not been identified. For TOS, we have now identified several new compounds that are markers for oils that cause disease, and that are candidates for animal and toxicological testing to determine whether some of the compounds may be etiologic agents. These compounds are products of aniline and triglycerides and come in the form of fatty-acid anilides, di-esters of DPAP (14 structures), mono-esters of MPAP (2 structures), and di-ester amides or DNAPs (9 structures).

TOS and EMS are strikingly similar (Kilbourne et al., 1988; Hertzman and Abaitua-Borda, 1993). This clinical similarity suggests that a common biological pathway was affected by both the TOS-associated case oils and the L-tryptophan case samples, and that this commonality may provide an answer to what caused the diseases. Recently, researchers from the Mayo Clinic have proposed a possible link between TOS and EMS (Philen and Hill, 1993). They reported the in vitro conversion of (N-phenylamino)alanine, an EMS-associated contaminant, to 3-N-phenylamino-1,2-propane diol. The latter compound forms the basic skeleton of the DPAPs, MPAPs and DNAPs.

Apart from the clinical similarities between TOS and the EMS, both diseases seem to damage the immune system and elicit an immune response. People with TOS and EMS still suffer from their disease in a chronic phase that includes peripheral neuropathy, hepatopathy, scleroderma and pulmonary hypertension. Many times their disease will include myalgias and arthralgias. In this chronic phase, the disease takes on the appearance of a rheumatic illness, particularly of scleroderma (Alonso-Ruiz et al., 1986; Philen and Hill, 1993). It has been suggested that scleroderma and related disorders may be triggered by exposure to some environmental agent(s) when a pre-existing susceptibility is present in the victim (Plotz et al., 1993). Certain host factors (e.g., age, immunogenetic background) may have played a role in determining the susceptibility of the victim to the illness. The study of TOS and EMS may shed light on the etiology and pathogenesis of not only these diseases but of a whole host of other illnesses.

Perhaps the greatest challenge in identifying the causes of TOS and EMS lies in the fact that no suitable animal models have been identified that simulate TOS or EMS. All compounds that are identified as possible etiologic agents must be evaluated as toxic agents. With the lack of an animal model it will be very difficult, if not impossible, to determine whether a compound is the actual etiologic agent. It is imperative that research on these diseases continues until suitable animal models, and perhaps a link between TOS and EMS have been identified and the etiologic agents of both diseases are known.

Acknowledgments

We thank Dr. Angel Messeguer for providing synthesized standards of the various di-esters and di-ester amides. The authors would also like to thank the continued support of the Government of Spain, which has allowed us to continue this research. This work was supported in part through an international collaborative agreement with the Fondo de Investigacion Sanitaria of the Government of Spain.
Use of trade names is for identification only and does not constitute endorsement by the Public Health Service or the U.S. Department of Health and Human Services.

References

Alonso-Ruiz A, Zea-Mendoza AC, Salazar-Vallinas JM, Rocamora-Ripoll A and Beltrán-Gutierrez J (1986) Toxic oil syndrome: a syndrome with features overlapping those of various forms of scleroderma. Sem. Arthritis Rheum. 15:200-12.

Benson LM, Mayers AN, Gleich GJ and Naylor S (1995) LC-APCI-MS/MS for the Analysis of PAP Biotransformation Products. Conference Proceedings of 43rd American Society for Mass Spectrometry Conference on Mass Spectrometry and Allied Topics, May 21-26, 1995 Atlanta, GA USA.

Bernert JT, Kilbourne EM, Akins JR, Posada de la Paz M, Meredith NK, Abaitua-Borda I, and Wages S (1987) Compositional analysis of oil samples implicated in the Spanish toxic oil syndrome. J. Food Sci. 52:1562-1569.

Ferrer M, Galceran M, Sanchez-Baeza F, Casas J and Messeguer A (1993) Synthesis of aniline derivatives with potential toxicological implications to the Spanish toxic oil syndrome. Liebigs Ann. Chem. 507-511.

Guitart R and Gelpi E (1991) Chemical composition of TOS related oils. In Toxic oil syndrome: current knowledge and future perspectives. (Grandjean P, Tarkowski S, Eds.) 42:95-138, World Health Organization Regional Publications, England.

Hertzman PA and Abaitua-Borda I (1993) The toxic oil syndrome and the eosinophilia-myalgia syndrome: pursuing clinical parallels. J. Rheum. (20)10:1707-1710.

Hill RH, Schurz HH, Posada de la Paz M, Abaitua-Borda I, Philen RM, Kilbourne EM, Head, SL, Bailey SL, Driskell WJ, Barr JR and Needham LL, (1995) Possible etiologic agents for toxic oil syndrome (TOS): fatty acid esters of 3-(N-phenylamino)1,2-Propanediol. Arch. Environ. Contam. Toxicol. 28:259-264.

Kamb ML, Murphy JJ, Jones JL, Caston JC, Nederlof K and Horney LF (1992) Eosinophilia-myalgia syndrome in L-tryptophan-exposed patients. J. Am. Med. Assoc. 267:77-82.

Kilbourne EM, Bernert JT, Posada de la Paz M, Hill RH, Abaitua Borda I, Kilbourne B and Zack MM (1988) Chemical correlates of pathogenicity of oils related to the toxic oil syndrome epidemic in Spain. Am. J. Epidem. 127:1210-1227.

Pestane A and Munoz E (1982) Anilides and Spanish Toxic Oil Syndrome. Nature 298:608.

Philen RM and Hill RH (1993) 3-(Phenylamino)alanine - A link between eosinophilia-myalgia syndrome and toxic oil syndrome? Mayo Clin Proc 68:197-200.

Plotz, PH, Leff, RL and Miller, FW (1993) Systemic sclerosis and related syndromes. In Primer on the Rheumatic Diseases (Schumacher, H.R.; Klippel, J.H.; Koopman, W.J.; Eds.) pp 118-127, Arthritis Foundation, Atlanta, GA.

Posada de la Paz M, Castro M and Kilbourne EM (1987) Toxic-oil syndrome case reports associated with the ITH oil refinery in Sevilla. Food Chem. Toxicol. 25:87-90

Posada de la Paz M, Philen RM and Abaitua-Borda I (1991) Manufacturing processes at two French rapeseed oil companies: possible relationships to toxic oil syndrome in Spain. Food Chem. Toxicol. 29:797-803.

Schurz HH, Hill RH, Posada de la Paz M, Philen RM, Abaitua Borda I, Bailey SL, Needham LL (1996) Products of aniline and triglycerides in oil samples associated with the toxic oil syndrome. Chem. Res. Toxicol.9(6):1001-1006.

Tabuenca JM (1981) Toxic allergic syndrome caused by ingestion of rapeseed oil denatured with aniline. Lancet 2:567-8.

Vazquez-Roncero A, Maestro Duran R and Gutierrez R (1984) New aniline derivatives in oils related to the toxic oil syndrome: I. fatty esters of 3-Phenylamino-1,2-propanediol. Grasas y Aceties 35:15-21.

Ventura Diaz M (1982) Conventional analysis of the available samples of purportedly toxic oils. Grasas y Aceities 33:73-77.

Immunological Aspects of the Toxic Oil Syndrome

Carlos Lahoz, Victoria del Pozo, Soledad Gallardo, Blanca Cárdaba, Aurora Jurado, Isabel Cortegano, Ana del Amo, Ignacio Arrieta and Pilar Palomino
Immunology Department. Fundación Jiménez Díaz, Av. Reyes Católicos 2, E-28040 Madrid, Spain

Introduction

The Toxic Oil Syndrome (TOS), which occurred in Spain in 1981, was caused by the ingestion of rapeseed oil denatured with 2% aniline, sold illegally as edible oil. The number of unaffected individuals exposed to the toxic oil is unknown, but more than 20,000 people suffered from the disease.

From a clinical point of view, TOS is a multisystemic disease with three consecutive phases. The acute phase (one to two months) was characterized by fever, rash, pulmonary oedema, myalgia and eosinophilia, the intermediate phase with myalgia, hepatopathy and sicca syndrome. Two to four month later, chronic sequelae involved scleroderma-like skin lesions, dysphagia, sicca syndrome, pulmonary hypertension, peripheral neuropathy, weight loss, hepatopathy (Kilbourne et al., 1983; Kilbourne et al., 1991). The pathologic findings in TOS showed primary endothelial injury, with cell proliferation and perivascular inflammatory infiltrates. About a thousand people died and approximately 60% of total patients developed some clinical features of the intermediate phase.

Due to the clinical course of the disease, immunopathological mechanisms have been suspected but a direct connection was never demonstrated. To analyze this possibility, we used three different approaches: First, we determined several immunological parameters in the sera of patients with TOS and without the disease, using a case-control design. Second, we have developed a mRNA extraction procedure from paraffin-embedded lung tissues to analyze cytokine profiles in samples from autopsied TOS patients with pulmonary involvement. We analyzed mRNA expression for different cytokines, namely IL-1, IL-2, IL-4, IL-5, IFN-γ, GM-CSF, CD25 and CD23, using an RT-PCR technique. Third we also analyzed the possible mechanisms of cytotoxicity induced by the toxics.

Our data support the implication of the immune system in TOS, with a possible activation of T cells and release of cytokines, that could explain some of the clinical findings of the disease.

Immunological Data in the Acute Phase of the Disease

Immunological findings on TOS patients obtained during the onset of the disease were difficult to interpret due to different reasons: The number of investigated patients was too small, studies were not carried out under blind conditions, without an adequate experimental design, and finally, sera from patients at different clinical phases were used (Lahoz et al., 1992). Our purpose was to investigate the immunological mechanisms in the acute and intermediate phase of TOS using a case-control design, under blind conditions.

The criteria used for case selection were those described by Kilbourne et al. (1988): peripheral eosinophilia of 500 cells/mm^3 or more, with cough, dyspnea, or pulmonary infiltrate was suggestive of an acute phase of TOS. Sera from May to August 1981 were selected from the sera bank created during the firsts months of the epidemic with samples from TOS patients and from non-TOS patients but with some characteristics similar to TOS. The final number of samples analyzed were n=126 of which there were 98 cases and 28 controls (Gallardo et al., 1992).

Total and Specific Immunoglobulins

No significant differences in IgG, IgM or IgA serum levels were found between controls and TOS patients. Total IgE levels of TOS patients were higher (276.64 ± 847 UI/ml) than those of controls (113.03 ± 177 UI/ml), but the difference did not reach statistical significance.

In regard to IgG and IgE antibodies against oleylanilide and other related products, all tested sera (n=126) were negative for specific IgG antibodies against oleylanilide or related products such as aniline, linoleylanilide and 3-phenyl-aminopropane-1,2-diol. No specific IgE antibodies were found in the sera tested.

The incidence of autoantibodies against smooth muscle, mitochondria, nuclei, reticulin, gastric parietal cells, centromeres, cytoskeletons, nucleoli, lymphocytes, cardiolipin and extactable nuclear antigens (ENA's) was within the range found in the normal population both in incidence and titers (data not shown).

No autoantibodies against type I-V collagen were found.

Cytokines in Sera from Acute Phase Patients

Table 1 shows the results obtained for sIL-2R (soluble IL-2 receptor) and sCD23. (soluble low affinity receptor for IgE). sIL-2R levels were significantly increased in TOS patients as compared to controls (1810 U/ml versus 892 U/ml, $p<0.0001$). Although sCD23 levels in TOS patients were increased three-fold as compared to controls (157.1 versus 50.1), statistical significance was not reached as several controls had high sCD23 levels.

High levels of GM-CSF (Granulocyte Macrophage-Colony Stimulating Factor) were detected in TOS patients (n=61), and controls (n=23): 432 ± 205 pg/ml and 410 ± 192 pg/ml respectively. No statistically significant difference was found.

Table 1 sIL-2R and sCD23 values in sera from TOS acute phase patients

	Cases (n=98)	Controls (n=28)	Wilcoxon[a]
	Median[b]	Median[b]	p
sIL-2R	1810 (900-3200)	892 (343-3200)	0.0001
sCD23	157.1 (0-800)	50.1 (0-544.3)	NS[c]

[a] Wilcoxon rank-sum test was used to analyze differences in variables
[b] Data represent median values (10-90 percentiles) of U/ml
[c] NS: Non significant

No IL-4 (Interleukin 4) and TNF (Tumor Necrosis Factor) were detected in all the tested sera and IL-6 (Interleukin 6) levels were not significantly different between TOS patients and controls (only one TOS patient and none of control sera tested gave positive results).

The most interesting finding in this study is the strong statistical significance in the sIL-2R serum levels in TOS patients compared to controls ($p<0.0001$). sIL-2R is elevated in several autoimmune processes: multiple sclerosis (Greenberg et al., 1988), systemic lupus erythematosus (Tokano et al., 1989), rheumatoid arthritis (Silverman et al., 1991) and primary Sjögren's syndrome (Manoussakis et al., 1989). Recently, it has been reported that sIL-2R levels are correlated with blood eosinophilia and lung function (Lassalle et al., 1991). This data suggests a possible link between T cell activation, eosinophils, and impairment of lung function in inflammatory processes. This cascade of events possibly occurred in TOS patients.

Specific IgG and IgE antibodies against oleylanilide and related toxic products were not detected. In previous studies, specific IgE antibodies against oleylanilide were described in the sera of patients in the acute phase of TOS (Tricas, 1984). Discrepancies may be due to several factors such as the origin, nature and method for antigen purification, and also there were technical differences.

Several groups investigated the possibility that autoimmune mechanisms were involved in the pathogenesis of TOS. TOS has been suggested to be similar to graft-versus-host disease (GVHD) in many aspects (Gleichmann H and Gleichmann E, 1984, Kammüller et al., 1984, Goldman et al., 1991). (Cyclic) phenylthiourea derivatives, possibly formed in adulterated rapeseed oil, have been hypothetized to be related with the causative factor of TOS, due to structural similarities with other autoimmune inducing drugs (Kammüller et al.,1988, 1989).

The negative results obtained in the present study may be due to the fact that sera were obtained in the acute phase of the disease, that is to say too early for autoantibodies to be detected.

Cytokine Profile in Pulmonary Tissues

The cellular infiltrates in lung tissue found in necropsies are composed mainly by lymphocytes (73% of cellular infiltrates), as has been described by Aguilera (1989), and Martinez Tello and Téllez (1992). Due to previous data we suspected a T cell activation; therefore, we wanted to know which type of T cell could be involved. For this puropose we analyzed the cytokine profile. Two subsets of T lymphocytes (CD4) have been described, depending on the cytokine profile that they produce: Th1 and Th2. Thus, Th1 produces γ-IFN and interleukin 2 (IL-2) and Th2 cells produce IL-4 and IL-5 among others (Kapsenberg et al., 1991, 1992, Romagnani 1991, Scott and Kaufmann 1991).

Taking into account the findings in patients at the acute phase of TOS (elevated sIL-2R and IgE and eosinophilia, among others), we postulate that T cell activation might have happened. The secretion of Th2 cytokines could explain the eosinophilia (IL-5) and the increased levels of IgE and sCD23 (IL-4).

In order to analyze which subset of T cells (Th1 or Th2) were implicated, lung specimens from patients who died in the early and chronic phases of TOS were obtained in order to isolate total RNA from paraffin embedded tissues. Control specimens were taken from patients who died by processes not related to immunological diseases, and whose lungs were not affected.

mRNA was extracted from necropsies of TOS patients who had pulmonary involvement and died in the period 1981 to 1983, following the method described by De Andrés, et al. (1995). We analyzed mRNA expression for IL-1, IL-2, IL-4, IL-5, IFN-γ, and GM-CSF cytokines and CD25 (receptor for IL-2) and CD23 (low affinity receptor for IgE), using a reverse transcriptase polymerase chain reaction technique (RT-PCR).

The final number of paraffin-embedded lung tissue from TOS patients who met the case definition and who died from 1981 to 1983 was 34, with 47% being from the acute phase, and 53% from the chronic phase.

Methods of mRNA extraction have been analyzed on different paraffin-embedded specimens dated from 1981. Comparison of different methods was done: the isothiocyanate-guanidinium method, using a cesium chloride gradient and the direct isolation by the oligo dT magnetic beads. Best results in terms of quality and quantity of messenger RNA have been obtained by using the oligo dT beads; therefore this method has been selected for messenger RNA analysis.

In order to discard false positive results PCR was monitored by Southern blot. Controls and patients were evaluated using Fisher's exact test, $p<0.05$ was considered significant

As is shown in Fig.1, when we compared the profiles for Th1 and Th2 cytokines in lung specimens from TOS patients, we found a dominant Th2 versus Th1 response ($p< 0.001$). However in the control group no prevalence of Th2 versus Th1 was found ($p= 0.11$).

Neither IL-1α, GM-CSF, CD25 nor CD23 had significant differences when comparing TOS cases and control group, although 43% of the TOS specimens analyzed had a positive signal to CD23 compared to only 20% for the comparison group. In the case of IL-1α, all specimens tested were negative. Results obtained for CD25 and GM-CSF were very similar between controls and patients.

We also analyzed in the TOS group whether the cytokine profile was different in the acute and chronic phases. We observed that the percentage of Th1 was increased in the acute phase with respect to the chronic phase (30% versus 14%) but this diference is not significant. The Th2 response is slightly decreased in the acute phase of TOS with respect to the chronic phase, but again the difference was not significant.

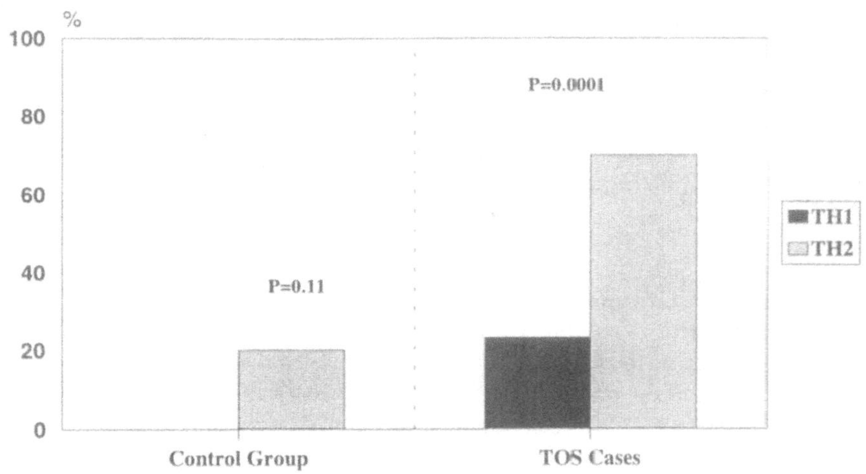

Fig. 1 Percentage of lung specimens showing a Th1 (mRNA for IL-2 and γ-IFN) and Th2 (mRNA for IL-4 and IL-5) profiles in TOS patients (n=34) and in the control group (n=15)

The results suggest that TOS may be a T cell-dependent process involving primarily Th2 cells. These cells produce IL-5 which could explain the eosinophilia found in acute patients because IL-5 is an eosinophilopoietic factor. Another cytokine produced by Th2 cells is IL-4, and this cytokine together with sCD23 has been implicated in IgE synthesis, that was elevated in some patients. In addition, positive immunohistochemical localization of IL-4 in the skin of patients with EMS and TOS has been demonstrated (Kaufman et al., 1994). We studied CD23 messenger RNA because it is implicated in the regulation of IgE. Although the percentage of TOS specimens positive for CD23 was elevated with respect to the comparison specimens (43% and 20%, respectively) this finding was not statistically significant. Eosinophilia Myalgia Syndrome (EMS) presents many features in common with TOS such as: severe myalgia, multisystem involvement and high levels of peripheral blood eosinophils in their acute phases; additionally, EMS appears to have the same potential for long term sequelae as TOS (Kilbourne et al., 1991). The final pathogenic patways are probably similar in the two syndromes, both of them related to the use of consumer products (Mayeno et al., 1992, WHO 1991, 1992). The indirect evidence for the T cell activation found in TOS has also be reported in EMS associated with high levels of sIL-2R in the sera of seven EMS patients (McKinley et al., 1993).

No animal model for TOS is available until now. Several reports have been published (Bell et al. 1992; Blasco and Ruiz-Gutierrez 1992), and the conclusions reached are similar to those found in the acute phase of TOS patients (Lahoz et al., 1983). High IgE levels and abnormal levels of T lymphocytes subsets were found in mice treated with anilide.

Mechanisms of Cellular Dammage by TOS Related Toxic Agents

Previous studies performed in our laboratory with some TOS related products led us to investigate the solubility and optimal concentration of these products in cell cultures using dose-response curves. We found that some of these products were toxic to cells in a way different from necrosis. These results led us to perform cytotoxicity studies with the TOS associated products which were found in case oils, such as 3-(N-phenylamino)-1,2-propanediol (PAP), 1,2-di-oleyl ester of PAP (DEPAP), and MEPAP, the monooleyl ester of PAP. We also tested triolein and aniline. Finally we included in our analysis 3-(phenylamino)-L-alanine (PAA), an EMS associated product due to its chemical similarity to PAP, and in order to elucidate the mechanisms implicated in both diseases.

We report here the cytotoxic effect of DEPAP on cultured human lymphocytes and demonstrate that these toxicants may exert their cytotoxicity, at least in part, by triggering programmed cell death.

Discrimination between viable, apoptotic and necrotic lymphocytes by flow cytometry

Cultured cells with and without each product were washed after different time periods, adjusted to 0.5 x 106 cells and treated with 500 µl of a solution containing 100 µg/ml of propidium iodide, 0.05% nonidet P-40 (to permeate the cell membrane) and 0.02% RNase. After one hour of incubation, the stained DNA was analyzed by flow cytometry.

Fig. 2 shows a dose-response curve obtained by culturing peripheral blood lymphocytes with different doses (10, 20 and 40 µg/ml in panels A, B and C respectively) of DEPAP after 3 days of incubation.

Changes in size and granularity can be observed using flow cytometry. Apoptotic, necrotic and viable cells differ in light scatter patterns. Apoptotic cells give lower forward scatter (FS) and higher side scatter (SS) values than viable cells, reflecting the smaller size (by water loss) and different nucleus/cytoplasm consistency (by chromatin condensation and nuclear fragmentation). Apoptotic cells also show a low DNA stainability with PI, resulting in a distint, quantifiable region (peak H) left to the Go/G1 diploid (D) peak, appearing as a broad hypodiploid DNA peak, easily discriminable from the narrow peak of lymphocytes with normal DNA content (diploid D, peak) in red fluorescence channels. The hypodiploid peak (H), which is due to DNA degradation, increases with the concentration of DEPAP. At higher concentrations or longer culture times, the diploid peak (D) disappears, increasing the hypodiploid (H) and necrotic (N) peaks, respectively.

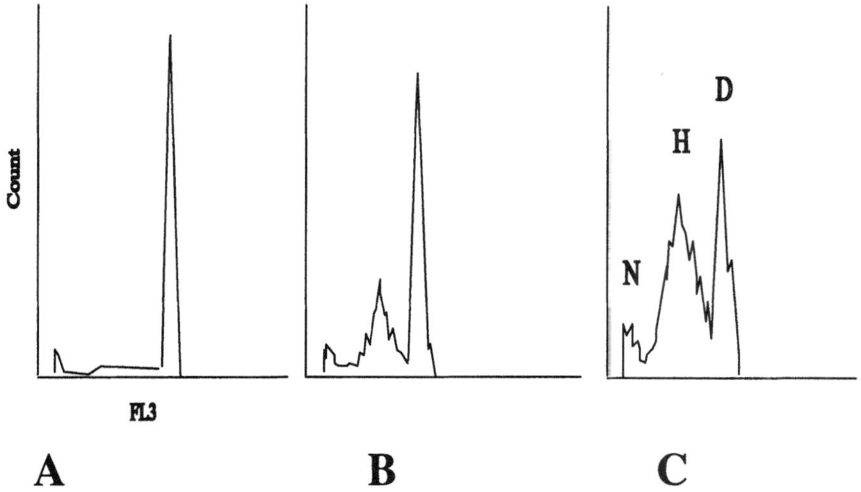

Fig. 2 Dose response curve with different concentrations: 10, 20 and 40 μg/ml in panels A, B and C of DEPAP incubated with peripheral blood lymphocytes for three days. D, H and N represents diploid, hypodiploid and necrotic peaks, respectively.

Cells cultured with the rest of the products had similar percentages in these determinations as the cellular control (with and without 0.2% ethanol). None of the results, except DEPAP, gave significant differences as regard to controls.

Cytotoxicity may induce autoimmunity by exposing autoreactive lymphocytes to abnormal forms of self material released during premature cell death, but it is also likely that subtoxic concentrations of some reactive drug metabolites, through the capacity to participate in oxidation-reduction reactions, may act at the lymphocyte membrane to cross-link accessible receptors involved in signal transduction. Thus, many agents which induce apoptosis are either oxidants or stimulators of cellular oxidative metabolism, for example low doses of H_2O_2. On the other hand, some authors have hypothesized the relationship between sclerodermia, autoimmunity and the generation of free radicals by chemical agents. Aniline derivates found in TOS case oils have been described to increase the production of reactive oxygen metabolites (ROM) from phagocytic cells, demonstrating a possible role for oxygen radical mechanisms in the pathophysiology of TOS.

Our results support our hypothesis that DEPAP causes cytotoxicity in cells and tissues by inducing apoptosis, and these inducer mechanisms could have implications in the development of autoimmunity, which we believe plays a role in TOS pathogenesis.

References

Aguilera B (1989) Lesiones vasculares en el Síndrome Tóxico por aceite de colza. Doctoral Thesis, Madrid Universidad Autónoma 26-41

Bell SA, Hobbs MV, Rubin RL (1992) Isotype-restricted hyperimmunity in a murine model of the toxic oil syndrome. J. Immunol. 148: 3369-3376

Blasco R, Ruiz-Gutierrez V (1992) Effect of fatty acid anilides on immune responses of Swiss mice. Clin. Exp. Immunol. 89: 131-135

De Andrés B, del Pozo V, Gallardo S, de Arruda-Chaves E, Cárdaba B, Martin-Orozco E, Posada M, Palomino P, Lahoz C (1995) Improved method for mRNA extraction from paraffin-embedded tissues. Biotechniques 18(1): 42-43.

Gallardo S, del Pozo V, Cárdaba B, de Andrés B, Martin-Orozco E, Fernandez JC, Tramón P, Posada M, Abaitua I, Palomino P, Lahoz C (1994) Immunological basis of toxic oil syndrome (TOS). Toxicology 93:289-299

Gleichmann H, Gleichman E (1984) GVHD a model for the Spanish Toxic Oil Syndrome. Lancet. i: 1174-1175.

Goldman M, Druet P, Gleichmann E (1991) Th2 cells in systemic autoimmunity: insights from allogeneic diseases and chemically-induced autoimmunity. Immunol. Today 12(7): 223-227

Greenberg SJ, Marcon L, Hurwitz BJ, Waldmann TA, Nelson D I (1988) Elevated levels of soluble IL-2 receptors in Multiple Sclerosis. New Eng. J. Med. 319: 1019-1020

Kammüller M E, Penninks A H, Seinen W (1984) Spanish Toxic Oil Syndrome is a chemically induced graft-versus-host-like epidemic. Lancet. i: 1174-1175

Kammüller M E, Bloksma N, Seinen W (1988) Chemically induced autoimmune reactions and Spanish toxic oil Syndrome. Focus on hydantoins and related compounds. J. Toxicol. 26: 157-174

Kammüller ME., Bloksma N, Seinen W. (1989) Autoimmunity and toxicology. Immune dysregulation induced by drugs and chemicals. In: Kammüller M.E., Bloksma, N. and Seinen, W. (eds), Autoimmunity and toxicology. Immune dysregulation induced by drugs and chemicals. Elsevier, Amsterdam, pp 4-34

Kapsenberg ML, Wierenga EA, Bos JD, Jansen HM (1991) Functional subsets of allergen-reactive human $CD4^+$ T cells. Immunol. Today 12(11): 392-395

Kapsenberg ML, Jansen HM, Bos JD, Wierenga E A (1992) Role of type 1 and type 2 T helper cells in allergic diseases. Curr. Opin. Immunol. 4: 788-793

Kaufman LD, Gruber BL, Gómez-Reino JJ, Miller F (1994) Fibrogenic growth factors in the eosinophilia-myalgia syndrome and the toxic oil syndrome. Arch. Dermatol 130: 41-47

Kilbourne EM (1983) Clinical epidemiology of toxic oil syndrome. New. Engl. J. Med. 309: 1408-1414

Kilbourne EM, Bernert JT, Posada M et al. (1988) Chemical correlates of the pathogenicity of oils related to the TOS epidemic in Spain. Am. J. Epidemiol. 127: 1210-1227

Kilbourne EM, Posada M, Abaitua I, et al. (1991) Toxic Oil Syndrome: A current clinical and epidemiologic summary, including comparisons with the eosinophilia-myalgia syndrome. J. Am. Coll. Cardiol. 18: 711-717

Lahoz C, Tricas L, Vela C, Lauzurica P, Gurbindo C, Garcia R (1983) Hyper IgE, eosinophilia and immunological hyperreactivity due to ingestion of adultered rapeseed oil (Toxic Oil Syndrome). Eur. J. Respir. Dis. 62: 415-418

Lahoz C, Rose NR, Robinson CJG (1992) Immunology. In: Nadal, J and Tarkowski, S (Eds), "Toxic Oil Syndrome: current Knowledge and future perspectives". WHO Regional Publications, European series, n° 42, pp. 143-152

Lassalle P, Sergant M, Delneste Y, et al. (1991) Levels of soluble IL-2 receptor in plasma from asthmatics. Correlation blood eosinophilia, lung functions, and corticosteroid therapy. Clin. exp. Immunol. 87: 266-271

Martínez Tello FJ, Téllez Y (1992) Pathology In: Nadal, J. and Tarkowski, S. (Eds), "Toxic Oil Syndrome: current Knowledge and future perspectives". WHO Regional Publications, European series, n° 42, pp. 39-65

Manoussakis MN, Papadopoulos GK, Drosos AA, Motsopoulos HM (1989) Soluble interleukin 2 receptor molecules in the serum of patients with autoimmune disease. Clin. Immunol. Immunopathol. 50: 321-32

Mayeno AN, Belongia EA, Fengling Lundu SK, Gleich GJ (1992) 3-(Phenylamino)alanine, a novel aniline-derived aminoacid associated with the Eosinophilia-Myalgia Syndrome: a link to the Toxic Oil Syndrome?. Mayo Clin. Proc. 67:1134-1139

McKinley KL, Harati Y, Schneider LW (1993) Chronic immune activation in the eosinophilia-myalgia syndrome. Muscle & Nerve 16: 947-951

Romagnani S. (1991) Human Th1 and Th2 subsets: doubt no more. Immunol. Today. 12(8): 256-257

Scott P, Kaufmann HE (1991) The role of T-cell subsets and cytokines in the regulation of infection. Immunol. Today. 12(10): 346-348

Silverman ED, Laxer R, Nelson DL, Rubin L A (1991) Soluble interleukin 2 receptor in juvenile rheumatoid arthritis. J. Rheumatol. 18: 1398-1402

Tokano Y, Murashima A, Takasaki Y, Hashimoto H, Okomura K, Hirose S (1989) Relation between soluble interleukin 2 receptor and clinical findings in patients with systemic lupus erythematosus. Annals of Rheumatic Diseases 48: 803-809

Tricas, L. (1984) "Immunological hyperreactivity in the acute phase of TOS' Doctoral Thesis, University of Zaragoza

WHO (1991) "Toxic oil syndrome and eosinophilia-myalgia syndrome. Pursuing parallels in pathogenesis". Report on a WHO meeting. Washington, DC., WHO Regional Office for Europe, Copenhagen, 22 pp.

WHO (1992) Toxic oil syndrome and eosinophilia-myalgia syndrome. Clinical aspects". Report on a workshop. Stony Brook, New York. WHO Regional Office for Europe, Copenhagen, 19 pp

Altered Gene Expression and Immunotoxicity

(Chair: H. Van Loveren, The Netherlands)

Regulation and Mechanisms of Apoptosis in T Lymphocytes

Diels J. van den Dobbelsteen, C. Stefan I. Nobel, Andrew F.G. Slater and Sten Orrenius
Institute of Environmental Medicine, Division of Toxicology, Karolinska Institutet, Box 210, S-171 77, Stockholm, Sweden.

Apoptotic cell death is now widely recognised as being a distinct process of importance in both normal physiology and pathology. It is characterised by morphological and biochemical changes in most cell compartments (Kerr et al., 1972; Wyllie et al., 1980). The microscopically observed nuclear alterations such as margination and condensation of chromatin and nuclear fragmentation are accompanied by sequential degradation of the DNA, first into fragments of approximately 700, 300 and 50 kilobasepairs and then into small oligonucleosomal fragments which are multiples of about 200 basepairs. This multi-step process can be readily visualised by appropriate electrophoresis of DNA (Walker et al., 1995). Most cell types also show a remarkable shrinkage presumably due to the loss of water and electrolytes and the cytoplasm gradually becomes more electron dense. The plasma membrane remains remarkably intact, though it blebs and the phospholipids redistribute. The specific translocation of phosphatidylserine to the outer leaflet of the plasma membrane has been shown to play a very important role in the recognition of apoptotic cells by phagocytes (Martin et al., 1995). In addition, other membrane changes have been reported to contribute in the specific recognition and removal of the apoptotic bodies that remain after fragmentation of the dead cell (Savill et al., 1993). In this final stage of the process, the dying cells do usually not spill their contents into the extracellular space and consequently the potential for inflammation is limited.

The current hypothesis for apoptotic cell death is that stimulation of the proteolytic activity of a member of the family of interleukin-1-β-converting enzyme (ICE)-like proteases is a crucial event in the initiation of the above mentioned biochemical and morphological changes. The activation of this specific class of proteases is thought to be a crucial event in the process (Martin and Green, 1995 and references therein). The activation of these proteases is associated with the specific cleavage of proteins involved in recognition and repair of DNA damage (poly(ADP-ribose)polymerase (= PARP), DNA-dependent protein kinase and U1-ribonucleoprotein), the attachment of DNA to the nuclear scaffold (lamin B1) and cytoskeletal proteins, such as fodrin and actin

(Casciola-Rosen et al, 1996; Martin et al., 1995; Ashkenas and Werb 1996). It appears that in the early stages of apoptosis the proteases selectively cut their substrates in order to disrupt the cellular metabolism in a controlled way.

Apoptosis in the Immune System

Some of the most widespread examples of the role of apoptosis in normal tissue development are found within the immune system (McConkey et al., 1996). The stem cells derived from the bone marrow migrate into the thymus and undergo a negative and positive selection that will be survived by approximately 10% of the population. A major part of the immature thymocytes, which either recognise self-derived polypeptides or are not able to recognise non-self antigens in a MHC restricted manner have to be deleted during thymic education. Similarly, B lymphocyte development requires apoptosis to delete for example autoreactive B cells. In the response to virally infected and transformed cells, lymphocyte-mediated cytotoxicity is a critical component of the immune response. Cell-mediated killing also underlies graft rejection following transplantation (Nagata and Golstein, 1995). These responses involve at least two general types of cells, the $CD8^+$ cytotoxic T lymphocyte (CTLs) and the Natural Killer (NK) cells. The killing by both cell types is affected by a cell-cell contact without effects on bystander cells (another hallmark of apoptosis). The very rapid kinetics of cell death observed in the target cells suggests that CTLs possess a uniquely efficient mechanisms to induce apoptosis. In fact, recent advances have shown that the CTL-induced lysis normally proceeds through the utilisation of two separate pathways (Kägi et al., 1994; Nagata and Golstein, 1995). The first mechanism comprises the formation of pores in the target cell membrane by secreted perforins. The second mechanism requires the expression of Fas Ligand (FasL) and Fas Receptor (Fas, APO-1 or CD95) on the cell membranes of the CTL and target cell, respectively. In the cell-cell contact the presence of FasL on the CTLs surface causes trimerisation of Fas on the target cell which initiates the activation of the proteolytic cascade triggering the apoptotic cell death. There is accumulating evidence that the Fas/APO-1/CD95 system not only plays a critical role in CTL killing, but is also involved in several other cell death inducing mechanisms existing in the immune system (Dhein et al., 1995; Bellgrau et al., 1995).

The Fas/FasLigand System

The Fas/FasL system is widely involved in the regulation of immune function. Fas is a member of the Tumour Necrosis Factor (TNF) receptor family of cytokine receptors. The extracellular domain of the receptor is linked by a transmembrane domain to the intracellular domain containing the so-called death domain. This domain shares large homology with a similar domain in the p60 TNF receptor and is essential for transmitting the apoptotic signal. FasL binding induces trimerisation of the Fas receptor such that the three clustered death domains form a binding site for several associating proteins. This results in a complex called the Death Inducing Signalling Complex (DISC). Two proteins identified as part of the DISC are Fas Associating protein with Death Domain (FADD or MORT1) and FADD-Like ICE (FLICE or MACHα). One part of FADD binds to the Death Domain of Fas whereas the other part provides the domains that are able to associate with FLICE. This second protein contains a pro-enzyme with an ICE-like structure coupled to a FADD binding domain. It is speculated that this pro-enzyme can be released and (autocatalytically?) activated into an ICE-like protease that in turn activates other ICE-like proteases (for more details see Boldin et al., 1996 or Muzio et al., 1996). One of the ICE-like proteases that gets activated is CPP32 (also called Yama or Apopain) which is capable of specifically cleaving PARP. Activation of CPP32 requires the processing of two 32 kD proenzymes into a homodimeric active protease consisting of two 17 kD and two 10 kD subunits (Schlegel et al., 1996).

The widespread expression of Fas in B and T lymphocytes as well as non-immune cells suggests an important role in regulation of immune function by cell death (Debatin, 1996). In addition to the involvement in CTL killing as described above, there is clear evidence for a role for the Fas/FasL system in the mechanism of immuneprivilege in the eye. As the eye can not tolerate destructive inflammatory responses, FasL is expressed in several compartments of the eye in order to kill any Fas carrying inflammatory cells (Griffith et al., 1995). A similar mechanism has been reported for the maintenance of the immuneprivilege of the testis (Bellgrau et al., 1995).

Normally the expression of FasL is very low and restricted (for example to tissues requiring immuneprivilege). However, activation of B and T lymphocytes in the periphery leads to an upregulation of FasL in those cells (Dhein et al., 1995; Brunner et al., 1995). These activated FasL expressing T cells can kill themselves (autocrine suicide), neighbouring T cells (fratricide) or other Fas carrying cells. In the peripheral immune system the consequence of activation of the Fas/FasL system during an ongoing immune response will be the modulation or down regulation of the response.

Too much Fas-mediated apoptosis and hyperactivity of the Fas/FasL system has been observed in HIV-infection and this may contribute to the T cell depletion that characterises the associated disease. The spontaneous apoptosis of $CD4^+$ and $CD8^+$ cells from HIV infected patients is strongly increased after T cell receptor triggering (Ameisen et al., 1995; Gougeon et al., 1995). The T cell receptor triggering induces autocrine cell death and this is greatly accelerated by viral proteins such as gp120 and HIV-1 Tat (Westendorp et al., 1995). In addition it has been shown that both Fas and FasL levels are increased in the patients T cells (Debatin, 1996). Thus, the Fas/FasL system plays an important role in the physiology and pathology of the immune system.

Redox Changes in Fas-Induced Apoptosis

As the Fas/FasL system is an important mechanism of inducing apoptosis our group has been focusing on biochemical changes during apoptotic cell death induced by anti-Fas/APO-1 antibody (which is mimicking the effect of FasL). One of our areas of interest is the oxidative change that occurs during apoptosis. Various reports have shown that as cells undergo apoptosis they gradually become more oxidised. The decreasing levels of reduced glutathione (GSH), α-tocopherol and protein thiols together with the down-regulation of antioxidant enzymes such as catalase, Mn and Cu/Zn superoxide dismutase, DT-diaphorase and thioredoxin evidence this change in redox state (Slater et al., 1995; Bustamante et al., 1995; Beaver and Waring, 1995; Briehl et al., 1995). Based on the use of redox sensitive dyes, several groups have observed an increased formation of reactive oxygen species (ROS) during apoptosis (Greenlund et al., 1995; Torres-Roca et al., 1995). However, it is still under debate whether or not oxidative stress is a critical part of the apoptotic process. The requirement for ROS to induce and execute apoptosis has been challenged by reports of ongoing apoptosis in a low oxygen environment (Muschel et al., 1995; Jacobson et al., 1995; Shimizu et al., 1995; Jacobsen, 1996).

Recent work suggests that apoptotic oxidative stress is not necessarily mediated by extensive formation of ROS. JURKAT T lymphocytes induced to undergo apoptosis with anti-Fas/APO-1 antibody were observed to rapidly lose their main cytosolic antioxidant GSH (50% decrease in GSH level approximately 60 min after adding the antibody). The reduced tripeptide was not oxidised but instead excreted into the medium of the cells. As this GSH efflux could be blocked by bromosulfophthalein and dibromosulphophthalein, known inhibitors of the hepatocyte canalicular membrane GSH transporter, a specific export rather than non-specific leakiness through the plasma membrane is proposed to be responsible (Van den Dobbelsteen et al., 1996). We have observed that stimulation of GSH efflux is present in several other cell types undergoing apoptosis. It provides a mechanism by which the cell can enhance its oxidative tone without requiring an increase in the production of ROS. The exact role of the GSH efflux in the apoptotic process is still unknown. We have found that other hallmarks of apoptosis such as nuclear fragmentation, DNA degradation, plasma membrane blebbing and phosphatidylserine exposure are not inhibited by blocking the extrusion of GSH (Van den Dobbelsteen et al.,

1996). We speculate that the export of GSH helps in the degradation of the apoptotic remnants by making the debris more vulnerable to proteolysis and stimulating phagocytic cells. In addition, the excreted GSH may be of nutritional benefit to normal bystander cells.

The Opposing Effects of Dithiocarbamates on Apoptosis

Dithiocarbamates comprise one class of molecules that have been widely used in studies in the redox regulation of apoptosis (Bessho et al, 1994; Wolfe et al., 1994; Nobel et al., 1995). They are known to exert both pro-oxidant and antioxidant effects in cell-free and biological systems. These usually water-soluble molecules have various agricultural (fungicides and insecticides), clinical (alcohol aversion therapy) and industrial applications. Chemically they are defined by possession of a $(R_1)(R_2)N\text{-}C(S)\text{-}S\text{-}R_3$ functional group. They readily form salts with various metals, and their thiol groups can also be oxidised to the corresponding disulphide.

When thymocytes are incubated with pyrrolidine dithiocarbamate (PDTC) or diethyldithiocarbamate (DDC) for longer incubation periods (in excess of 6 h) typical features of apoptosis, such as cell shrinkage, chromatin condensation and fragmentation, can be observed. Although these compounds can inhibit the rapid thymocyte apoptosis induced by other agents (Bessho et al., 1994; Wolfe et al., 1994; Nobel et al., 1995), in long term experiments they induce apoptosis themselves. We observed that this toxicity was inhibited by the inclusion of non-cell membrane permeable metal chelators in the incubation medium, suggesting that the toxicity of PDTC was related to its ability to chelate metals (present in the fetal calf serum) in a lipophilic, membrane-permeable complex (Nobel et al., 1995). This was confirmed by the detection of an 8-fold increase in intracellular copper content of PDTC-treated thymocytes. As it is well known that mild oxidative stress can induce apoptosis in different cell types (Lennon et al., 1991; Dypbukt et al., 1994), a copper-dependent oxidative stress probably signals the cell to undergo apoptotic cell death.

In contrast to the ability of inducing apoptotic cell death in long term incubations, we and others have found that in short term incubations both PDTC and DDC inhibit apoptosis induced by a variety of agents (Nobel et al., 1996). It has been argued that this inhibitory effect of dithiocarbamates is due to their radical scavenging properties (Wolfe et al., 1994; Bessho et al., 1994; Nobel et al., 1995). However, when rat thymocytes were incubated with micromolar concentrations of PDTC, we detected a considerable oxidation of intracellular glutathione. This PDTC-dependent glutathione oxidation was associated with inhibition of apoptosis induced by either the glucocorticoid methylprednisolone or the topoisomerase II inhibitor etoposide (Nobel et al., 1995). We have recently found that dithiocarbamates can by oxidised in the cell to their corresponding disulphide in a copper dependent oxidation. The disulphide disulfiram (DSF) was, on a molar basis, a more potent inhibitor of apoptosis than either PDTC or DDC, even though it is a poorer radical scavenger. This suggests that dithiocarbamate inhibition of apoptosis may relate to oxidation of critical thiols rather than a general scavenging of oxygen radicals.

The inhibitory effect on apoptosis by short term incubation with dithiocarbamates is not unique to thymocytes. Using the human JURKAT T lymphocyte cell line treated with anti-Fas/APO-1 antibody, we have observed that both PDTC and DSF exert a time and concentration-dependent inhibition of apoptotic cell death (unpublished data). The inhibition by DSF is associated with reduced processing of the 32 kD proenzyme of CPP32, an ICE-like protease critically involved in Fas-mediated apoptosis in these cells (Schlegel et al., 1996). Furthermore DSF directly inhibits CPP32 activity in enzyme assays using pre-activated protein. This direct inhibition of CPP32 is removed in the presence of DTT, indicating that it requires the presence of a disulphide bond. We propose that dithiocarbamates and thiuram disulphides impose a copper-dependent oxidative stress on cells, that by altering the glutathione and protein thiol redox states prevents the apoptotic machinery from functioning. We hypothesise that this inhibition operates via redox inactivation of one or more ICE-like proteases.

Thus, these results obtained with dithiocarbamates demonstrate that oxidative stress can have opposing effects on the induction of apoptosis in cells of the immune system. Whereas in short term incubations thiol oxidation leads to inactivation of critical members of the ICE-like family of proteases, in long term incubations the effects of the oxidative stress are gradually overcome and the mild oxidative stress actually triggers the apoptotic program.

Conclusions

With our studies on the metabolism of GSH and the opposing effects of dithiocarbamates on T cell apoptosis we have made an attempt to better understand the interaction between oxidative stress and apoptosis. The presence of an efflux of the cell's major antioxidant GSH during apoptosis provides a non-oxidative mechanism by which the reducing environment of the cell is lost. An increase in oxidative damage to proteins and lipids would then result even in the absence of an increased formation of oxidants. This may explain the seemingly contradictory finding that even in the absence of an increased formation of oxidants there can be increased oxidative damage to proteins and lipids in apoptotic cells.

The opposing effects that thiol oxidising molecules such as dithiocarbamates have on apoptotic cell death, confirm that the machinery involved in regulation and execution of apoptosis can be easily disturbed. Toxicants altering the cellular redox state can easily interfere with apoptosis and are therefore of potential hazard to organ systems that rely on the tight regulation of apoptosis such as the immune system.

Acknowledgements

This work was supported by fellowships from the European Science Foundation and Karolinska Institute (to DD) and a grant from the Swedish Medical Research Council.

References

Ameisen JC, Estaquier J, Idziorek T and De Bels F (1995) Programmed cell death and AIDS: significance, perspectives and unanswered questions. Cell Death Diff 2: 9-22

Ashkenas J and Werb Z (1996) Proteolysis and the biochemistry of life or death decisions. J Exp Med 183: 1947-1951

Beaver JP and Waring P (1995) A decrease in intracellular glutathione concentration precedes the onset of apoptosis in murine thymocytes. Eur J Cell Biol 68: 47-54

Bellgrau D, Gold D, Selawry H, Moore J, Franzusoff A and Duke RC (1995) A role for CD95 ligand in preventing graft rejection. Nature 377: 630-632

Bessho R., Matsubara K, Kubota M, Kuwakado K, Hirota H, Wakazono Y, Lin YW, Okuda A., Kawai M, Nishikomori R and Heike T (1994) Pyrrolidine dithiocarbamate, a potent inhibitor of NFκB activation, prevents apoptosis in human promyelocytic leukemia HL-60 cells and thymocytes. Bioch Pharmacol 48: 1883-1889

Boldin MP, Goncharov TM, Goitsev YV and Wallach D (1996) Involvement of MACH, a novel MORT1/FADD-interacting protease, in Fas/APO-1- and TNF receptor-induced cell death. Cell 85: 803-815.

Briehl MM, Cotgreave IA and Powis G (1995) Downregulation of the antioxidant defense during glucocorticoid-mediated apoptosis. Cell Death Diff 2: 41-46

Brunner T, Mogil RJ, LaFace D, Yoo NJ, Mahboubi A, Echeverri, Martin SJ, Force WR, Lynch DH, Ware CF and Green DR (1995) Cell-autonomous Fas (CD95)/Fas-ligand interaction mediates activation-induced apoptosis I T-cell hybridomas. Nature 373: 441-444

Bustamante J, Slater AFG and Orrenius S (1995) Antioxidant inhibition of thymocyte apoptosis by dihydrolipoic acid. Free Rad Biol Med 19: 339-347

Casciola-Rosen L, Nicholson DW, Chong T, Rowan KR, Thornberry N, Miller DK and Rosen A (1996) Apopain/CPP32 cleaves proteins that are essential for cellular repair: a fundamental principle of apoptotic cell death. J Exp Med 183: 1957-1964

Dhein J, Walczak H, Bäumler C, Debatin KM and Krammer PH (1995) Autocrine T cell suicide mediated by APO-1/(Fas/CD95). Nature 373: 438-441
Dypbukt JM, Ankarcrona M, Burkitt M, Sjöholm Å, Ström K, Orrenius S and Nicotera P (1994) Different prooxidant levels stimulate cell growth, activate apoptosis, or produce necrosis in insulin-secreting RINm5F cells. J Biol Chem 269: 30553-30560
Gougeon ML (1995) Does apoptosis contribute to CD4 T cell depletion in human immunodeficiency virus infection? Cell Death Diff 2: 1-8
Greenlund LJS, Deckwerth TL and Johnson EM, Jr. (1995) Superoxide dismutase delays neuronal apoptosis: A role for reactive oxygen species in programmed neuronal death. Neuron, 14: 303-314
Griffith TS, Brunner T, Fletcher SM, Green DR and Ferguson TA (1995) Fas ligand-induced apoptosis as a mechanism of immuneprivilege. Nature 370: 1189-1192
Jacobson MD and Raff MC (1995) Programmed cell death and Bcl-2 protection in very low oxygen. Nature 374: 814-816
Jacobson MD (1996) Reactive oxygen species and programmed cell death. Trends in Bioch Sci 21: 83-86
Kerr JFR, Wyllie AH and Currie AR (1972) Apoptosis: a basic biological phenomenon with wide ranging implications in tissue kinetics. Br J Cancer 26: 239-257
Lennon SV, Martin SJ and Cotter TG (1991) Dose-dependent induction of apoptosis in human tumour cell lines by widely diverging stimuli. Cell Proliferation 24: 203-204
Martin SJ and Green DR (1995) Protease activation during apoptosis: death by a thousand cuts? Cell 82: 349-352
Martin SJ, O'Brien GA, Nishioka WK, McGahon AJ, Mahoubi A, Saido and Green DR (1995) Proteolysis of fodrin (non-erythroid spectrin) during apoptosis. J Biol Chem 270: 6425-6428
Martin SJ, Reutelingsberger CPM, McGahon AJ, Rader JA, Van Schie RCAA, LaFace DM and Green DR (1995) Early redistribution of plasma membrane phosphatidylserine is a general feature of apoptosis regardless of the initiating stimulus: inhibition by overexpression of Bcl-2 and Abl. J Exp Med 182: 1545-1556
McConkey DJ, Zhivotovsky B and Orrenius S (1996) Apoptosis - Molecular mechanisms and biomedical implications. Mol Asp Med 17: 1-110
Muschel RJ, Bernhard EJ, Garza L, McKenna WG and Koch C (1995) Induction of apoptosis at different oxygen tensions: Evidence that oxygen radicals do not mediate apoptotic signaling. Cancer Res 55: 995-998
Muzio M, Chinnaiyan AM, Kischkel FC, O'Rourke K, Shevchenko A, Ni J, Scafidi K, Bretz JD, Zhang M, Gentz R, Mann M, Krammer PH, Peter M and Dixit VM (1996) FLICE, a novel FADD-homologous ICE/CED-3-like protease, is recruited to the CD95 (Fas/APO-1) Death Inducing Signaling Complex. Cell 85: 817-827
Nagata S and Golstein P (1995) The Fas death factor. Nature 267: 1449-1455
Nobel CSI, Kimland M, Lind B, Orrenius S and Slater AFG (1995) Dithiocarbamates induce apoptosis in thymocytes by raising the intracellular level of redox-active copper. J Biol Chem 270: 26202-26208
Savill JS, Fadok V, Henson P and Haslett C (1993) Phagocyte recognition of cells undergoing apoptosis. Immunol Today 14: 131-136
Schlegel J, Peters I, Orrenius S, Miller DK, Thornberry NA, Yamin TT and Nicholson DW (1996) CPP32 is the ICE-like protease in Fas-mediated apoptosis. J Biol Chem 271: 1841-1844
Shimizu S, Eguchi Y, Kosaka H, Kamiike W, Matsuda H and Tsujimoto Y (1995) Prevention of hypoxia-induced cell death by Bcl-2 and Bcl-xL. Nature 374: 811-813
Slater AFG, Nobel CSI, Maellaro E, Bustamante J, Kimland M and Orrenius S (1995) Nitrone spin traps and a nitroxide antioxidant inhibit a common pathway of thymocyte apoptosis. Biochem J 306: 771-778

Torres-Roca JF, LeCoeur H, Amatore C and Gougeon ML (1995) The early intracellular production of reactive oxygen species mediates apoptosis in dexamethasone-treated thymocytes. Cell Death Diff 2: 309-319

Van den Dobbelsteen DJ, Nobel CSI, Schlegel J, Cotgreave IA, Orrenius S and Slater AFG (1996a) Rapid and specific efflux of reduced glutathione during apoptosis induced by anti-Fas/APO-1 antibody. J Biol Chem 271: 15420-15427

Van den Dobbelsteen DJ, Nobel CSI, Samuelsson A, Orrenius S and Slater AFG (1996b) Glutathione metabolism during apoptosis. In: Oxidative stress, Cancer, AIDS and Neurodegenerative diseases. Marcel Dekker, New York, in press

Westendorp MO, Frank R, Ochsenbauer C, Stricker K, Dhein J, Walczak H, Debatin KM and Krammer PH (1995) Sensitisation of T cells to CD95-mediated apoptosis by HIV-1 Tat and gp120. Nature 375: 497-500

Wolfe JT, Ross D and Cohen GM (1994) A role for metals and free radicals in the induction of apoptosis in thymocytes. FEBS Lett 352: 58-62

Wyllie AH, Kerr JFR and Currie AR (1980) Cell death: The significance of apoptosis. Int Rev Cytol 68: 251-306

Immunoregulatory Genes and Immunosuppression by Glucocorticoids

Nathalie Auphan*, Joseph A. Didonato, Arno Helmberg, Caridad Rosette and Michael Karin.
Department of Pharmacology, Center for Molecolar Genetics, School of Medicine, University of California San Diego, La Jolla, CA 92093-0636, USA.
* Present address: Centre d'Immunologie INSERM-CNRS de Marseille-Luminy, Case 906, F-13288 Marseille cedex 09, France.

Introduction

Glucocorticoids (GCs) are physiological inhibitors of inflammatory responses and are widely used as immunosuppressive and anti-inflammatory agents. GCs induce (i) lymphocyte apoptosis (Cohen, 1989, Galili, 1983) and (ii) inhibit synthesis of lymphokines (Arya et al., 1984, Culpepper and Lee, 1985, Knudsen et al., 1987, Wang et al., 1993, Zanker et al., 1990) and cell surface molecules (Von Knebel et al., 1990) required for immune function. In spite of the widespread use of GCs, the molecular mechanisms that underlie their therapeutic effects are poorly understood.

GCs induce target gene expression through binding to a specific receptor (GR) which functions as a ligand-activated transcription factor (Beato, 1989). Therefore, it was suggested that GCs-dependent apoptosis is mediated by induction of so-called "lysis genes". However, such genes have not yet been identified.

Another relevant mechanism that could mediate the immunosuppressive effects of GCs was suggested by the ability of ligand-activated GR to repress gene expression through transcriptional interference with other transcription factors, most notably AP-1 (Jonat et al., 1990, Yang-Yen et al., 1990). Indeed, analysis of the *cis* elements through which GCs inhibit IL-2 production failed to identify GR binding sites (GREs) and implicated sites recognized by AP-1 as mediators of this effect (Northrop et al., 1992). Such findings raised the possibility that induction of T cell apoptosis may also be mediated by interference with transcription factors, such as AP-1, required for cell survival rather than direct induction of "lysis genes". To test this hypothesis, we introduced into a GR-deficient T cell leukemia cell line, expression vectors encoding either wild type (wt) GR or a mutant derivative that is defective in gene activation but fully competent in interference with AP-1 activity. We found that clones expressing this mutant GR were fully sensitive to glucocorticoids, both in respect of inhibition of IL-2 production and in undergoing apoptosis.

A number of lymphokine encoding genes which are repressed by GCs are not transcriptionally controlled by AP-1 raising the possibility that GCs may interfere with others transcription factors. IL-2 promoter contains *cis* elements required for maximal induction during T cell activation, which are recognized by AP-1, NF-κB, NF-AT and octamer binding proteins (OBP). We investigated which of these factors, in addition to AP-1, mediate the inhibitory effect of GCs. We found that GCs are potent inhibitors of NF-κB through induction of the IκBα inhibitory protein, which traps activated NF-κB in inactive cytoplasmic complexes.

Materials and Methods

Cell Culture Conditions: Jurkat cells were cultured in RPMI 1640, containing 10% fetal calf serum, 10 mM Hepes and 50 μM β-mercaptoethanol.

Isolation of stable transfectants: 10^7 Jurkat cells were electroporated with 20 μg of PvuI-linearized pRc/βact-GR-wt or pRc/βact-GR-LS7 plasmids described in Helmberg et al. (1995). G418-resistant clones were examined by thymidine incorporation assays in the presence or absence of dexamethasone (dex).

Transient Transfections: 10^7 Jurkat cells were electroporated with 10 μg of reporter plasmid. After overnight culture, cells were stimulated for 8 hours in different conditions, lysed and luciferase activity was measured and normalized to the protein concentration of extracts.

DNA fragmentation assay: 10^6 Jurkat cells were cultured overnight in presence of 10^{-8} M dex. Cells were lysed in 400 μl of DNA isolation buffer (10 mM Tris-HCl pH 8.0, 100 mM NaCl, 25 mM EDTA, 0.5% SDS and 0.5 mg/ml proteinase K). After a 5 hours incubation at 55°C, DNA was ethanol precipitated, resuspended in 50 μl of TE, resolved on a 2% agarose gel and visualized by ethidium bromide.

Electrophoretic Mobility Shift Assay (EMSA): Nuclear extracts were prepared as described (Devary et al., 1993). 5 μg of each extract were incubated with 20,000 cpm of ^{32}P-labeled κB, Oct-1, or AP-1 (TRE) binding site probes and analyzed on non-denaturing polyacrylamide gels.

RNA Extraction and Analysis by Northern Blot: Total cytoplasmic RNA was extracted and 20 μg samples were separated on 1 % agarose gel, transferred to a nylon membrane, and analyzed by hybridization to *MAD-3* (IκBα) and p65 (RelA) complementary DNA probes.

Protein Extraction and Analysis by Western Blot: Cells were lysed and 40 µg of each lysate was separated by SDS-polyacrylamide gel electrophoresis (PAGE), transferred to a polyvinylidene difluoride membrane (Millipore), and immunoblotted with either anti-IκBα or anti-p65 and donkey anti-rabbit immunoglobulin conjugated to peroxidase as described (Devary et al., 1993).

Results

Transcriptional Repression and Activation by GR mutant: Mutant LS7 contains two adjacent amino acid substitutions in the second half of the second zinc finger, replacing Pro493 and Ala494 by Arg and Ser, respectively (Fig. 1) (Godowski, et al., 1989). Transient transfections were performed to determine the relative ability of the receptor mutant to repress AP-1 activity (Fig. 2a) in Jurkat cells. AP-1 activity, as measured by -73Col-Luc expression (containing a truncated collagenase promoter previously defined as AP-1-dependent (Deng and Karin, 1993)), was not detected in non-activated Jurkat cells, but was strongly induced upon activation with TPA and the Ca^{2+}-ionophore ionomycin. All transfected GR types markedly suppressed this induction in response to the synthetic glucocorticoid dexamethazone (dex).

Several stably transfected cell clones were isolated that constitutively expressed either wild-type or mutated GR. GRE-mediated transactivation in GR-wt, GR-LS7 and in parental Jurkat cells is shown in Figure 2b. GRE-mediated transactivation was determined by a MMTV-Luc reporter containing two consensus GREs. As previous results in other cell type (Godowski et al., 1989), LS7 exhibited minimal transactivating potential.

Dex-mediated apoptosis of GR-expressing Jurkat cells: DNA fragmentation by internucleosomal cleavage, a characteristic of apoptosis, was detected after a 40 hours incubation with dex in GR-wt and GR-LS7 expressing clones but not in the parental cells (Fig. 2). Thus, dex inhibits proliferation of GR expressing cells by inducing apoptosis, independently of GRE-mediated transactivation.

Dex-mediated inhibition of NF-κB activation requires induction of a short-lived inhibitory protein: Because of its role in the activation of lymphokine genes and other immunoregulatory genes (Grilli, Chiu et al.,1993), we examined whether NF-κB activity is inhibited by GCs.

Inhibition of NF-κB activation by dex was observed in GR-wt transfected Jurkat cells, but not in the parental GR^- subclone (Fig 3a). Therefore, the inhibitory effect of dex is mediated through the GR.

Inhibition of NF-κB activity required 45 to 60 min of dex treatment as well as new protein synthesis, as it was prevented by cycloheximide (Fig.3b). Upon dex removal, but in the continued presence of TPA, NF-κB binding activity reappeared after 30 min and peaked at 60 min. However, if dex was left on the cells for 5 hours, NF-κB activity remained repressed (data not shown, see (Auphan et al., 1995)). These results suggest that inhibition of NF-κB is mediated by a short-lived dex-induced protein.

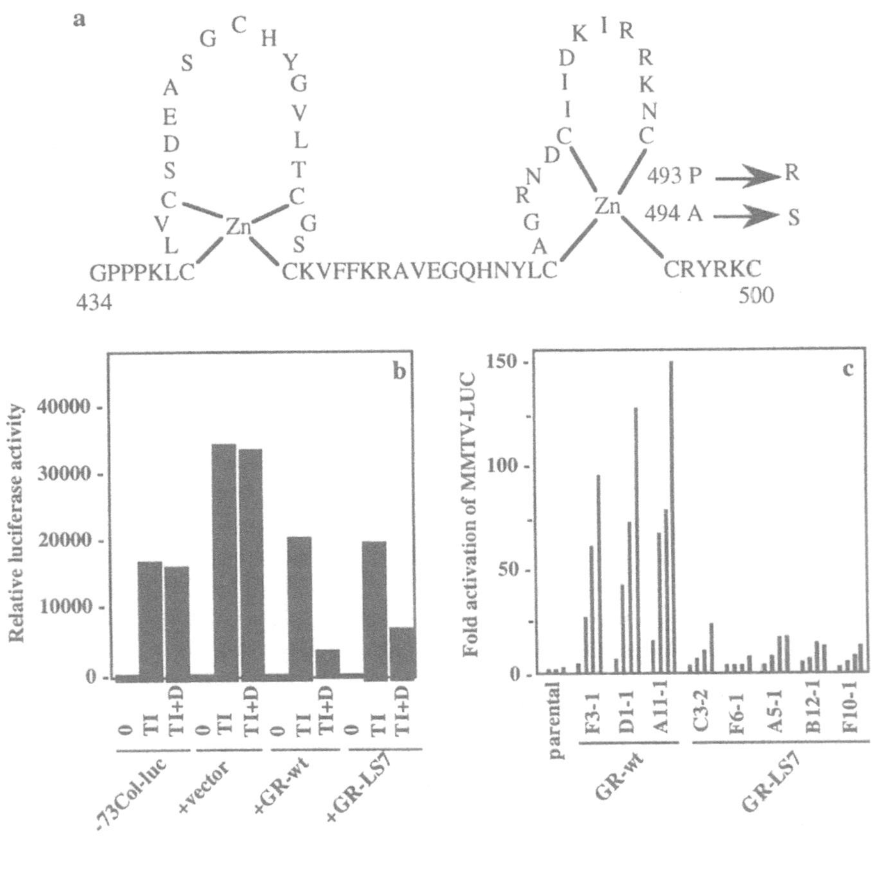

Fig. 1: Transcriptional repression and activation by GR mutant. **a)** Primary struture of the Zn-finger region of wt and LS7-GR (Godowski et al, 1989); **b)** dex-dependent repression of the collagenase promoter in Jurkat cells. Cells were cotransfected with - -73Col-Luc and the respective GR expression vectors or the empty expression vector, grown overnight and treated with TPA (T) plus ionomycin (I) in absence or presence of dex (D). **c)** GR-mediated transactivation in parental and stably transfected Jurkat cells. Different transfected clones and the parental Jurkat cells were elctroporated with MMTV-Luc reporter. Cells were divided in five samples and cultured either in medium alone or in presence of $3x10^{-10}$, $1x10^{-9}$, $3x10^{-9}$ and 10^{-8} M dex. For each clone, the four bars indicate luciferase activity in dex treated cells divided by luciferase activity in cells cultured in medium alone.

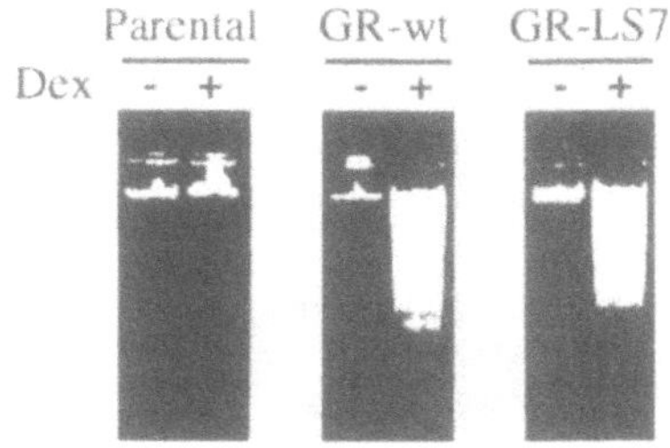

Fig. 2: DNA fragmentation associated with glucocorticoid-induced apoptosis. Parental, GR-wt and GR-LS7 cells were incubated with 10^{-8} M dex. After overnignt culture, DNA was extracted and run on a 2% agarose gel in presence of ethidium bromide.

Dex treatment increases abundance of IκBα (*MAD-3*): A possible mediator of the inhibitory effect of dex is IκBa, encoded by the *MAD-3* gene (Haskill et al., 1991), as it is a short-lived protein whose degradation is further enhanced by various NF-κB inducers (Brown et al., 1993, de Martin et al., 1993, Henkel et al., 1993). These inducers also cause IκBα resynthesis by stimulating *MAD-3* transcription, a feed-back response that terminates NF-κB activation. Immunoblotting experiments showed that treatment with TPA led to a partial disappearance of IκBα after 30 min (Fig. 4). However, in cells treated with TPA plus dex, the amount of IκBα was the same as in unstimulated cells. None of these treatments affected p65 abundance.

Dex induces expression of the gene encoding IκBα *(MAD-3)*: Because dex did not prevent IκBα degradation at early time points (Fig. 5), the increase in IκBα abundance in dex-treated cells is probably the result of increased IκBα synthesis. Northern blot analysis verified this assumption and indicated that dex treatment of GR-wt Jurkat cells, but not the GR$^-$ parental clone, led to 7-fold induction of *MAD-3* (IκBα) mRNA. Stimulation of either GR-wt or GR$^-$ Jurkat cells with TPA resulted in 6-fold induction of *MAD-3* mRNA after 1 hour. Treatment with both TPA and dex resulted in a synergistic induction (23-fold) but only in cells expressing GR-wt. None of these treatments had a substential effect on expression of p65 (RelA) mRNA. Induction of IκBα expression by TPA and dex is most likely the result of increase transcription because it was inhibited by actinomycin D. Indeed, at the same time, Scheinman et al. demonstrate that dex increases the transcription rate of the MAD-3 gene (Scheinman et al., 1995).

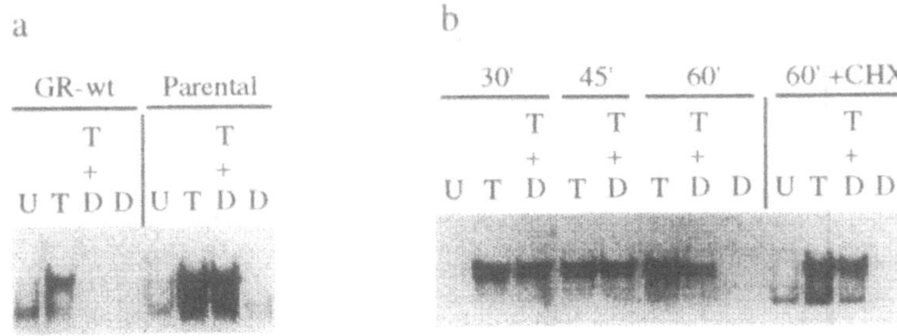

Fig. 3: Inhibition of NF-κB activation requires both the GR and the synthesis of a short-lived inhibitory protein. a Parental and GR-wt transfected Jurkat cells were either unstimulated (U) stimulated for 1 hour as indicated (T:TPA; T+D: TPA+dex; D:dex). b GR-wt Jurkat cells were stimulated for different times as indicated. When mentioned, cells were pretreated with cycloheximide (CHX, 10μg/ml) for 30 min prior stimulation with TPA and /or dex. For a and b, nuclear extracts were examined for NF-κB binding activity.

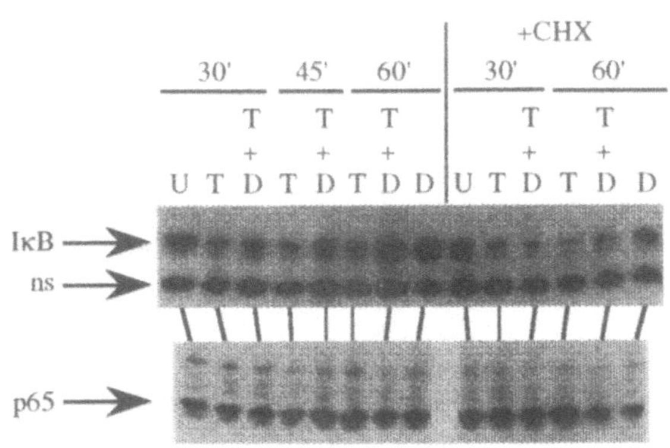

Fig. 4: Dex treatment increases the abundance of IκBα. GR-wt Jurkat cells were treated for different lenghts of time as indicated and lysates were analyzed by immunoblotting with either anti-IκBα or anti-p65 serum. When mentioned, cells were pretreated with cycloheximide (CHX, 10μg/ml) for 30 min prior to stimulation with TPA and /or dex.

Conclusion

Since the classical mode of action of the ligand activated GR is induction of gene transcription, it has been assumed for many years that T cell apoptosis is mediated by induction of "lysis genes". However, no functional glucocorticoid-inducible "lysis genes" were identified. The work described here strongly suggests that an alternative mechanism is involved. Indeed, a variant GR that is severely compromised in its ability to activate GRE-mediated gene transcription, but fully capable of interfering with AP-1 activity, confers full glucocorticoid sensitivity when expressed in GR-deficient Jurkat cells. Analysis of five different clones expressing this mutant receptor LS7 indicates that they all undergo apoptosis in response to dex. Likewise, expression of retinoic acid receptor α, which shares only its AP-1 interfering effect with the GR, was found to confer retinoic acid-mediated apoptosis on these cells (data not shown, see Helmberg et al., 1995). Thus, GCs are unlikely to cause apoptosis in Jurkat cells by direct induction of "lysis genes". Another hypothesis consistent with our findings is that GCs induce apoptosis by inhibiting the expression of genes needed for survival (Raff, 1992) which still have to be identified, this repression being likely mediated through transcriptional interference between the activated GR and transcription factors such as AP-1.

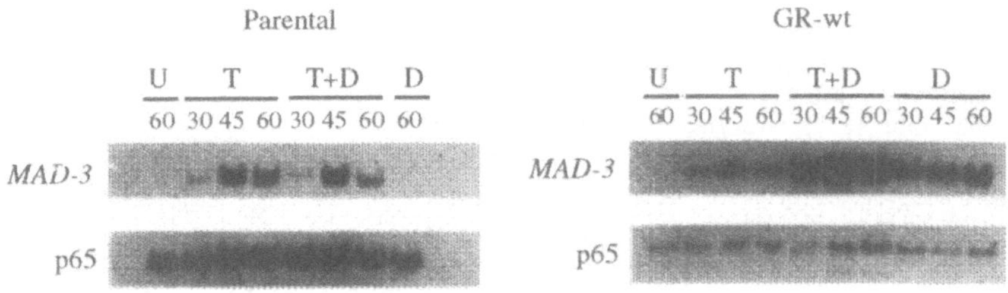

Fig. 5: Dex induces expression of the gene encoding IκBα (*MAD-3*). Parental and GR-wt Jurkat cells were treated for different lenghts of time (in minutes), as indicated and total RNA was analyzed by hybridization to *MAD-3* and p65 complementary DNA probes.

However, not all the repressive functions mediated by GCs can be explained by interference of its receptor with AP-1. Indeed, here we define a novel and simple mechanism through which GCs repress NF-κB activity, which does not rely on direct protein-protein interaction between the activated GR and any NF-κB constituent. Rather, the inhibition is based on induction of *MAD-3* mRNA resulting in the fast reappearance of IκBα in activated cells treated with dex, which then lead to the retention of p65 in the cytoplasm (result not shown, see Auphan et al., 1995).

NF-κB plays a central role in the induction of a large number of important immunoregulatory genes including those encoding IL1, IL2, IL3, IL6, IL8, TNFα, IFNγ, GM-CSF (Grilli et al., 1993). Several of these genes are also regulated by AP-1, which synergizes with NF-κB. Because GCs inhibit both AP-1 and NF-κB activities, albeit through different mechanisms, it is not surprising that they repress the expression of a wide variety and a whole spectrum of immunoregulatory genes.

References

Arya SK, Wong-Staal F and Gallo RC (1984) Dexamethazone-mediated inhibition of human T cell growth factor and γ-interferon messenger RNA. J. Immunol. 133: 273-276.

Auphan N, DiDonato JA, Rosette C, Helmberg A and Karin M (1995) Immunosuppression by glucocorticoids: inhibition of NF-κB activity through induction of IκB synthesis. Science 270: 286-290.

Beato M (1989) Gene regulation by steroid hormones. Cell 56: 335-344.

Brown K, Park S, Kanno T, Franzoso G and Siebenlist U (1993) Mutual regulation of the transcriptional activator NF-κB and its inhibitor, IκB-α. Proc. Natl. Acad. Sci. 90: 2532-2536.

Cohen JJ (1989) Lymphocyte death induced by glucocorticoids. Anti-inflammatory steroid action, basic and clinical aspects. I. Academic press. 110-131.

Culpepper JA and Lee F (1985) Regulation of IL-3 expression by glucocorticoids in cloned murine lymphocytes. J. Immunol. 135: 3191-3197.

de Martin R, Vanhove B, Cheng Q, Hofer E, Csizmadia V, Winkler H and Bach FH (1993) Cytokine-inducible expression in endothelial cells of an IκBα-like gene is regulated by NFκB. EMBO J. 12: 2773-2779.

Deng T and Karin M (1993) JunB differs from c-Jun in its DNA-binding and dimerization domains, and represses c-Jun by formation of inactive heterodimers. Genes Devel. 7:479-490.

Devary Y, Rosette C, DiDonato J and Karin M (1993) NF-κB activation by ultraviolet light not dependent on a nuclear signal. Science 261: 1442-1445.

Galili U (1983) Glucocorticoid induced cytolysis of human normal and malignant lymphocytes. J. Steroid Biochem. 19: 483-490.

Godowski PJ, Sakai DD and Yammamoto KR (1989) Signal transduction and transcriptional regulation by glucocorticoid receptor. DNA-protein interactions in transcription. 95. I. A. R. Liss. 197-210.

Grilli M, Chiu JJ and Lenardo MJ (1993) NF-κB and Rel: participants in a multiform transcriptional regulatory system. Int. Rev. Cyt. 143: 1-62.

Haskill S, Beg AA, Tompkins SM, Morris JS, Yurochko AD, Sampson-Johannes A, Mondal K, Ralph P and Balwin Jr AS (1991) Characterization of an immediate-early gene induced in adherent monocytes that encodes IκB-like activity. Cell 65: 1281 1289.

Helmberg A, Auphan N, Caelles C and Karin M (1995) Glucocorticoid-induced apoptosis of human leukemic cell is caused by the repressive function of the glucocorticoid receptor. EMBO J. 14: 452-460.

Henkel T, Machleidt T, Alkalay I, Kronke M, Ben-Neriah Y and Baeuerle P (1993) Rapid proteolysis of IκB-α is necessary for activation of transcription factor NF-κB. Science 365: 182-185.

Jonat C, Rahmsdorf HJ, Park KK, Cato A, Gebel S, Ponta H and Herrlich P (1990) Antitumor promotion and antiinflammation: Down-modulation of AP-1 (Fos/Jun) activity by glucocorticoid hormone. Cell 62: 1189-1204.

Knudsen PJ, Dinarello CA and Strom TB (1987) Glucocorticoids inhibit transcriptional and post-transcriptional expression of interleukin 1 in U937 cells. J. Immunol. 139: 4129-4134.

Northrop J, Crabtree GR and Mattila PS (1992) Negative regulation of interleukin 2 transcription by the glucocorticoid receptor. J. Exp. Med. 175: 1235-1245.

Raff M (1992) Social controls on cell survival and cell death. Nature 356: 397-400.

Scheinman R, Cogswell P, Lofquist A and Baldwin AJ (1995) Role of transcriptional activation of IκBα in mediation of immunosuppression by glucocorticoids. Science 270: 283-286.

Von Knebel Doeberitz M, Koch S, Drzonek H and Zur Hausen H (1990) Glucocorticoid hormones reduce the expression of major histocompatibility class I antigens on human epithelial cells. Eur. J. Inmmunol. 20: 35-40.

Wang Y, Campbell HD and Young IG (1993) Sex hormones and dexamethasone modulate interleukin-5 gene expression in T lymphocytes. J. Steroid Biochem. and Mol. Biol. 44: 203-210.

Yang-Yen H-F, Chambard J-C, Sun Y-L, Smeal T, Schmidt TJ, Drouin J and Karin M (1990) Transcriptional interference between c-Jun and the glucocorticoid receptor: mutual inhibition of DNA binding due to direct protein-protein interaction. Cell 62: 1205-1215.

Zanker B, Walz G, Wieder KJ and Strom TB (1990) Evidence that glucocorticoids block expression of the human interleukin-6 gene by accessory cells. Transplantation 49: 183-185.

Molecular Aspects of UVB-Induced Immunosuppression

Johan Garssen, Rob J. Vandebriel and Henk van Loveren
Laboratory for Pathology and Immunobiology, National Institute of Public health and the Environment, P.O. Box 1, NL-3720 BA Bilthoven, The Netherlands

Abstract

Ultraviolet light can affect the immune system locally as well as systemically leading to an impaired resistance to neoplastic cells and/or infections. Prior to the biological effect, UVB must be absorbed by a chromophore in the skin where it will give a signal that can lead to an altered immune response in the skin or elsewhere. These altered immune responses may be constituted by alteration in among others: cytokine profile, growth factors and costimulatory signals. Several hypotheses about the identity of the photoreceptor have been put forward. One photoreceptor in the skin is urocanic acid (UCA), that can isomerize from the *trans*- to the *cis*-isomer. The *cis*-isomer has immunosuppressive properties. Another photoreceptor is DNA that also efficiently absorbs UV wavelenghts. After absorption the structure of the DNA molecule is altered. This alteration might lead to gene activation responsible for the immunotoxic outcome (altered gene expression). It has been demonstrated that the formation of DNA photoproducts by UV light is associated with the activation of many genes. Several studies indicate that UV-induced DNA damage, in the form of cyclobutyl pyrimidine dimers plays a role in UV-induced suppression of the immune system locally as well as systemically. In mice that were injected with liposomes containing the excision repair enzyme T4 endonuclease UVB-induced dimers were removed more efficiently as compared to control mice. In these mice UV-induced immunosuppression was prevented. Pilot studies by Kripke et al. indicated that the release of IL-10 and TNFα that are both induced by DNA damage might be involved. In preliminary studies with mice that were deficient with respect to DNA repair lower doses of UV were needed for the induction of immunosuppression as compared to their normal littermates. These studies indicate that altered gene expression plays a pivotal role in UVB-induced immunosuppression. In addition to a role for UCA and DNA in UV-induced immunosuppression it is postulated recently that signal transduction (EGF-receptor mediated upregulation of phospholipase A2) and transcription factors (NFκB, p91) are involved in UV-induced immunomodulation.

Introduction

Ultraviolet radiation, a component of sunlight, has been recognized by photobiologists, dermatologists, and oculists as a potential hazard for human health because of its genotoxic, carcinogenic and *immunotoxic* properties. Its effects on human health include the induction of skin cancers, ocular damage and impairment of immunity to certain infections. A few decennia ago it was demonstrated that UV photons can affect the activity of the immune system through interactions with the skin. This means that UV not only changes normal cells into cancer cells but also permits the outgrowth of the UV-transformed cells by depressing the immune system. An intriguing question is what interactions between UV radiation and the skin initiates alterations in immune function in the exposed skin and systemically, i .e. in other places than the exposed skin.

During the last 20 years many studies have been performed in order to investigate the immunosuppressive activities of UVB in laboratory animals and in human volunteers. In particular effects of UVB radiation on resistance to tumours and skin associated infections have been examined. In addition, effects of UVB radiation on immune parameters such as contact hypersensitivity and delayed-type hypersensitivity (both type IV hypersensitivity reactions), mixed lymphocyte reactions, mixed skin lymphocyte reactions, antigen presentation and numbers and function of Langerhans cells have been studied intensively.

The antigenicity of murine tumours which are caused by UVB radiation was one of the first items to be investigated (Kripke, 1974). These antigenic tumours were rejected when transplanted into normal mice, and accepted in UVB-exposed mice, indicating a systemic UVB-induced immunosuppression. This suppression was transferable by lymphocytes from the spleen of UV-irradiated donor mice (Daynes and Spellman, 1977).

The first indications that the immune system also plays an important role in the incidence of skin tumours *in humans* came from studies in which patients with organ transplants were exposed to sunlight in combination with therapeutic immunosuppression. It is clear that this combination can increase the incidence of malignancies especially at sun-exposed sites. Allograft recipients (taking immunosuppressive drugs) are at greater risk of developing skin malignancies after sunlight exposure than the general population, probably as a result of a decrease in "immune surveillance" due to immunotherapy in combination with an additional but crucial UV-induced immunosuppression. These patients are particulary prone to developing skin cancers in areas of the world where UVB exposure is high, for instance Australia. Whether immunosuppression in these areas is more severe compared to areas where UVB exposure is low remains to be analyzed (Bouwes-Bavinck, 1995).

Limited clinical data on immunosuppressive activity in patients with UV-related skin cancers suggest that the immunosuppressive activity of UVB radiation in humans is similar to that in mice. Exposure of human skin to suberythemal doses of UVB followed by application of a contact sensitizer to the irradiated site resulted in a decreased contact hypersensitivity (CHS) response in

40% of the normal subjects and in 90% of the patients with (sunlight induced) skin cancer. In the persons who failed to respond, immunological "tolerance" was demonstrated in the majority of the skin cancer patients but not in the controls (Yoshikawa et al., 1990). It is remarkable that skin type does not determine the potential immunosuppressive effects of UVB radiation; black volunteers were equally susceptible as caucasian volunteers to UVB induced suppression of CHS (Vermeer et al., 1991).

The effect of UVB radiation on immunity against certain skin infections has also been studied. *Herpes simplex* virus has been widely used as an experimental model of virus infection in mice. The immune response against this virus was suppressed by suberythemal doses of UVB. Antigen presentation of HSV by epidermal cells was impaired and DTH responses to the virus were decreased. People who are exposed to burning doses of sunlight (e.g. during outdoor activities) can show reactivation of the persistent latent form of *Herpes simplex* virus (cold sores). The hypothesis has been put forward that UVB can activate the virus and in addition can impair the immune system in the skin enabling the virus to replicate in the skin and cause the clinical lesions (Norval et al., 1994). Other skin-associated infections that might be influenced by UV exposure are *Leishmania*, which is a parasitic protozoon with an initial cutaneous phase, *Mycobacterium leprae*, and *Candida albicans* (a commensal yeast), that can cause thrush in the oral cavity and chronic skin disease. Indications regarding these skin-associated infectious diseases were in majority found in animal studies but also in human studies. Recently, Armstrong reviewed the relation between the thickness of the stratospheric ozone layer and human health, and demonstrated that already in the early 1950's, there was some evidence, that increased exposure to sunlight was related to an increased incidence of several skin-associated infectious diseases such as staphylococcal impetigo (Armstrong , 1994).

Because UV can induce systemic effects on the immune system it might be suggested that also the resistance to systemic infections and tumours is affected. There is increasing evidence that UVB radiation can affect the immune response to systemic, non-skin associated infections. For instance, a widely used bacterial infection model in mice is *Mycobacterium bovis* causing tuberculosis. These bacteria cause major problems in tropical and subtropical countries; it may be hypothesized that this is influenced in part by the high level of UVB radiation in these areas. UVB exposure, before but also during infection, was shown to reduce the DTH against mycobacteria in BALB/c and C3H mice (Jeevan et al., 1989, 1992). The total number of bacteria was increased in spleen and lymph nodes after UVB exposure, which indicates a systemic effect of UVB radiation. Supernatants of UVB exposed epidermal cells can induce similar immunosuppressive effects *in vivo*, indicating that the systemic effects by UVB radiation are mediated by soluble compounds released by the epidermis. Also the pathological effects of a murine leukaemia virus, LP-BM5 were enhanced by *in vivo* UVB exposure, suggesting that UVB radiation enhances the immunosuppressive effects of a retroviral infection. In HIV-infected individuals, UVB radiation might enhance the virus induced immunosuppression and activate viral replication. Thus, HIV infected persons are at higher risk for deleterious effects of UVB radiation. In several countries HIV patients are therefore advised to stay out of the sun. However, UVB exposure of HIV positive psoriasis patients did not give any clinical problems. In addition to studies regarding systemic infections in mice by Jeevan et al., we demonstrated recently that the resistance to *Listeria monocytogenes*, *Trichinella spiralis*, and *Cytomegalovirus* infectons in the rat are significantly suppressed. All these studies indicate that especially the cellular immune response (i.e. T cells) is suppressed by UV exposure and that this suppression might be crucial for the impairment of host resistance (Goettsch et al., 1994,1996).

Since a few years many research groups have focused on the mechanisms that are responsible for the systemic effects of UV light regarding immunomodulation. There is consensus in the literature that for UV to have a biological effect, it must be absorbed by a chromophore in the skin that translates the energy into a biochemical signal. This signal will in turn lead to an increase or decrease of cellular components such as cytokines, growth factors, and costimulatory signals that can modulate the immune response. There are different candidates for acting as a UV-photoreceptor that are not mutually exclusive. UV can cause biological effects by altering the DNA structure directly. In addition, UV can be absorbed by cellular components other than DNA, such as lipids in cell membranes, which generate oxidative DNA damage. In addition, lipid peroxidation reactions can trigger intracellular signalling pathways, which in turn may lead to gene activation. A completely different photoreceptor is urocanic acid (UCA). Already in the early eighties it was demonstrated that UV can isomerize UCA. UCA is located predominantly in the stratum corneum where it accumulates because no urocanase is present in the skin to catabolize it. On UV-irradiation the *trans*-isomer is converted to the immunosuppressive *cis*-isomer.

UCA could be involved in photoimmunosuppression through its ability to photosensitize the formation of pyrimidine dimers, i.e. a UV specific DNA damage, indicating that the DNA- and UCA-mediated effects might be linked (Mohammed et al., 1994). Recently, Pouwels et al. (1994) demonstrated that pyrimidine radical-cation can cycloadd to a neutral pyrimidine to form the cyclobutane dimer. However, Terrian et al. (1996) demonstrated that any reversal of photoimmunosuppression by photoreactivation cannot significantly involve cleavage of UCA-DNA cyclobutane adducts pointing to a limited role for UCA-DNA adducts in photoimmunosuppression.

Role of Urocanic Acid in UV-induced immunosuppression

UCA is a major absorber of UV-light in the skin and it is known to be located predominantly in the stratum corneum where it is synthesized from histidine by deamination. It represents about 0.7% of the dry weight of the epidermis

(Tabachnik, 1957). It was first proposed that UCA has an important function as a natural 'sun-screen' to protect against sunburn (Zenisek et al., 1955) and secondly that it may act as a natural photoprotecting agent against damage to DNA in the skin (Morison, 1985). In the early eighties it was postulated that *trans*-UCA acts as a receptor for UVB and that the resulting *cis*-isomer may mediate immunosuppression (De Fabo and Noonan, 1983). After exposure to sunlight, photons are absorbed by *trans*-UCA and the immunosuppressive *cis*-isomer is formed (figure 1). Analysis of the action spectrum of suppression of contact hypersensitivity (CHS) by UVB radiation and the absorption spectrum of *trans*-UCA acid revealed that they were similar. In histidase-deficient mice, which have less than 10% of wild type amounts of skin UCA, UVB-induced immunosuppression was very limited (De Fabo et al., 1983), while in histidine (precursor of UCA) fed mice that had increased levels of UCA, UVB-induced immunosuppression was more severe than in control mice (Reilly and De Fabo, 1991). *Cis*-UCA can penetrate the deeper skin, and can be detected in urine and in the blood (Kammeyer et al, 1994). Another approach was to expose *trans*-UCA *in vitro* to UVB radiation and inject mice with the UV-irradiated UCA.

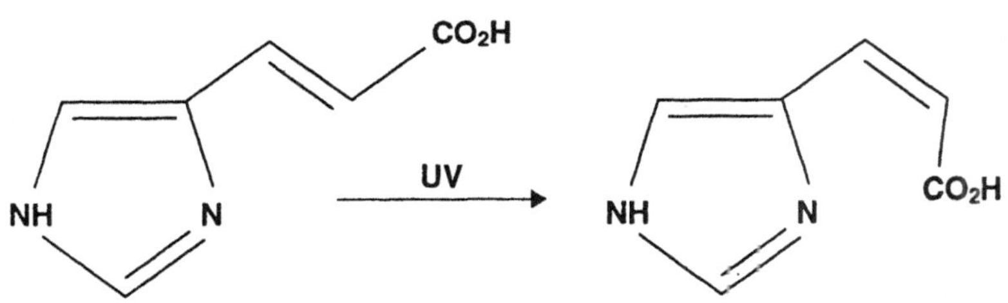

Figure 1: Photoisomerization of urocanic acid

It has been shown that this UV-UCA causes suppression of DTH responses to HSV (Ross et al., 1986, 1988a,b). The cellular target for *cis*-UCA is still unknown, but there are several suggestions. There is evidence that *in vivo cis*-UCA treatment alters the immune response of animals. However, *in vitro* exposure to *cis*-UCA does not have these effects in any case. Thus, *cis*-UCA may induce a cascade of events, ultimately leading to immunosuppression (Noonan et al., 1988). Production of suppressor factors that can decrease IL-1 production by epidermal cells, and can reduce HLA-DR expression by antigen-presenting cells may be involved in this cascade leading to immunosuppression (Rasanen et al., 1987). The release of TNF-α by keratinocytes may be induced by *cis*-UCA, leading to immunosuppression (Kurimoto and Streilein, 1992). Noonan and De Fabo suggested in recent experiments that skin fibroblasts would be the main target for *cis*-UCA, and that those fibroblasts would release factors after cAMP modulation induced by cis-UCA (Palaszynskin et aL, 1992). In HSV infection experiments it was demonstrated that structural analogues of *cis*-UCA, including histamine, could suppress DTH reactions (Norval et al., 1989). Histamine receptor antagonists could prevent the generation of UVB- or *cis*-UCA-induced immunosuppression, indicating a role for histamine-like receptors (Matheson and Reeve, 1991; Norval et al., 1990). In other experiments it was studied whether *cis*-UCA plays a role in suppression of the resistance to an oral *Trichinella* infection induced by UVB exposure as was found earlier in other studies earlier (Goettsch et al., 1994). These studies indicated that *cis*-UCA plays at least a partial role in the inhibition of resistance to an oral *Trichinella spiralis* infection induced by suberythemal UVB exposure, indicating that *cis*-UCA might play a role in systemic effects of UV (unpublished observation).

Role of DNA damage in UV-induced immunosuppression

UV-induced DNA damage and repair: DNA can absorb UV radiation (Sage, 1993). The best-studied photolesions are induced by 254 nm, i.e. part of the UVC wavebands. This is close to 260 nm which is the absorption maximum of DNA. Wavelengths absorbed by DNA lead to the induction of: 1) cyclobutane pyrimidine dimers (CPD)(Beukers and Berends, 1960; Wang, 1960; Setlow and Carrier, 1966), 2) pyrimidine (6-4) pyrimidone photoproducts (Lippke et al., 1981; Rycyna and Alderfer, 1985), 3) 8,8-adenine dehydrodimer (Gasparro and Fresco, 1986), 4) a purine product (Gallagher and Duker, 1989), and 5) a product at TA sequences (Bose and Davies. 1984). UVB irradiation, which is thought to be the main wavelength-range that is responsible for UV-induced immunosuppression, results mainly in pyrimidine dimers and (6-4) photoproducts. These DNA lesions comprise about 95% of the UVB-induced DNA lesions (figure 2). Both cyclobutane pyrimidine dimers and (6-4) pyrimidine pyrimidone photoproducts seem to be mutagenic in mammalian cells (Mitchell and Nairn, 1989) and might play a crucial role in photocarcinogenesis and photo-immunosuppression.

There are at least four different mechanisms described that may play a role in repair of UV induced DNA damage.

- i: Photorepair mechanisms. In many organisms photorepair entails direct monomerization of CPD by an enzyme that is activated by visible light. Nevertheless, this mechanism has never been found to occur in adult human skin (Li et al, 1993) nor in the skin of adult hairless mice that are frequently used for UV-carcinogenesis studies (Ley et al, 1977).
- ii: Post-replication repair mechanisms. Photoproducts in the DNA molecule are bypassed during replication after which the gaps in the daughter strands are filled by a DNA polymerase (Kaufmann, 1989).

- iii: Some observations suggest that somatic recombination acts as a mechanism to remove DNA damage to allow DNA replication.

- iv: Probably the most important repair system for UV-induced DNA damage is the nucleotide-excision repair mechanism. During this repair process damaged DNA fragments are recognized, followed by incisions of the damaged DNA strand at two sites at some distance from the lesion. The damaged DNA (within an oligomer) is subsequently removed. The gap is filled using the intact opposite strand as a template. The importance of this mechanism is the protection of cells against malignant conversion. In Xeroderma pigmentosum patients the nucleotide-excision repair mechanism is disturbed leading to a high mutability of cells and a 1000-fold higher chance of getting skin cancers (Kraemer et al., 1987).

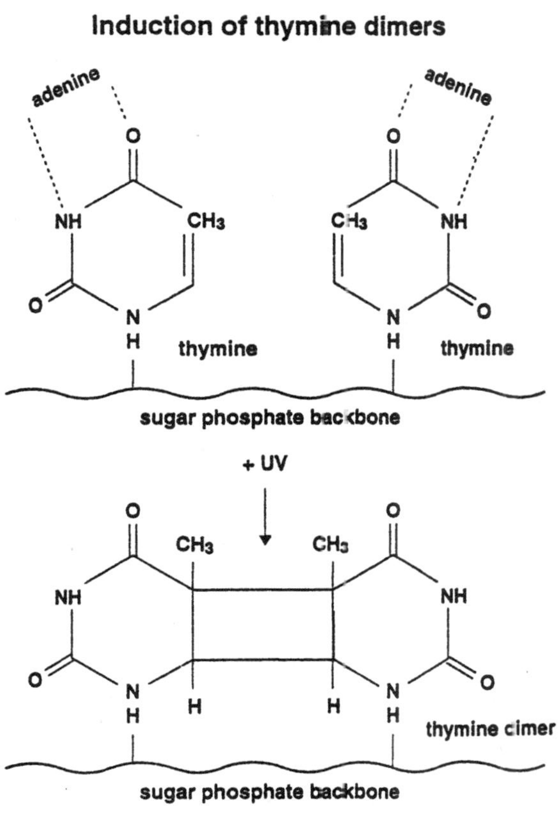

Figure 2:

Thymine dimer induction

UV-induced DNA mutations: The occurrence of DNA damage can ultimately lead to mutations. This implies that the efficiency of DNA repair is inversely proportional to the mutation frequency. In the case a mutation occurs in genes that play a role in the regulation of cell cycling and/or differentiation, the normal growth might be disregulated. Such genes are divided into proto-oncogenes and tumor-suppressor genes. It is suggested that the *ras*-proto-oncogene may be mutated by UV light leading to mutations in that particular gene. Mutations in ras oncogenes are found in 10-40% of human and mouse skin cancers. An example of a tumor-suppressor gene in many tumors (animal and man) is P53. This gene codes for a protein, acting as a transcription factor, that is able to block progression of the cell cycle in the G1-phase in case DNA damage has been induced by e.g. UVB. Blocking the progression in that phase of the cell cycle prevents unrepaired DNA from entering the DNA replication phase which can lead to mutations (Kastan et al., 1991). The P53 gene plays a crucial role during apoptosis (programmed cell death) of cells with damaged DNA. It is evident that a dysfunctioning P53 gene facilitates mutations in proto-oncogenes and/or tumor-suppressor genes and thus facilitates progression towards malignancy. In human and murine skin cancer induced by UV light it is demonstrated that mutations in the P53 gene may play a crucial role. The mutations are in general C→Tand CC→TT mutations. These mutations are unique for UV exposure.

Mutations can lead to an altered function of the cells as described above. For that reason if cells that play a role (directly or indirectly) in the immune system are mutated by UV light this may lead to an altered immune response (*altered gene expression*) (immunotoxicity). Several studies indicate that certain genes are activated or inactivated leading to e.g. an altered cytokine profile. UVB exposure can regulate the synthesis and secretion of GM-CSF, IL-1, Ll-6, IL-8, IL-10, TNF-α, IL-15, and probably many other cytokines.**DNA damage in photoimmunology:** The first studies indicating that UV induced DNA damage may play a role in UV-induced immunosuppression were performed using the *Monodelphis domestica*. Cells from this marsupial contain a photolyase DNA repair enzyme that is activated to repair cyclobutyl pyrimidine dimers by visible light, i.e. photoreactivating light. In these animals the suppression of contact hypersensitivity by UVB was prevented due to photorepair of UVB-induced dimers (Applegate et al., 1989).

More recently it was demonstrated that DNA repair enzymes encapsulated in liposomes can remove DNA damage from UV-exposed mouse skin in vivo. Liposomes containing the excision repair enzyme T4 endonuclease V (T4N5) have been shown to increase the rate of dimer removal when applied to UV-exposed mouse skin immediately following irradiation (Yarosh et al., 1990). This procedure prevented UV-induced suppression of type IV hypersensitivity to antigens applied or injected at a distant site, whereas liposomes containing heat-inactivated enzyme had no effect (Kripke et al., 1992). Systemic immunosuppression induced by PUVA (8-methoxypsoralen plus UVA) was not prevented by this liposome treatment indicating that the effect of T4N5 containing liposomes was specific for UVB-induced DNA damage (Yarosh et al., 1994). There are several hypotheses dealing with the mechanisms that are involved in of UVB-induced DNA damage causing systemic immunosuppression. Recently we demonstrated that UVB can induce dimers in the skin UV-dose dependently and that even in lymph nodes that drain the skin, cells can be detected that contain these dimers. This indicates that cells that contain UVB specific DNA lesions can migrate from the skin to lymph nodes.

This might be the cause for the systemic effects of UVB exposure. We propose that these cells are Langerhans cells (Sontag et al., 1995). If these cells are Langerhans cells it might be suggested that the UVB induced DNA lesions are involved in altered antigen presentation by these Langerhans cells. It was shown recently that if these mice were treated with the liposomes as descibed above the number of DNA lesion containing cells in the lymphnodes was significantly reduced. Even the impaired antigen presenting function of lymph node cells was significantly restored by the liposome treatment (Vink et al., 1996). Other studies indicated that cytokines released by keratinocytes are involved in systemic UV-induced immunosuppression. It was proposed that DNA damage induced the release of these cytokines. Recently it was demonstrated by Kripke et al. (unpublished observations) that treatment of UV-exposed keratinocytes with T4N5 liposomes inhibited the secretion of IL-10 which might be one of the crucial mediators for UVB-induced systemic immunosuppression. If DNA lesions, i.e. double strand breaks, were induced by liposomes containg the restriction enzyme HindIII immunosuppression was induced as well. This latter immunosuppression was correlated with increased IL-10 and TNF-α production in the skin. For this reason it is not unrealistic to propose that UV-induced immunosuppression is mediated by DNA damage via the secretion of cytokines such as IL-10 and TNF-α. An alternative hypothesis is that UVB affects co-stimulatory molecules such as ICAM-1. The adhesion molecule ICAM-1, by virtue of its capacity to serve as a ligand for the LFA-1 molecule on leucocytes, participates in the generation and maintenance of a variety of immune responses in a.o. the skin. Krutmann and Trefzer (1993) demonstrated that UVB can modulate ICAM-1 expression in the skin and that this might have a major impact on immunity. It is well known that the constitutive expression of ICAM-1 in human skin cells is low but proinflammatory cytokines such as interferon-γ can upregulate ICAM-1 mRNA and protein expression. The interferon-γ induced upregulation of ICAM-1 can be inhibited by UVB exposure in a dose dependent manner. Very recently it was demonstrated that cyclobutane pyrimidine dimer formation is involved in the inhibition of interferon-γ induced upregulation of ICAM-1. This was found in animal studies as well as in humans. Treatment with liposomes containing the photorepair enzyme photolyase, which especially removes dimers from cellular DNA after exposure to photoreactivating light, prevents the UV-induced effect on ICAM-1. Therefore it is clear that UVB induced DNA damage represents a necessary intermediate in UVB-induced suppression of interferon-γ mediated ICAM-1 upregulation (Krutmann et al., 1996). It is even suggested that susceptibility/resistance toward UVB-induced immunomodulation might be defined at the level of DNA repair enzyme activity.

Signal transduction and transcription factors elicited by UVB: It is known that UVB can upregulate IL-6 expression. Recently Simon et al. (1994) demonstrated that in these studies cytosolic extracts of UVB-exposed human epidermal cells free of nuclei were capable of activating NFκB (nuclear factor B). This NFκB activation can be modulated by scavenging of oxygen free radicals (Simon et al., 1994).

One of the first changes in the skin due to UVB exposure is skin redness. Especially prostaglandins are involved in the induction of erythema. The responsible enzyme for the UVB-induced synthesis of prostaglandin E2 (PGE2) is cytosolic phospholipase A2. The UVB-signalling mechanism responsible for upregulation the expression of this enzyme is under investigation now. Recent studies indicate that phospholipase A2 is activated via an epidermal growth factor -receptor (EGF-receptor) mediated pathway (Gresham et al., 1996). PGE2, produced via EGF-receptor mediated activation of phospholipase A2, is known to upregulate the production of IL-10 and downregulate IL-12 leading to a shift from Th1 toward Th2 cell mediated immune reactions which is a well known feature of UVB-induced immunomodulation (Ullrich, 1995).

Conclusions

It is clear that exposure to ultraviolet light can induce skin cancer and that at least two mechanisms are involved in the proces of photocarcinogenesis, i.e. DNA damage/mutations (genotoxicity) and immunomodulation (immunotoxicity). In the last decade it has been demonstrated by several groups that UV can affect the immune system locally (exposed skin) as well as systemically. Even the immunological resistance to non-skin-associated infections can be impaired by UV exposure indicating that systemic effects occur. Several hypotheses have been put forward dealing with the mechanisms responsible for the systemic immunomodulation. One of the targets of UV radiation within the skin is UCA, which may modulate immune responses, another is DNA. Several studies indicated that DNA damage, due to absorption of UV energy, and mutation can lead to immunosuppression (locally and systemically). Due to an altered gene-expression, cytokine profiles and the expression of adhesion molecules are affected either induced directly or by altered transduction of signals. This may lead to an inhibiton of Th1 cell function and preferential activation of Th2 cells. The exact mechanism responsible for immunomodulation induced by the UVB induced altered gene expression remains to be analyzed.

References

Applegate LA, Ley RD, Alcalay J, Kripke ML (1989) Identification of the molecular target for the suppression of contact hypersensitivity by ultraviolet radiation. J Exp Med 117:1117-1131

Armstrong BK (1994) Stratospheric ozone and health. Int J Epidemiology 23:873-885

Berg R JW (1994) UV-induced damage in relation to skin cancer induction in hairless mice. PhD thesis, Utrecht, The Netherlands, ISBN 90-393-0594-3.
Beukers R, Berends W (1960) Isolation and identification of the irradiation product of thymine. Biochem. Biophys. Acta 41:550-551
Bose S, Davies RJH (1984) The photoreactivity of T-A sequences in oligodeoxyribonucleotides and DNA. Nucleic Acids Res 20:7903-7914
Bouwes Bavinck JN, Robertson I, Wainwright RW, Green A (1995) Excessive numbers of skin cancers and premalignant lesions in an Australian heart transplant recipient. Br Heart J 74:468-470
Daynes CK and Spellmann CW (1977) Evidence for the generation of suppressor cells by ultraviolet radiation. Cell Immunol 31:182-187
De Fabo EC, Noonan FP (1983) Mechanism of immune suppression by ultraviolet irradiation in vivo. I. Evidence for the existence of a unique photoreceptor in skin and its role in photoimmunology. J Exp Med 157:84-98
De Fabo EC, Noonan FP, Fisher MS, Burns J, Kacser H (1983) Further evidence that photoreceptor mediating UV-induced systemic immunosuppression is urocanic acid. J Invest Dermatol 80:319
Gallagher PE, Duker NJ (1989) Formation of purine photoproducts in a defined human DNA sequence. Photochem Photobiol 49:599-605
Gasparro F, Fresco J (1986) Ultraviolet-induced 8,8-adenine dehydrodimers in oligo-and polynucleotides. Nucleic Acids Res 10:4239-4251
Goettsch W, Garssen J, Deijns A, De Gruijl FR, Van Loveren H (1994) UV-B exposure impaires resistance to infection by Trichinella spiralis. Environmental Health Perspectives 102 (3):298-301
Goettsch W, Garssen J, De Klerk A, Herremans MPT, Dortant P, De Gruijl FR, Van Loveren H (1996) Effects of ultraviolet-B exposure on the resistance to Listeria monocytogenes in the rat. Photochemistry and Photobiology 63: 672-679.
Gresham A, J Masferrer, X Chen, Leal-Khouri S, Pentland AF (1996) Increased synthesis of the high molecular weight cytosolic phospholipase A2 mediates early UV-induced prostaglandin E2 in human skin. Am J Physiol. In press.
Jeevan A, Kripke ML (1989) Alteration of immune responses to mycobacterium bovis BCG in mice exposed chronically to low doses of UV radiation. Cell Immunol 145:2087-2091
Jeevan A, Ullrich SE, Dizon V, Kripke ML (1992) Supernatants from ultraviolet-irradiated keratinocytes decrease the resistance and delayed-type hypersensitivity response to Mycobacterium bovis bacillus Calmette-Guerin and impair the phagocytic ability of macrophages. Photodermatol Photoimmunol Photomed 9:255-263
Kammeyer MBM, Theunissen MA, de Rie JD, Bos JD (1994) Retention of increased *cis*-urocanic levels in human body upon UV-B exposure. Photochem Photobiol 59S:27S
Kastan MB, Onyekwere O, Sidransky D, Vogelstein B, Craig RW (1991) Participation of p53 protein in the cellular response to DNA damage. Cancer Res 51:6304-6311
Kaufmann WK (1989) Pathways of human cell post-replication repair. Carcinogenesis 10:1-11
Kraemer KH, Lee MM, Scotto J (1987) Xeroderma pigmentosum. Arch Dermatol 123:241-250
Kripke ML, Cox PA, Alas LG, Yarosh DB (1992) Pyrimidine dimers in DNA initiate systemic immunosuppression in UV-irradiated mice. Proc Natl Acad Sci 89:7516-7520
Kripke ML (1974) Antigenicity of murine skin tumors induced by ultraviolet light. J Natl Cancer Inst 53:1333-1336
Krutmann J, Trefzer U (1993) Regulation of keratinocyte ICAM-1 expression by UV irradiation. Semin Immunopathol 13:333-344

Krutmann J, Ahrens C, Roza L, Arlett CF (1996) The role of DNA damage and repair in ultraviolet B radiation-induced immunomodulation: relevance for human photocarcinogenesis. Photochem Photobiol 63:394-396

Kurimoto I, Streilein JW (1992) cis-Urocanic acid suppression of con#tact hypersensitivity induction is mediated via tumor necrosis factor-α. J Immunol 148: 3072-3078

Ley RD, Sedita BA, Grube DD (1977) Absence of photoreactivation of pyrimidine dimers in the epidermis of hairless mice following exposure to ultraviolet light. Photochem Photobiol 27:483-485

Li YF, Kim S-T, Sancar A (1993) Evidence for lack of DNA photoreactivating enzyme in humans. Proc Natl Acad Sci (USA) 90:4389-4393

Lippke JA, Gordon LK, Brash DE, Haseltine WA (1981) Distribution of UV light-induced damage in a defined sequence of human DNA: detection of alkaline sensitive lesions at pyrimidine nucleoside-cytidine sequences. Proc Natl Acad Sci (USA) 78:3388-3392

Matheson MJ, Reeve VE (1991) Effect of anti-histamine cimetidine on UV radiation tumourigenesis in the hairless mouse. Photochem Photobiol 53: 639-642

Mitchell DL, Nairn RS (1989) The biology of (6-4) photoproduct. Photochem Photobiol 49:805-819

Mohammed T, Tessman I, Morrison A, Kennedy MA, Simmonds SW (1994) Photosensitized inactivation of infectious DNA by urocanic acid, indoleacrylic acid and rhodium complexes. Photochemistry Photobiol 59:189-196

Morison H (1985) Photochemistry and photobiology of urocanic acid. Photodermatology 2:158-165

Noonan FP, De Fabo EC, Morrison H (1988) *Cis* urocanic acid, a product formed by ultraviolet B irradiation of the skin, initiates an antigen presentation defect in splenic dendritic cells in vivo. J Invest Dermatol 90:92-99

Norval M, Simpson TJ, Ross JA (1989) Review article, urocanic acid and immunosuppression. Photochem Photobiol 50:267-275

Norval M, El-GHorr A, Garssen J, Van Loveren H (1994) The effects of ultraviolet light irradiation on viral infections. British Journal of Dermatology 130:693-700

Norval M, Gilmour JW, Simpson TJ (1990) The effect of histamine receptor antagonists on immunosuppression induced by the cis-isomer of urocanic acid. Photodermatol Photoimmunol Photomed 7:243-248

Norval M, Gibbs NK, Gilmour J (1995) The role of urocanic acid in UV-induced immunosuppression: recent advances (1992-1994). Photochem Photobiol 62: 209-217

Palaszynski EW, Noonan FP, De Fabo EC (1992) Cis urocanic acid down regulates the induction of adenosine 3'5'-cyclic monophosphate by either trans-urocanic acid or histamine in human dermal fibroblasts. Photochem Photobiol 55:165-171

Pouwels PJW, Hartman RF, Rose SD, Kaptein R (1994) CIDNP evidence for reversibility of the photosensitized splitting of pyrimidine dimers. J Am Chem Soc 116:6967-6968

Rasanen L, Jansen CT, Reanala T, Morrison H (1987) Stereospecific inhibition of human epidermal cell IL-1 secretion and HLA-DR expression by cis-urocanic acid. Photodermatol 4: 182-186

Reilly SK, De Fabo EC (1991) Dietary histidine increases mouse skin urocanic acid levels and enhances UV-B induced immune suppression of contact hypersensitivity. Photochem Photobiol 53:431-438

Ross JA, Howie SEM, Norval M, Maingay J, Simpson TJ (1986) Ultraviolet-irradiated urocanic acid suppresses delayed-type hypersensitivity to Herpes simplex in mice. J Invest Dermatol 87:630-633

Ross JA, Howie SEM, Norval M, Maingay J (1988) Induction of suppression of DTH to herpes simplex virus by epidermal cells exposed to UV-irradiated urocanic acid *in vivo*. Viral Immunol 1:191-198

Ross JA, Howie SEM, Norval M, Maingay J (1988) Systemic administration of urocanic acid generates suppression of the DTH response to herpes simplex virus in a murine model of infection. Photodermatol 5:9-14

Rycyna R, Alderfer J (1985) UV irradiation of nucleid acids: formation, purification and solution conformational analysis of the '6-4 lesion' of dTpdT. Nucleic Acids Res 13:5949-5963

Sage E (1993) Distribution and repair of photolesions in DNA: genetic consequences and the role of sequence context. Photochem Photobiol 57:163-174

Setlow RB, Carrier WL (1966) Pyrimidine dimers in ultraviolet irradiated DNAs. J Mol Biol 17:237-254

Simon M, Aragane Y, Schwarz A, Luger T, Schwarz T (1994) UVB light induces nuclear factor B activity independently from chromosomal DNA damage in cell-free cytosolic extracts. J Invest Dermatol 102:422-427

Sontag Y, CLH Guikers, AA Vink, De Gruijl FR, Van Loveren H, Garssen J, Roza L, Kripke ML, Van Der Leun JC, Van Vloten WA (1995) Cells with UV-specific DNA damage are present in murine lymph nodes after in vivo UV irradiation. J Invest Dermatol 104:734-738

Tabachnik J (1957) Urocanic acid, the major acid soluble UV absorbing compound in guinea pig epidermis. Arch Biochem Biophys 70:295-298

Terrain LD, Kuhl CN, Tessman I, Morrison H (1996) On the roles of urocanic acid in photoimmunosuppression: Attempted photorepair of urocanic acid-DNA cyclobutane adducts with DNA Photolyase. Photochemistry and Photobiology 63 (6):898-900

Ullrich SE (1995) The role of epidermal cytokines in the regulation of cutaneous immune reactions and ultraviolet radiation-induced immune suppression. Photochem Photobiol 62 (3):389-401)

Vermeer M, Schmieder GJ, Yoshikawa T, Van Den Berg JW, Metzman MS, Taylor JR, Streilein JW (1991) Effects of ultraviolet B light on cutaneous immune responses of humans with deeply pigmented skin. J Invest Dermatol 97:729-734

Vink AA Strickland FM, Bucana C, Cox PA, Roza L, Yarosh DB, Kripke ML (1996) Localization of DNA damage and its role in altered antigen presenting cell function in UV-irradiated mice. J Exp Med, In press

Wang SY (1960) Photochemical reactions in frozen solutions. Nature 188:844-846

Yarosh DB, Tsimis J, Yee V (1990) Enhancement of DNA repair of UV damage in mouse and human skin by liposomes containing a DNA repair enzyme. J Cosmet Chem 41:85-92

Yarosh DB, Bucana CD, Cox P, Alas L, Kibitel T, Kripke ML (1994) Localization of liposomes containing a DNA repair enzyme in murine skin. J Invest Derm 103:461-468

Yoshikawa T, Rae V, Bruins-Slot W, Van Den Berg JW, Taylor JR, Streilein JW (1990) Susceptibility to effects of UVB radiation on induction of contact hypersensitivity as a risk factor for skin cancer in humans. J Invest Derm 95:530-536

Zenisek A, Kral JA, Hais IM (1955) Sunscreening effect of urocanic acid. Biochim Biophys Acta 18:589-591

The Use of Ecotoxicology and Human Toxicology in the Regulation of Chemical Safety in the European Union

(Chairs: M. Pugh, Ireland,
and J.V. Tarazona, Spain)

Options for the Regulation and Control of the Environmental Impact and Human Health Consequences of Chemicals in the European Union

Robin Mason,
Nuffield College, Oxford University, OX1 1NF, U.K..

Introduction

This paper gives a general introduction to the economic principles of regulation of environmental hazards, and then assesses three different approaches to chemical regulation. In doing so, it reviews the current European Union approach to the regulation of hazardous chemicals, and considers the recent Green Paper (1993) proposals on the use of a civil liability system for environmental regulation. The conclusions that emerge from this discussion are: (i) *ex post* systems (for example, liability rules) are inadequate for the regulation of long-term risks; (ii) *ex ante* regulation should be based on the chemical characteristics which give rise to harm; (iii) any regulatory approach must be assessed carefully for the incentives for strategic behaviour that it creates. The paper concludes by making suggestions for the regulation of chemicals in the E.U. which reflect these general principles.

Economic Principles for the Regulation of Environmental Hazards

Pollution of the environment by man-made chemicals is a growing and widespread concern. Synthetic chemicals are an essential part of modern industry; and so world-wide sales of these products has doubled every 7-8 years, exceeding US$750 billion in 1990 (see Macauley et al., 1992). Major industrial accidents have highlighted, however, the potential for disaster posed by the large volumes and variety of chemicals used in industrial processes. Every environmental compartment is at risk - water (due to contamination by agricultural chemicals, for example), the atmosphere (by e.g. chlorofluorocarbons, or CFCs), and the organic chain (by e.g. heavy metals such as cadmium and lead).

This section examines the reasons why private decision-making may lead to excessive degradation of the environment. The source of the problem is *market failure*: some significant economic factor goes 'unpriced', so that economic activity is undertaken without consideration of its full impact. The most important form of market failure for the issue of environmental degradation is the existence of *externalities*. An externality arises when the decisions of some economic agents (individuals, firms, governments) (whether in production, in consumption, or in exchange (affect other economic agents, and are not included in the priced system of commodities. In this case, victims receive no compensation from injurers for any damage caused. An alternative way of expressing this problem is to say that economic agents consider only their private costs when making decisions; they do not consider the total, or social, costs of their actions. It is the gap between private and social costs that gives rise to the externality.[1]

This gap is illustrated in figure 1. The quantity of a polluting activity undertaken is shown on the horizontal axis; the (marginal) benefits (e.g. increased agricultural productivity from pesticide use) and costs (damage to the

Figure 1: Social and Private Marginal Benefit and Cost Schedules

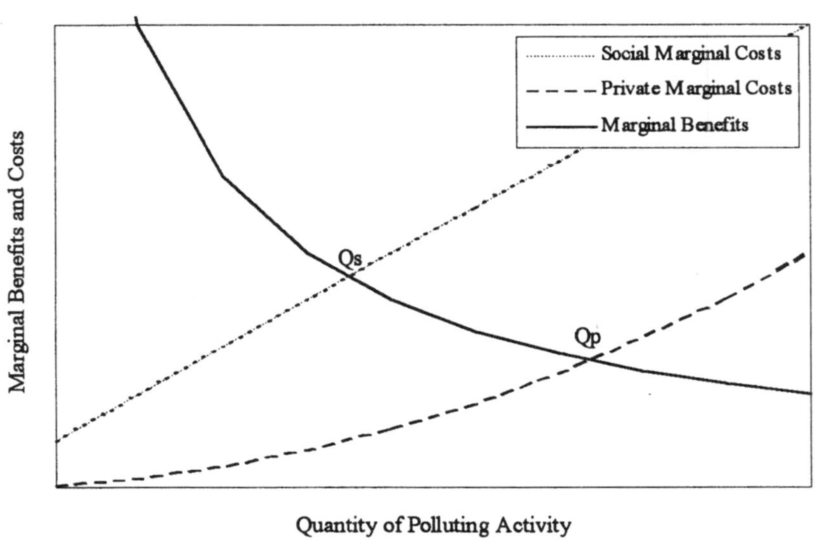

[1] To be precise, marginal, rather than total, costs are the decisive factor

environment and human health consequences) of the activity appear on the vertical axis. For the purposes of the illustration, suppose private and social marginal benefits from the activity are equal; they are shown by the downward sloping curve in the figure. Private individuals and a benevolent social planner are distinguished only by the costs that they face. Marginal costs are the upward sloping curves[2]; private costs are less than social costs, and so are represented by the lower of the two cost curves. The level of polluting activity undertaken is determined by the intersection of the marginal benefit and the appropriate marginal cost curve. As the figure shows, the private amount of polluting undertaken (Q_P) exceeds the socially optimal amount (Q_S).

The market failure in this case is the non-existence of a market for a certain commodity (the externality itself. This leads to a distortion in the market's allocation of resources, so that in the presence of externalities, the market outcome does not maximise society's welfare. Regulation is required to correct the market failure and restore social efficiency. Three possible approaches are suggested by the analysis: (i) impose a tax on the activity (equal to the difference between social and private marginal costs at the optimal quantity Q_S); (ii) control the amount of the polluting activity undertaken (e.g. specify that activity levels cannot exceed Q_S); (iii) use the legal (court) system to deter excessive pollution. Each of these three systems will be discussed in the next sections.

Before embarking on the discussion, it is important to note that the previous analysis, summarised in figure 1, contains (at least) two implicit assumptions. First, it is a static analysis, and therefore considers only short-term or acute costs of chemical use. Secondly, it is concerned only with the total *quantity* of a chemical used; it fails to recognise that the *characteristics* of chemicals are also important. These are not always reasonable assumptions. In many cases, the problem facing society is not the acute toxicity of chemicals, or the absolute quantities used, but their ability to accumulate in natural resources. This point is highlighted by the troubles caused by DDT and other chlorinated hydrocarbon pesticides developed in the 1960s. It was considered at the time that the relative stability of these chemicals was a desirable characteristic (the compounds could be applied at relatively low rates of toxicity, with fewer applications required per year. Problems soon arose, however, as the pesticides accumulated at increasing rates while moving up the biological chain. The ultimate waste sink for these persistent chemicals turned out to be man himself. By the 1970s, human milk in many countries contained levels of DDT concentrations exceeding the World Health Organisation's guidelines for DDT in cow's milk (see Conway and Pretty, 1991, for further details). Table 1 shows that many pesticides which are currently causing concern due to their high concentrations in drinking water are comparable to aspirin in terms of their LD_{50}. It is the ability of these chemicals to accumulate in the environment which presents the regulatory challenge.

The three regulatory approaches suggested by the analysis will now be assessed for their effectiveness in controlling long-term chemical damage.

---

[2] Decreasing marginal benefits and increasing marginal costs are standard assumptions in economics; see Varian, 1992.

Table 1: Toxicity of Selected Agricultural Chemicals

Insecticides	Herbicides	Known Comparison
	Super-toxic Chemicals: $LD_{50} < 10$ mg/kg	
Endrin Several organophosphates	No herbicides	Deadly nightshade
	High Toxicity: $LD_{50} < 50$ mg/kg	
Many organochlorines Most organophosphates	No herbicides	Nicotine, cyanide, vitamin D
	Moderate Toxicity: $LD_{50} < 500$ mg/kg	
Most organochlorines e.g. DDT Many organophosphates	Most herbicides e.g. paraquat	Barbiturates
	Slight Toxicity: $LD_{50} < 5$ g/kg	
No organochlorines Some organophosphates	Most herbicides e.g. atrazine, diuron	Aspirin
	Low Toxicity: $LD_{50} > 5$ g/kg	
		Compare salt at 30 g/kg

The Use of Liability Systems to Regulate Chemicals

In theory, a system of liability should address fully the problems caused by externalities. Courts will act to identify the cause of environmental harm, assess the responsibility of each agent involved, quantify damages to victims, and apportion costs; in this way, victims will be compensated fully for any damage or injury suffered, and injurers will face the full (social) cost of their actions. Moreover, a decentralised liability system allows a more efficient dissemination of information than could be achieved in practice by any government (see Landes and Posner, 1987, and Shavell, 1987, for further discussion). This argument underlies the European Commission's recent conclusion that "civil liability is a useful legal instrument for recovering the costs of restoring environmental damage as well as for its prevention and enforcement functions" (see the E.C. Green Paper, 1993, p. 27).

There are three primary reasons why *ex post* liability systems fail to live up to this ideal. First, the transactions costs (lawyers' fees; the cost of delays caused by lengthy court proceedings; etc.) can be very large. For example, recent estimates suggest that the total transactions costs of the U.S. Comprehensive Environmental Response Compensation and Liability Act (CERCLA, also known as the Superfund) could exceed US$44 billion, out of a total cost of US$100 billion (see Menell, 1991). Secondly, attribution of liability requires that the court be able to identify the responsible party. Often, however, there are many injurers (particularly if the damage relates to the environment, where hundreds of firms may be involved in the contamination of a waste disposal site, for example), whose responsibilities may differ. In addition, identification may be hampered by delays between actions and damages, further muddying the waters of causation. It may also be difficult to establish a clear relationship between action and damage, for example because of scientific uncertainty. All of these factors limit the court's ability to assign the correct liability to the correct party (see Swanson, 1991). Finally, liability can effectively be truncated because of limited liability and bankruptcy laws; this is known as the 'judgement proof' problem, in which responsible parties are unable to pay fully the amount for which they have been found legally liable. As the *ex post* liability of the polluting company may be far smaller than the actual harm done, injurers will not consider the full social costs of their actions. The result will be that injurers will be encouraged to engage in risky activities to a socially excessive extent; they will take too little care in undertaking these activities; and they will have inadequate incentives to purchase liability insurance (see Schwartz, 1985, Roe, 1986, Shavell, 1986, Ringleb and Wiggins, 1990, Hansmann and Kraakman, 1991, and Mason and Swanson, 1996, for further development of these arguments).

This is not to negate the value of a liability system altogether; far from it. The key point is that liability systems on their own will not be sufficient when chemicals are accumulative and damages are long-term. *Ex ante* regulation will be needed to supplement liability systems.

Ex ante Regulation of Chemicals

There are two generic *ex ante* regulatory systems: those that are quantity-based, and those that operate via prices. In an ideal situation (perfect information and competition, with zero administration costs), the two are entirely equivalent; see Baumol and Oates (1988). The world is not ideal, however, and the two approaches can have very different effects. This section reviews how quantity-based and price-based systems fare in the regulation of long-term chemical damage.

It is important at the outset to distinguish between how *ex ante* systems should operate in principle, and how they are run in practice. More often than not, quantity restrictions are imposed on specific products, and taxes are charged on

Figure 2: GUS Indices of Herbicides (taken from Vighi and Zanin, 1994)

The GUS index indicates the ability of an agricultural chemical to accumulate in groundwater. It is defined by GUS = $t_{½}(4 - \log K_{OC})$, where $t_{½}$ is the half-life in soil, and K_{OC} is the octanol/water partition coefficient for non-ionic molecules. Substances with high GUS-values have a high leaching potential.

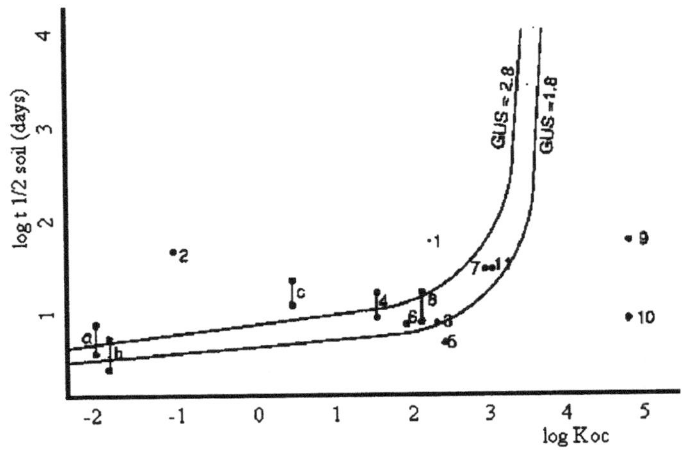

the quantity sold of a particular product. Neither method will be effective, since both permit substitution, either between products or between characteristics of a product. To illustrate the first point, figure 2 shows the use of herbicides in Italy before and after its ban on the use or sale of atrazine, a herbicide used in maize cultivation which violated E.C. standards for drinking water in 1990 (see Bergman and Pugh, 1994). The figure gives the GUS indices for 14 herbicides. Chemicals labelled with numbers are herbicides used in Italy before the banning of atrazine (labelled '1' in the figure). The chemicals labelled with letters are three herbicides that have been registered for use in Italy after the imposition of the ban on atrazine. The GUS indices of two of these products are greater than those of the previous substitutes for atrazine (linuron and terbutylazine, labelled '7' and '11' respectively); the GUS index of the third is comparable to previous levels. Figure 2 shows clearly, therefore, that the atrazine ban has only effected substitution to similar chemicals.

Similarly, taxation of output alone will simply cause an increase in the unregulated characteristic when output and the characteristic are substitutes, so that a decrease in one can be compensated for by an increase in the other. Classic examples (to step away from chemicals for a moment) are the taxation of bricks in England during the Industrial Revolution, and the house-front tax common in many towns. In the former, a uniform tax was imposed on each brick, with the consequence that bricks became very large. In the latter, a tax is charged on the length of pavement, or sidewalk, that fronts a house (the idea being, presumably, that this factor is related to the amount of public services consumed by the householder). In areas where such a tax is made, houses have become very narrow, very tall and very deep. The moral is clear: whenever possible, economic agents will substitute away from taxed factors.

Both arguments suggest that regulation must be based on the *total accumulative ability* of the chemical, and not simply on the basis of its name or sales. More generally, a comprehensive approach will involve regulating a vector of chemical characteristics, which will include not only persistence, but also ability to diffuse within environmental media, toxicity to animals, synergistic effects (for example, CFCs' action in ozone breakdown), etc. (see Swanson, 1996).

The discussion now turns to the specifics of quantity- and price- based systems.

Quantity-based Systems: The static analysis of figure 1 shows that chemicals can be regulated by requiring that the total use be limited to the quantity Q_S. The dynamic counterpart to this approach is to ban further use of a chemical once a critical concentration is exceeded. Examples include:

- the withdrawal from use of DDT in European countries in the 1970s;
- in the Federal Republic of Germany alone, three hundred of the (approximately) one thousand pesticides registered have been banned individually; a further 77 countries had used this type of regulation by 1993 (see Rolike, 1995);
- many European countries have responded to a European Commission directive (number 80/778/EEC), setting a maximum admissible concentration (MAC) for pesticides in drinking water, by banning the sale and use of particular chemicals;

The approach seems to have had limited success, however, and there is little evidence that chemical accumulation in the environment has abated; for example, the concentration of agricultural chemicals in groundwater continues to rise, despite regulatory efforts. Part of the failure of this approach can be explained by figure 2 (bans have been product-specific, and therefore achieved little more than substitution to other, equally harmful chemicals. In addition, it can be shown that this regulatory approach creates incentives for the manufacturers of chemicals to behave strategically, in order to disadvantage existing rivals (see Mason, 1996), to deter entry into the industry by potential rivals (see Mason and Polasky, 1994; and Mason, 1996), and to protect the value of patents on future products (see Mason, 1996). The factor underlying all three effects is that this form of regulation provides firms with a means to commit themselves or their industry to a certain course of action. This commitment ability can be of great value to firms.

The regulation of CFCs provides an excellent example of the third point. Du Pont, the major producer of CFCs 11 and 12 (chemicals used in refrigerators, foams and aerosols, but which have high ozone depleting potentials), opposed vigorously for many years any attempt to ban the use of these chemicals. Then, in 1988, it reversed its position, and become a major proponent of a world-wide ban. Industry sources suggest that the reason for this reversal lay in Du Pont's belief that it would be successful in winning the patent on a substitute from a similar class of chemicals, HCFCs. ICI and other European producers, on the other hand, have invested in a competing class of chemicals known as HFCs. They have therefore argued that production of HCFCs should be limited, on the grounds that they are ozone depleters (although to a smaller extent than CFCs).

(See National Wildlife Federation, 1989, for a fuller account.)

Price-based Systems: The idea that socially inefficient pollution can be controlled by taxation is not new to economics (Pigou suggested in 1946 that charging a tax on the polluting activity equal to the difference between social and private marginal costs would cause competitive resource exploitation to coincide with the optimum. (See figure 1.) In an ideal world, this would yield exactly the same outcome as quantity-based regulation. In reality, price-based systems appear to offer less opportunity for the sort of strategic behaviour decribed above. This robustness to anti-competitive behaviour constitutes one of the major advantages of a price-based system, and consequently, this form of regulation emerges as the most promising option for managing chemical hazards.

There are, of course, disadvantages to the use of *ex ante* taxes. First, substantial amounts of information are required to set tax rates optimally. (This is one of the main advantages of an *ex post* liability system (the 'rate' is set after the level of damage is determined.) Secondly, the administrative costs of tax systems tend to be high, and rise as the tax is more finely targeted; although by imposing the tax on manufacturers, of whom there are relatively few (compared to the number of users, for example), this factor is minimised. Finally, if there is uncertainty about the costs of designing products with lower accumulation abilities, and the marginal benefits to lower accumulation is very steeply sloped (with respect to decreases in accumulation), then taxation may lead to very inefficient outcomes (see Weitzman, 1974). Even with these reservations, however, some form of *ex ante* taxation is likely to be required.

Conclusions

Three general principles have emerged from the discussion. First, *ex post* liability systems will not be sufficient, on their own, to regulate long-term environmental and human health hazards from chemical use. Secondly, *ex ante* regulation must be designed carefully to avoid mere substitution from one product to another, or from one chemical characteristic to another. Finally, taxation of the relevant characteristics of chemicals appears to be the leading regulatory option, as it minimises the potential for strategic behaviour. This option has been adopted already by a few European countries; for example, Sweden imposes a charge per kg of active ingredient on pesticides (see OECD, 1989, for further details).

The discussion clearly can be nothing more than the first steps in a comprehensive analysis of the regulation of chemicals. Several important questions were not even addressed; for example, the optimal mix of *ex post* and *ex ante* systems; and the appropriate degree of regulatory differentiation across countries with e.g. different environmental sensitivities. These questions are the subject of ongoing research.

References

Baumol, W J and. Oates, WE (1988) The Theory of Environmental Policy, Cambridge; New York and Sydney: Cambridge University Press (Second edition).

Bergman, L and Pugh, DM (eds.)(1994) Environmental Toxicology, Economics and Institutions: The Atrazine Case Study, Kluwer Academic Publishing.

Commission of the European Communities, "Green Paper on Remedying Environmental Damage", Com (93) 47, March 1993.

Conway, G P and Pretty, JW (1991) Unwelcome Harvest: Agriculture and Pollution, London: Earthscan Publications.

Faure, M (1994)"The E.C. Directive on Drinking Water: Institutional Aspects". In: Bergman, L and Pugh, DM (eds.), Environmental Toxicology, Economics and Institutions, Kluwer, pp. 39-87.

Hansmann, H and Kraakman, R (1991) "Toward Unlimited Shareholder Liability for Corporate Torts", Yale Law Journal, Vol. 100(7), May 1991, pp. 1897-1934.

Landes, W M and Posner, RA (1987)The Economic Structure of Tort Law, Cambridge, Mass. and London: Harvard University Press.

Macauley, M K, Bowes, MD and Palmer, KL (1992) Using Economic Incentives to Regulate Toxic Substances, Resources for the Future, Washington.

Mason, C F and Polasky, S (1994) "Entry Deterrence in the Commons", International-Economic-Review, Vol. 35(2), pp. 507-25.

Mason, R A (1996) "The Regulation of Chemical Accumulation in the Environment", Mimeo, University of Oxford.

Mason, R A and Swanson, TM (1996) "Long-term Liabilities and the Choice of Liquidation", Geneva Papers on Risk and Insurance, Vol. 21, April 1996, pp. 204-223.

Menell, P S (1991) "The Limitations of Legal Institutions for Addressing Environmental Risks", Journal of Economic Perspectives, Vol. 5(3), pp. 93-113.

National Wildlife Federation (1989) "Du Pont Freons Products Division", Class discussion paper prepared by Forest Reinhardt (copies available from the author on request).

OECD (1989) Economic Instruments for Environmental Protection, Paris.

Opschoor, J B and Pearce, DW (eds.)(1991) Persistent Pollutants: Economics and Policy, Kluwer.

Ringleb, A H andWiggins, SN (1990) "Liability and Large-scale, Long-term Hazards", Journal of Political Economy, Vol. 98(3), pp. 574-595.

Roe, M (1986) "Corporate Strategic Reaction to Mass Tort", Virginia Law Review, Vol. 72(1), Feb. 1986, pp. 1-59.

Rolike, A (1996) "The Ban of a Single Pesticide", Geneva Papers on Risk and Insurance, Vol. 21, April 1996, pp. 224-239.

Schwartz, A (1985) "Products Liability, Corporate Structure, and Bankruptcy: Toxic Substances and the Remote Risk Relationship", Journal of Legal Studies, Vol. 14, Dec. 1985, pp. 689-750.

Shavell, S (1986) "The Judgement Proof Problem", International Review of Law and Economics, Vol. 6, pp. 45-58.

Shavell, S (1987) Economic Analysis of Accident Law, Cambridge, Mass. and London: Harvard University Press.

Swanson, T M (1991) "The Use of Liability-Rule Systems to Regulate the Environment: The Case for 'Preventative Policy". In: Opschoor, J B and Pearce, DW (eds.), Persistent Pollutants: Economics and Policy, Kluwer.

Swanson, T M (1996) "Optimal Policies for Regulating Persistent Chemicals", Mimeo, University of Cambridge.
Varian, H (1992) Microeconomic Analysis, New York and London: Norton (Third edition).
Vighi, M and Zanin, G (1994) "Agronomic and Ecotoxicological Aspects of Herbicide Contamination of Groundwater in Italy". In: Bergman, L and Pugh, DM (eds.), Environmental Toxicology, Economics and Institutions, Kluwer, pp. 111-139.
Weitzman, M L (1974) "Prices vs. Quantities" Review of Economic Studies, Vol. 41(4), Oct. 1974, pp. 477-91.

The Identification of Thresholds of Acceptability and Danger: The Chemical Presence Route

Michael Matthies[1], Frank Koormann[1], Geert Boeije[2] and Tom CJ Feijtel[3]
[1] University of Osnabrück, Institute of Environmental Systems Research, Osnabrück, Germany
[2] University of Ghent, Department of Applied Mathematics, Biometrics and Process Control, Ghent, Belgium
[3] Procter & Gamble N.V., European Technical Center, Strombeek-Bever, Belgium

Abstract

European Union chemical legislation requires the calculation of local and regional Predicted Environmental Concentrations (PEC) for the assessment of the exposure of new and existing chemicals to aquatic and terrestrial ecosystems. Current methods use local models for air, water and soil to estimate chemical concentrations close to the source and a generic multimedia 'unit world' approach to estimate regional PECs. These models assume generic environmental scenarios representing typical situations in European countries and do not account for the spatial heterogeneity and temporal variability in ecosystem characteristics, soil properties, river flow rates, chemical emissions, etc. The environmental and ecological complexity can best be represented in a Geographic Information System (GIS). By coupling a GIS with a fate simulation model the concentrations of substances in a specific environment are predicted more realistically. The GREAT-ER project (Geography-referenced Regional Exposure Assessment Tool for European Rivers) was launched to refine regional and local exposure assessments for down-the-drain chemicals by applying real, spatial-referenced datasets instead of generic or average values. A modular approach was developed consisting of a hydrological model and a waste-flow, river quality and fate model which are linked to a regional GIS-database. For the calibration and validation in two European study areas representative detergent chemicals (LAS, boron) are used. In a parallel study, high-volume intermediates discharged into the river Rhine are simulated.

Environmental Risk Assessment

Environmental risk assessment is characterized by both the probability of the presence of chemicals in ecosystems and of their adverse effects on ecosystem structure and function. To assess the environmental exposure the level of compounds at the location of the effects have to be determined. Exposure is usually predicted from emission rates and environmental modelling. Adverse effects on ecosystems have to be extrapolated from laboratory tests. Exposure modelling is a mechanistic approach based on the physical and (bio-) chemical processes of transport and transformation in air, water and soil. Effects assessment is a data-driven approach by applying extrapolation or assessment factors to acute or chronic test results with laboratory organisms to ecosystems (Van Leeuwen and Hermens, 1995). The objectives of exposure assessment are (Matthies, 1991):

- understanding of the governing processes and their variability
- determination of the chemicals presence route from the entry into the primary environmental compartment to the location of action
- recognition of the population at risk
- prediction of the concentration in the environmental medium of concern (PEC = Predicted Environmental Concentration).

European Union chemical legislation requires the calculation of generic Predicted Environmental Concentration (PEC) and Predicted No-Effect Concentration (PNEC) values in order to perform environmental risk assessment of chemicals available on the EU market (Fig. 1) (EEC, 1993a, 1993b, 1994a, 1994b; ECETOC, 1993, 1994a).

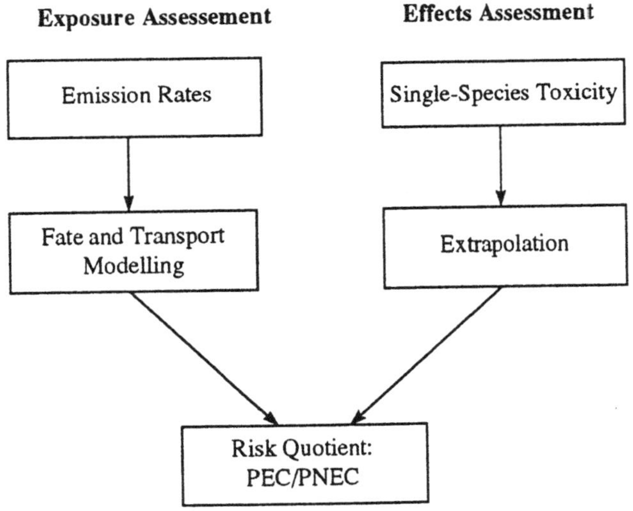

Fig. 1 EU Risk Assessment Process

Environmental Exposure Assessment of Chemicals

Over the last two decades a lot of experience was gained which enabled us to have a better understanding of the processes governing the fate of chemicals of various structure classes in aquatic and terrestrial ecosystems (Calamari, 1992; Trapp and Matthies, 1996). However, spatial and temporal scale have to be taken into account when the chemical presence route has to be determined. To cope with the spatial scale problem specific generic environments were defined. A common approach distinguishes between local and regional environments (Van Leeuwen and Hermens, 1995). Current techniques use local models for air, water, and soil to calculate chemical concentrations in the vicinity of the source and a generic multimedia 'unit world' approach to estimate regional PECs. Model predictions have to be validated to increase confidence in the general approach.

The European Science Foundation (ESF) supported the „Evaluation of Fate and Exposure Models" within the programme on „Environmental Damage Assessment" (Matthies, 1992). Field data obtained in various European countries were used to assess the influence of the environmental variability and the capability to cope with different environmental situations and scales. The transport of pesticides, persistent organic halocarbons as well as many other chemicals were predicted and the results were compared with measurements from field and laboratory experiments). The results give us further confidence in the basic concepts and their general applicability. Various computer programmes are now on the market which summarize this knowledge and experience, e.g. HAZCHEM (ECETOC, 1994b), USES (RIVM VROM & WVC, 1994), CemoS (Scheil et al., 1995; Trapp and Matthies, 1996), SAMS (Matthies et al., 1992). The ESF programme also addressed the question of model quality assurance. Criteria for good modelling practice were defined to select the most appropriate fate and exposure model (Wagner and Matthies, 1996; Vermeer and Wolff, 1996).

For *regional* assessments a Mackay-type multi-media model is currently being used. This method describes the mass transfer of chemicals between the different environmental media (soil, air, water, sediment), as well as removal from compartments. Multimedia models were evaluated by an international task force to ascertain their strengths and limitations and to recommend improvements to enhance model applicability and use (Cowan et al., 1995). The investigated models differ mainly in their intermedia transport expressions and parameters. Suggestions were made to improve models by including vegetation and the treatment of chemical fate in soils and sediments.

The methods to determine regional PECs use generic or average data per country and do not account for spatial and temporal variability in landscape characteristics, soil properties, river flows and/or chemical emissions. They can only be validated if the appropriate scale is selected together with the adequate data set. The simulation results offer no realistic fate data, and are merely applicable on a screening level (European Science Foundation, 1995).

Geography-Referenced Regional Exposure Assessment

The environmental and ecological complexity can best be represented in a Geographic Information System (GIS). By coupling a GIS with a fate simulation model the concentration of a chemical substance in a specific environment is predicted more realistically. The GREAT-ER project (Geography-referenced Regional Exposure Assessment Tool for European Rivers) was launched to refine regional and local exposure assessments for chemicals discharged into rivers by applying real, spatial-referenced datasets instead of generic or average values.

Geography-referenced regional exposure assessments are approached in a modular way (Fig. 2):

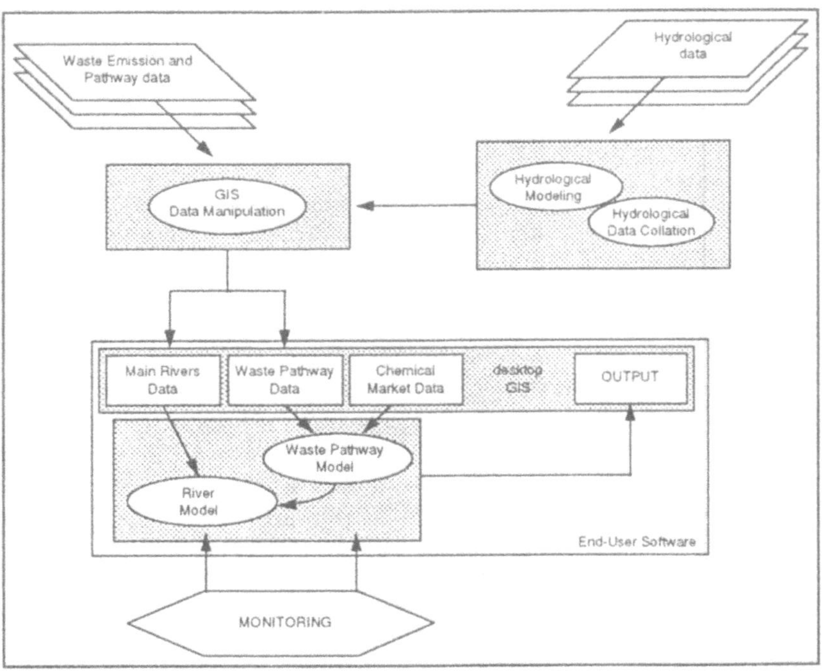

Fig. 2 Modular Approach of the GREAT-ER Project

- Waste Pathway and River Modelling

This module is used for the prediction of chemical emission, of chemical transformation during conveyance and treatment and of chemical fate along rivers.

- Hydrology

The hydrology module combines several hydrological databases with a hydrological model. It provides the GREAT-ER system with the required flows and river characteristics.

- GIS Data Manipulation

In the data manipulation module, input data sourced from several databases and from the hydrology module are transformed into appropriate GIS formats, including geographical segmentation.

- End-User Desktop GIS

In this module, access to and visualization of the data banks and model results is achieved, as well as the linking of the models with the data banks.

- Monitoring

A monitoring programme is executed in order to provide the environmental measurements required for model calibration and validation.

Waste Pathway and River Model

The purpose of this model is to provide the necessary data on all significant waste water discharges into rivers. Each main river is linked to its catchment area waste pathway model, which uses geographical and technical information about waste water emission, transportation and treatment infrastructure. 'Down-the-drain' chemicals can be routed to rivers (Fig. 3) via point discharges of treated or untreated domestic waste water.

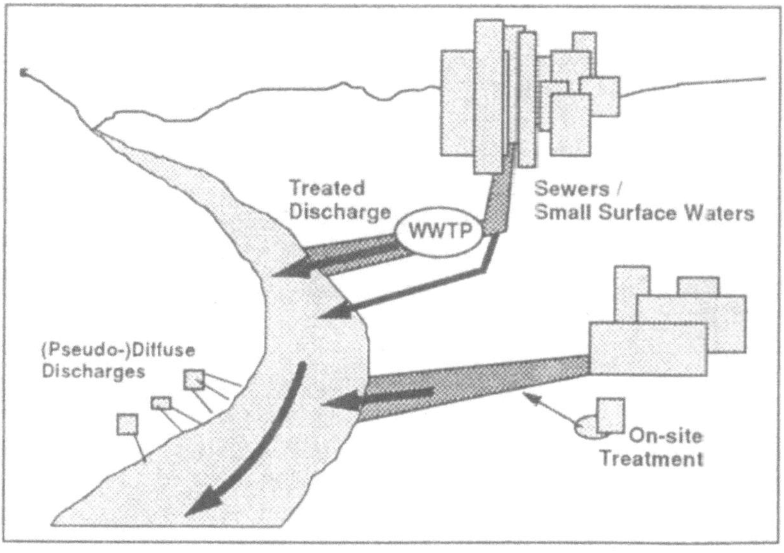

Fig. 3 'Down-the-drain' chemical routes to rivers

The objective of the river modelling is to predict the concentrations of chemicals along the main rivers. Fate models provide information about concentrations in various environmental media and about the mass balance of chemicals (Trapp et al., 1991). Natural rivers are not homogeneous in their hydrological regimes. Rivers are segmented into plug- flow stretches with relatively constant properties. For mathematical purposes, stretches can be divided further into completely mixed 'boxes' (Fig. 4). Each of the stretches can be modelled individually. A river stretch receives water from the previous (upstream) stretch and discharges water into the next (downstream) one. It may also receive (waste) water from discharge points, via a tributary or via runoff.

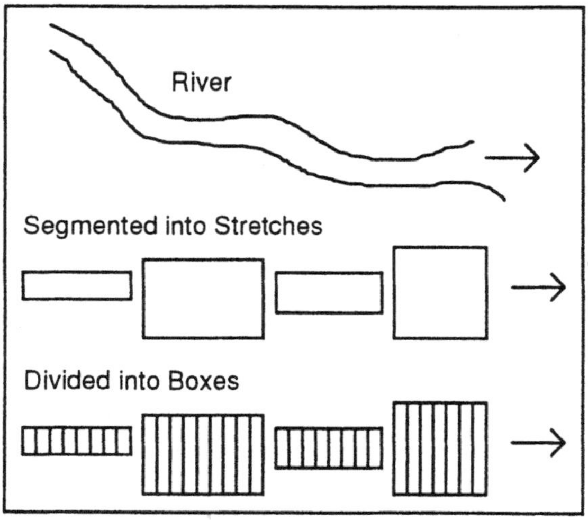

Fig. 4 River Segmentation

The main fate processes to be considered are: degradation, volatilization, sorption/desorption and sedimentation/resuspension. Other processes such as sediment diffusion or bio-accumulation may also be incorporated (Fig. 5).

Geographical Information Systems (GIS) are used for the storage of the necessary hydrological and waste pathway data in a format which is easily accessible by the models and which allows user-friendly interactions. A flexible data-driven approach is used, since the ideal grid shape and size are dependent on a desired level of detail and on data availability for specific regions.

The (standardized) GIS data banks, the waste pathway models and river models are integrated into one coherent simulation system. Such an integration process will result in an operational end-user system, which runs on a PC platform. The integration methodology is represented in Fig. 6.

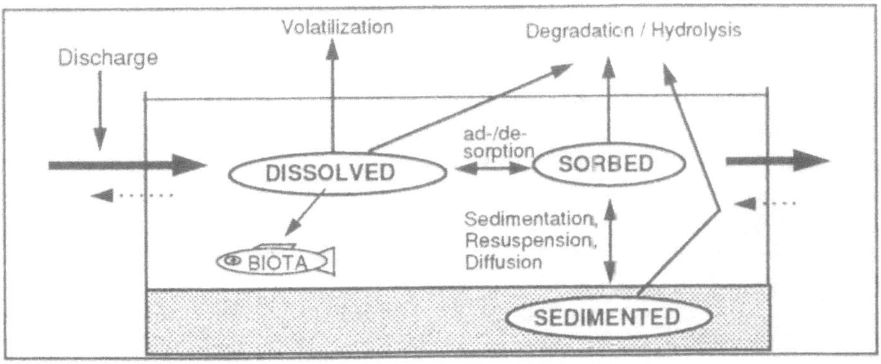

Fig. 5 Chemical fate processes

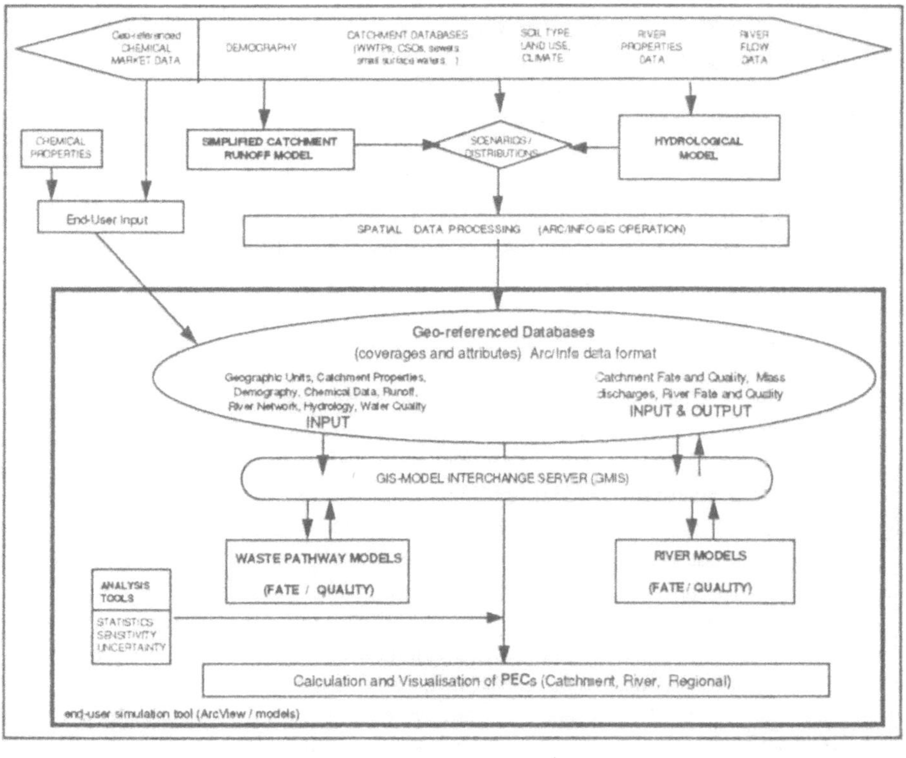

Fig. 6 Integration Methodology

System Application

Concentrations in water and sediment are to be calculated using the proposed simulation system, and then compared to measured values. Initially, monitoring results will be used to calibrate and fine-tune the modelling, to improve the predictive power. Two different compound classes are under investigation: (1) down-the-drain chemicals for which the detergent LAS serves as the representative compound and (2) intermediates which are only emitted at a few discharge points in Europe. However, the ultimate objective is to implement the GREAT-ER system for the entire European Union. In the first phase, the blueprint system will be applied to three pilot study areas, which will allow its refining and validation.

a) Down-the drain chemicals

One pilot study area for the fate simulation of LAS is situated in the northern part of Europe (UK, Yorkshire Ouse basin). For the southern European situation, a sub-catchment of the Po basin was chosen (in the Milan area). They differ in many respects, e.g. waste water treatment and discharges, topography, population density and distribution, etc. A monitoring programme will serve to test the reliability of the predictions (Hennes & Rapaport, 1989).

For the calibration, measured data representative for one point in time will be used. This implies the use of hydrological parameters in the model which are representative for the particular flow-regime at that time. The calibration for specific flow regimes can be extrapolated to other flow regimes. For validation, data will be measured throughout the year in order to obtain time-averages and distributions. In this case, the model will also use time-average hydrological data.

b) Intermediates in the River Rhine

Intermediates are produced in large quantities by chemical companies located along the River Rhine. Production volumes of 55 chemical intermediates were collected by ECETOC (1994a) and annual estimated release, location and capacity of the German companies were published. The river network of the Rhine and its main tributaries were divided into segments with one gauging station in each segment. The boundaries of the segments were chosen by the mouths of the tributaries or by barrage weirs assuming homogeneous hydrological properties within one segment (IKSR, 1991). Each segment receives only one discharge. The Rhine itself consists of 15 segments from Karlsruhe-Maxau to Rees; its tributaries, the Main, the Mosel and the Lippe consist of one, three and two segments, respectively. The mean flow and mean water level of the summer half-year were taken from DIHJ (1989a and 1989b). Flow velocities in each Rhine segment were estimated by the regression equations of Teuber and Wander (1987) or from rectangular cross-sections of the tributaries. A deterministic fate model assuming advection, dilution, volatilization, sedimentation and degradation was used to calculate downstream concentrations (Trapp and Matthies, 1996). All spatial-referenced data were stored in a database which is coupled to the desktop GIS ARCView. The results of a simulation run can immediately be visualized for the whole river network. Visualization requires the dynamic segmentation technique of ARC/INFO which provides ARCView with the geo-referenced linear objects.

As an example, the simulation of the high-volume intermediate vinyl chloride was selected. The basic substance properties, as well as the release data are given in Table 1. The results of the simulation are shown as a map of the Rhine valley in Fig. 7 and as a concentration plot in Fig. 8. Obviously, vinyl chloride is eliminated from the water body, mainly by volatilization. Since the wind speed determines the volatilization rate sensitively, the concentration profiles are plotted in Fig. 8 for an average of 4 m/s and a low value of 1 m/s. Both figures also show that the discharges of vinyl chloride cannot be considered separately. E.g. the concentration in the lower Rhine valley is determined by all discharges except that of BASF in Ludwigshafen. Vinyl chloride concentration of 0.2 µg/l was measured at the German-Dutch border (Rippen, 1995) which lie between the calculated concentrations of 0.1 µg/l and 0.28 µg/l for a wind speed of 4 m/s and 1 m/s, respectively. The great advantage of the GIS-coupled simulation is the simple analysis of all spatial parameters (hydrological, environmental, release, location, etc) and non-spatial data (substance data, production, etc) on the concentration profile along the river network and their graphical visualization.

Table 1 Data set for simulation of vinyl chloride

Substance properties (Rippen, 1995)		
CAS-No.		75-01-04
Molar Mass		62.5 g/mol
Henry's law constant (KAW)		0.77
LogKow		1.27
Koc		7.7 cm3/g
Degradation rate		0
Production data (ECETOC, 1994a)		
Production volume (1979)		1 500 000 t/a
Release via synthesis		0.000 7 %
Release via use		0.002 1 %
Total release		0.002 8 %
Production sites	Capacity	Estimated Release
Ludwigshafen (Rhine)	100 000 t/a	2.8 t/a
Knapsack (Rhine)	100 000 t/a	2.8 t/a
Cologne (Rhine)	225 000 t/a	6.3 t/a
Rheinberg (Rhine)	200 000 t/a	5.6 t/a
Marl (Lippe)	350 000 t/a	9.8 t/a

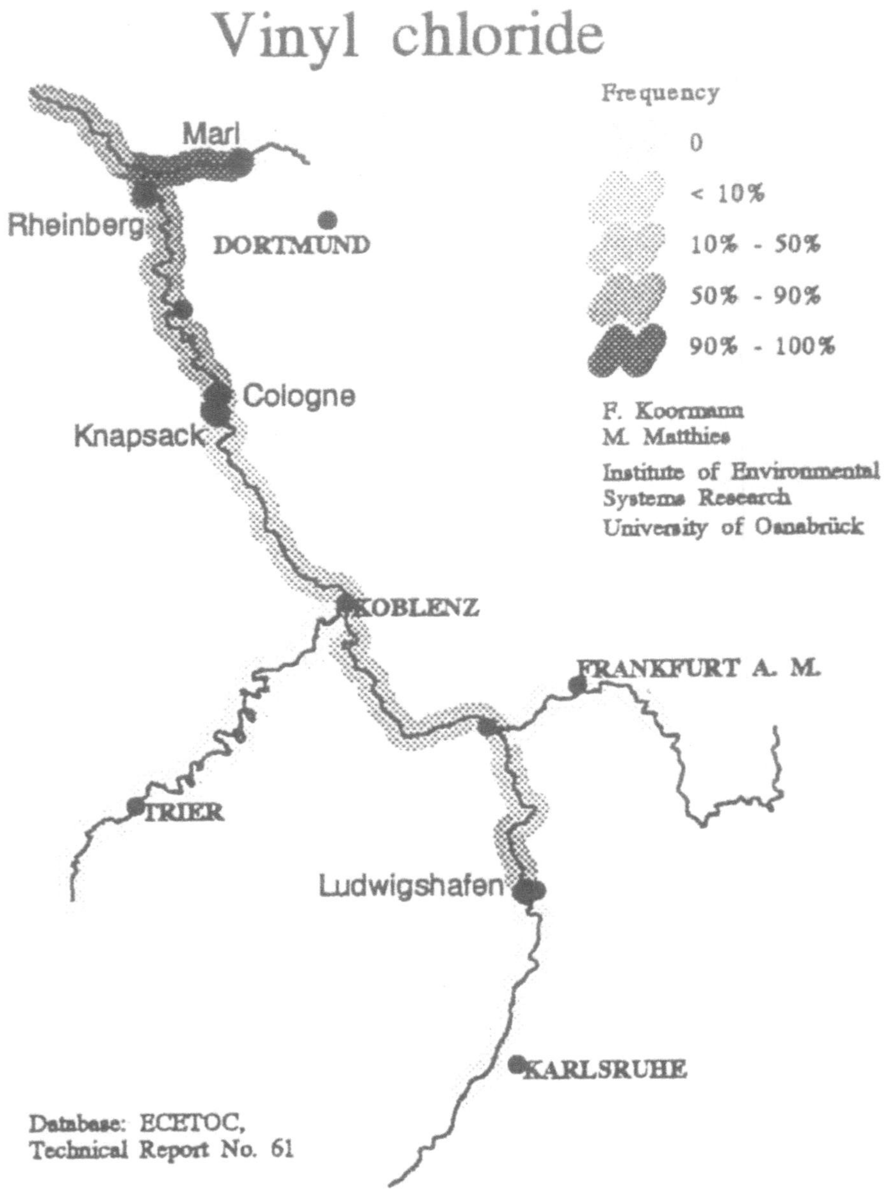

Fig. 7 Map of the average vinyl chloride concentration (summer 1989)

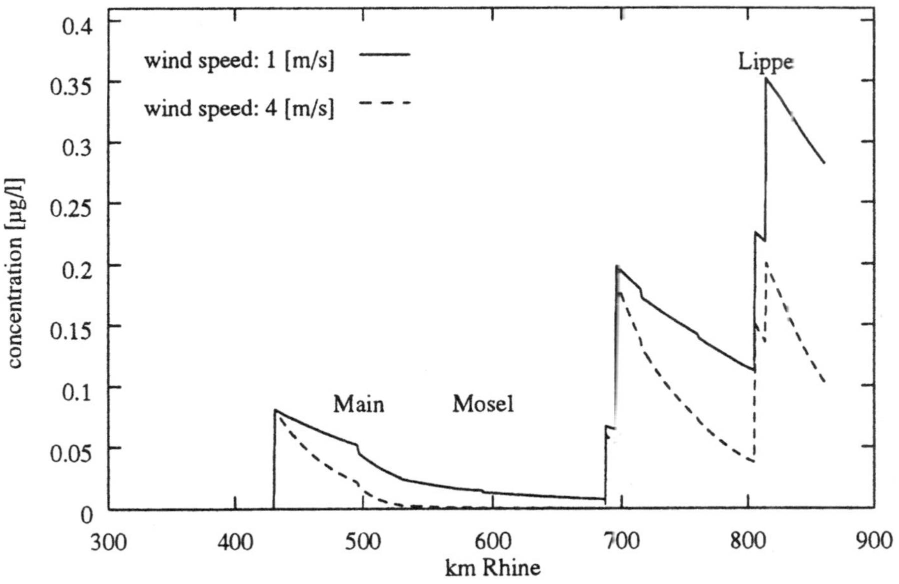

Fig. 8 Vinyl chloride concentration profile in the River Rhine (kilometres along the Rhine from Constance bridge left)

Conclusions

The objective of the new proposed regional exposure assessment is to solve the drawbacks associated with current regional PEC calculation methods. This project will allow redefinition of the regional PEC concept, using realistic region-specific characteristics instead of generic systems, thus allowing for a more accurate exposure prediction. Next to geographical or spatial variability, seasonal or temporal variability will also be accounted for.

As the objective is to use a region- and geography-specific approach for exposure assessments within EU chemical legislation (EEC, 1993a, 1993b, 1994a, 1994b), it is obvious that this project has to be carried out on a European level. Indeed, input information about river systems and catchment areas in all the member states is required. The modelling system will be available and applicable for all member states.

The immediate applicability of this project is therefore directed towards the environmental risk assessment of new and existing chemicals. The proposed exposure assessment system will mainly be applicable for the fate modelling of chemicals in rivers. However, possibilities for further expansion towards other compartments (eg. soil, air) are left open.

Acknowledgement

The financial support of CEFIC/ERASMUS is gratefully acknowledged.

References

Calamari D (Ed) (1991): Chemical Exposure Prediction. Lewis Publ Boca Raton

Cowan C, Mackay D, Feijtel TCJ, van de Meent D, di Guardo A, Davies J, Mackay N (1995): The Multi-Media Fate Model: A Vital Tool for Predicting the Fate of Chemicals. SETAC Press, Pensacola

Dhji (1989a): Deutsches Hydrologisches Jahrbuch, Rheingebiet Teil III: Mittel und Niederrhein mit Issel und Maasgebiet, Abflußjahr 1989, Landesamt für Wasser und Abfall Nordrhein-Westfalen, Düsseldorf

Dhji (1989b): Deutsches Hydrologisches Jahrbuch, Rheingebiet Teil II: Main und bayrisches Elbegebiet, Abflußjahr 1989, Bayrisches Landesamt für Wasserwirtschaft, München

ECETOC (1993): Environmental Hazard Assessment of Substances. Technical Report No 51

ECETOC (1994a): Environmental Exposure Assessment. Technical Report No 61

ECETOC (1994b): HAZCHEM, a Mathematical Model for Use in Risk Assessment of Substances. Special Report, No 8

EEC (1993a): Commission Directive 93/67/EEC of 20 July 1993 laying down the principles for assessment of risks to man and the environment of substances notified in accordance with Council Directive 67/458/EEC. Off J of the European Communities, L 227/9

EEC (1993b): Council Regulation (EEC) No 793/93/EEC of 23 March 1993 on the evaluation and control of the risks of existing substances. Off J of the European Communities, L 84/1

EEC (1994a): Commission Regulation of laying down the principles for the assessment of risks to man and the environment of existing substances in accordance with Council Regulation (EEC) No 793/93/EEC

EEC (1994b): Technical guidance documents in support of the Risk Assessment Directive (93/67/EEC) for substances notified in accordance with the requirements of Council Directive 67/548/EEC, Brussels

European Science Foundation (1995): Problems and Perspectives of Regional Fate Modelling of Xenobiotics. A Strategy for Enhancing the Role of Multimedia Mass Balance Models for the Stewardship of Chemicals. Proceedings of a workshop held at Hillerroed, Denmark, 30 Jan - 1 Feb 1995

Hennes EC, Rapaport RA (1989): Calculation and Analytical Verification of LAS Concentrations in Surface Waters, Sediment and Soil. Tenside Surfactants Detergents 26(2), pp 141-147

IKSR (1991): IKSR/KHR Expertengruppe: Rheinalarmmodell 2.0, Kalibrierung und Verifikation. Koblenz

Matthies M, Brüggemann R, Münzer B (1992): SAMS Screening Assessment Model System. OECD Paris and UBA Berlin

Matthies M (1994): Evaluation of Fate and Exposure Models. ESPR-Environ Sci & Pollut Res 1, pp 93

Matthies M (1991): Expositionsmodelle. UWSF-Z.Umweltchem Ökotox 3 (1), pp 37 - 41

Rippen G (1995): Handbuch Umweltchemikalien. Permanently updated. Ecomed, Landsberg/Lech

RIVM, VROM & WVC (1994): Uniform System for the Evaluation of Substances (USES), version 1.0. National Institute of Public Health and Environmental Protection (RIVM), Ministry of Spatial Planning and the Environment (VROM), Ministry of Welfare, Health and Cultural Affairs (WVC), The Hague, Distribution No 11144/150

Scheil S, Baumgarten G, Reiter B, Schwartz S, Wagner JO, Trapp S, Matthies M (1995): An Object-Oriented Software for Fate and Exposure Assessments. ESPR-Environ Sci & Pollut Res 2 (4), pp 238 -241

Teuber, Wander (1987): Fließzeiten im Rhein aus Flügelmessungen. Bundesanstalt für Gewässerkunde, Koblenz

Trapp S, Brüggemann R, Münzer B (1991): Exposure Analysis of the Phosphate Substitutes NTA and EDTA by Use of the Surface Water Model EXWAT. In: Wrobel LC, Brebbia CA (Eds) Water Pollution: Modelling, Measuring and Prediction. Elsevier Publ, pp 195-209

Trapp S, Matthies M (1996): Dynamik von Schadstoffen - Umweltmodellierung mit CemoS. Eine Einführung. Springer-Verlag, Heidelberg

Van Leeuwen CJ, Hermens JLM (1995): Risk Assessement of Chemicals: An Introduction. Kluwer Acad Publ Dordrecht

Veerkamp W, Wolff C (1996): Fate and Exposure Models in Relation to Risk Assessment. Developments and Criteria. ESPR-Environ Sci & Pollut Res 3 (2), pp 91 - 95

Wagner JO, Matthies M (1996): Guidelines for Selection and Application of Fate and Exposure Models. ESPR-Environ Sci & Pollut Res 3 (1), pp 47 - 51

The Identification of Thresholds of Acceptability and Danger: The Biological Route

Jose V. Tarazona
Division of Environmental Toxicology. CISA-INIA, E-28130 Valdeolmos, Madrid, Spain.

Abstract

The application of scientific concepts to management decisions is a clear need nowadays. The regulation of environmental pollutants requires the application of cost/benefit analysis. Human health and ecosystem status can be seriously affected and, therefore, are essential parts of the cost assessment. Ecotoxicology must provide the scientific basis for the prediction, control, monitoring and diagnosis of the expected/observed effects of chemical substances on the structure and function of ecosystems. The basic problem is to select which levels of risk/alteration can be accepted, taken into account the degree of uncertainty in the assessment. This paper summarizes some biologically based approaches that try to reduce the level of uncertainty and/or to improve the decisions on the basic thresholds: acceptable/unacceptable and safe/dangerous. The presentation will focus on two specific aspects. Starting with prediction related topics, the first aspect will regard on the evolution of Hazard and Risk Assessments and the biological developments that try to improve these evaluations. The discussion will cover ecotoxicological concepts included in the adopted criteria for different regulatory aspects, i.e. setting Water Quality Objectives, classifying potential hazards using the "inherent" toxicity of a substance, or predicting Non Effect Concentration in risk characterization procedures. The second aspect will cover the use of biologically based systems as complementary and even alternative tools in management decisions related to the control, monitoring and diagnosis of the environmental alterations produced by highly toxic chemicals on biological systems.

Introduction

Until relatively recently in humanity's history, pollution has been primarily a local problem. However, the explosion of human population and the industrialization of society have caused a tremendous increase in waste by-products coupled to the exponential growth in the production of goods and services. The "throwaway" attitude without considering potential consequences has resulted in major environmental disasters. The response includes the regulatory enforcement through governmental and international organizations. Public awareness of the extent of pollution will eventually force governments to undertake more effective environmental planning and adopt more effective antipollution measures. The scientific improvement of risk, and associated, management decisions are major goals for both Ecotoxicology and the environmental aspects of Human Toxicology. The basic problem is to select which levels of risk/alteration can be accepted, these threshold values must consider the degree of uncertainty in the assessment.

An excellent review on the state of the art of the toxicological knowledge on Human Health Risk Assessment has been recently published (Wilson et al., 1995). From a biological (toxicological) point of view, the fundamental problems in the human health risk assessment of toxic substances are related to the extrapolation of animal data to human effects (ultimate human relevance of animal studies) and the extrapolation of the high experimentally used doses to the low environmentally relevant concentrations, as e.g. distribution and elimination processes may be nonlinearly dependent on dose (Scheuplein, 1995). The essential methodologies to produce scientifically sound extrapolation models are physiologically based pharmacokinetics (PBPKs), biologically based dose-responses (BBDRs), and structure-activity relationships (SARs) (Vandenberg, 1995).

The ecotoxicological risk assessment, by contrast, presents different problems and potential solutions. This paper focuses on the state of the art of the environmental hazard and risk assessment and control of toxic pollutants, including the main problems and some new biologically based developments. These developments, when fully implemented, will produce a revolutionary change in the application of ecotoxicological thresholds for regulatory purposes.

Biological *versus* Chemicals tools in Ecotoxicology

The main aims of Ecotoxicology can be summarized in five keywords: Prediction, Control, Monitoring, Diagnosis and Restoration (Tarazona et al., 1995). These keywords focus on the type of information required by regulatory bodies to produce scientifically sound assessments on the effects of chemicals on biological systems, with the conservation of the structure and function of ecosystems as final goal. The *Prediction* of the effects of a new substance or technology must be followed by the *Control* of the prediction accuracy in terms of both emission/exposure and effect assessments. Different reasons, from the lack of control in the past to the uncertainty on the synergistic effects of environmental pollutants, sustain the need of continuous *Monitoring* programs on the status of potentially exposed population and ecosystems. The alterations, when observed, require specific cause-effect *Diagnosis*, using the basic concepts of Pathology, and the available technological efforts should be employed, after the appropriate cost/benefit evaluation, in the *Restoration* of the altered system.

Regulatory aspects related to Prediction, Control and Monitoring focus on the assessment of ecotoxicological thresholds: the border between what is considered safe (or at least acceptable) and what is considered dangerous, and therefore requires management decisions such as emission restrictions, diagnosis of specific effects, or restoration of the polluted environment. The basic instruments to solve the different questions arising from the above mentioned items are chemical and biological methods.

The effect of a substance on a living organism can only be established by biological methods, i.e. by bioassays. Toxicity is species specific and dose/effect relationships also depend on a myriad of physiological and environmental conditions. Regulatory ecotoxicology requires, obviously, a certain level of pragmatism. The extrapolation from a set of doses to a different one, the analogy in the toxicological response of certain species within taxonomic groups, the estimation of effects for intermediate environmental conditions using marginal data, or the extrapolation of effects between analogous and related compounds, are basic pieces for most assessments. All these extrapolations require scientific support. Therefore, and following with the same examples, research on dose/effect relationships, physiologically based toxicokinetics and toxicodynamics, speciation/bioavailability modeling, or structure-activity relationships, are basic aspects of ecotoxicology. Anyhow, the use of biological methods and data is essential: Toxicity can only be assessed by the direct observation of the effects, mostly under the experimental exposure of organisms (or suborganism structures) to the chemical, or by inference from real data observed for related organisms, chemicals, or environmental conditions. Therefore, in regulatory ecotoxicology, those items related to the prediction of effects have been always based on the use of biological methods.

By contrast, the control, monitoring, and even diagnosis of environmental effects have traditionally focused on the measurement of the pollutant concentration using chemical methodologies. From an ecotoxicological point of view, this approach is acceptable when it includes the subsequent toxicological interpretation, assessing which effects can be expected or related to the observed concentrations. However, this interpretation requires a large amount of toxicological information, which is restricted, actually, to a few priority pollutants. The lack of skillful and cost/effective bioassays must be included within the different reasons that explained this situation. The rapid development of biological methods in recent years has opened new possibilities and expectations. Therefore, biological methods are, nowadays, a new set of complementary/alternative possibilities for the control, monitoring and diagnosis of environmental effects.

Following this rationale, the paper will be divided in two parts. Hazard and Risk Assessment (HRA) is the essential aspect of threshold-based ecotoxicological prediction. Biologically based information has always been the essential tool for HRA; the state of the art, and particularly, the levels of uncertainty when setting thresholds and the potential solutions, will be considered in the first part. The second part will focus on the incorporation of biological methods to set ecotoxicologically-based thresholds for control and monitoring environmental problems.

Concepts and Developments for Environmental HRA.

Regulations on environmental risk assessment within the European Union entails a sequence of actions which is outlined below:

1. Assessment of effects, comprising
   a) Hazard identification: identification of the adverse effects which a substance has an inherent capacity to cause.
   b) dose (concentration)-response (effects) assessment: estimation of the relationship between dose, or level of exposure to a substance, and the incidence and severity of an effect, where appropriate.
2. Exposure assessment: estimation of the concentrations (doses) to which environmental compartments are, or may be, exposed.
3. Risk characterization: estimation of the incidence and severity of the adverse effects likely to occur due to actual or predicted exposure to a substance.

Different regulatory principles are associated to each action. According to Directive 67/548/EEC and its amendments, the identification of the hazard is followed by the classification and labelling of those substances identified as dangerous. Dose (concentration)-response (effects) assessments are the basis for setting Water Quality Objectives according to Directive 76/464/EEC. Risk characterization is covered by Directive 93/67/EEC for new notified substances, Commission regulation EC 1488/94 for existing substances, or Directive 91/414/EEC for pesticides.

In each case, thresholds of acceptability/danger are estimated using specifically designed protocols. These protocols are generally scheduled in three-step procedures: first establishing a set of scientifically sound criteria; second, developing agreed upon guidance schemes; and finally, studying the existing information (i.e. toxicity data) to select the most reliable and representative results. Each step produces different types of uncertainty as outlined in Fig 1.

A main epidemiological source for uncertainty is our incomplete understanding of ecological systems. Unlike human health risk assessments, which have a clear endpoint, ecological goals, i.e. ecosystem integrity or health

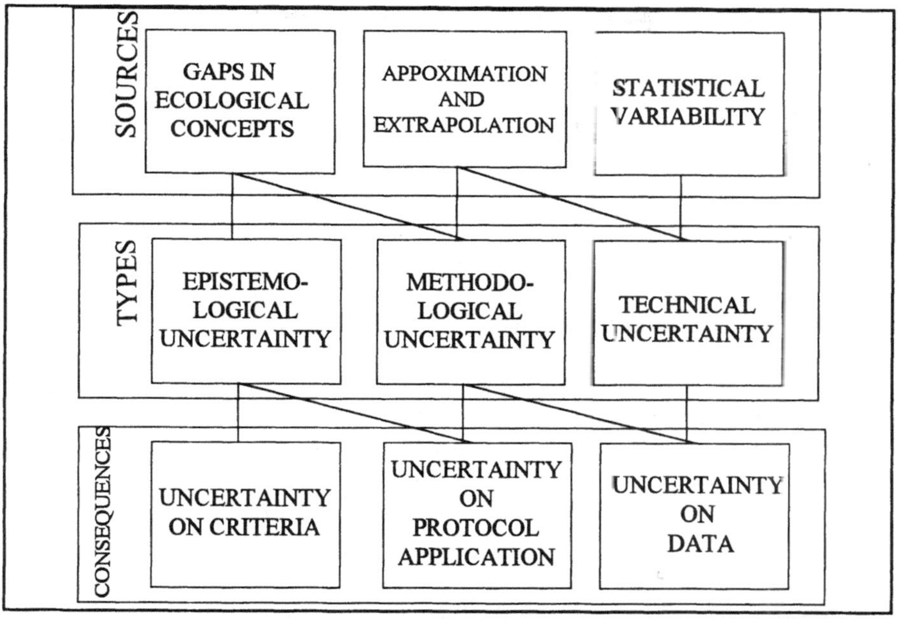

Fig. 1 Distribution of the uncertainty (modified from van Asselt et al., 1996)

are too vague (Renner, 1996). The aim of risk assessment should be the protection of the structure and function of ecosystems, allowing human development. However, this aim is, nowadays, reduced to less ambitious objectives: to estimate the probability that some undesired events will occur. Such events have been named ecological endpoints. Confidence in risk estimations remains constricted by the relative inability to identify critical endpoints (Bartell et al., 1992). When the selected methods are simple enough to produce clear information on the endpoints, i.e. single-species toxicity tests, the ecological relevance of these endpoints is uncertain. By contrast, when the methodology increases its ecological complexity, i.e. mesocosms studies, the estimations of relevant endpoints become equivocal. This uncertainty is challenged nowadays with the following solutions (Table 1):

a.-The application of protective factors selected by groups of experts according to the available information, widely used in the EU, i.e., when setting the Water Quality Objectives (Bro-Rasmussen et al., 1994) or to assess hazard quotients based on response concentration for the most sensitive species and the maximum environmental concentration in risk characterization (ECB, 1996).

b.-The use of probabilistic extrapolation methods, with special development in the USA, i.e. to estimate concentrations intended to be protective for a certain percentage -p.e. 95%- of the taxonomic groups (Harras, 1996) or to compare probability distributions of environmental concentrations with probability distribution of species response data for risk characterization (USEPA, 1992).

The first solution focuses on the most sensitive, but reliable, data, using application factors, mostly orders of magnitude, to move from acute to chronic effects, and from the most sensitive tested species to the most sensitive non-tested species (Bro-Rasmussen et al., 1994). Additional safety factors can be required due to lack of information or due to specific hazard potentials, i.e., genotoxicity. The second solution uses the whole available information set, and assumes that protecting a certain percentage of taxons produces the required ecosystem protection.

Both solutions use the precautionary principle as alternative for the lack of conceptual information on ecologically relevant endpoints. The evolution of ecology and the basic knowledge on the mechanisms of toxicity of each substance and/or group will enhance our capability to produce conceptual approaches to this problem. In this way, the use of biomarkers as sublethal endpoints in toxicity tests can be highly valuable. Both early-warning-biomarkers and effect biomarkers can produce information on which ecological events can be relevant for a particular substance.

Methodological uncertainties

At the present state of the art, only a limited number of endpoints, mostly related to survival, reproduction and growth, on a restricted number of species can be tested in a cost-effective way. However, the enhancement of our methodological capability has produced an exponential enhancement of the methodological possibilities (Table 1). The most significant points include: a) technological applications for the development of cost-effective tests, i.e. commercialization of toxkits (e.g. those developed by Prof. Persoone at the Gent University) or development of reliable *in vitro* tests (Castaño et al., 1996); b) increase in the number of potential species and particularly in the use of exposure conditions (i.e. Muñoz et al., 1996); and c) increase the potential for multi-species tests.

The first point enhances the possibility of regulators to require information from the industry (the use of miniaturized tests can reduce the cost of the bioassays by factors between 10 and 1000). The second point enhances the possibilities for a more randomized selection of test organisms and for obtaining information on more taxonomic groups and for more exposure routes (both aspects are essential when using probabilistic methods). The third point allows the experimental assessment of more ecologically relevant endpoints. An additional potential solution to reduce this source for uncertainty is the use of multivariate methods for data management. This point has been identified as essential for the regulatory use of multispecies methods (i.e. aquatic micro- and mesocosms or model terrestrial ecosystems).

Technological, including data analysis, uncertainties.

The third major component for the uncertainty of ecological risk assessment focuses on the reliability of the data. A toxicity test will generate different estimates of toxicity when performed by different laboratories or even by the same laboratory on different occasions. Within the explanations for data dispersion several aspects must be considered, including those related to the variability in testing conditions allowed by the guidelines, those related to the problems of a laboratory to run reliable assays, and those regarding on the relatively high random error of biological measurements.

Most guidelines allow for a certain degree of freedom in their experimental design. For example, the standardized guideline for the acute toxicity test on fish allows the use of different fish species, with different weights and physiological status (i.e. minnow adults or trout/carp alevins), a relatively wide range of water quality conditions (temperature, hardness, pH, etc.), and different exposure possibilities (from static to flow-through). Obviously, each laboratory will adapt this level of freedom to their particular characteristics, producing a large source for variability even for "similar" tests produced for the same regulatory objective.

Table 1: Sources of uncertainty in the threshold assessment and potential solutions.

Sources	Classical Solutions	Other solutions
Concepts: lack of definition in the ecological endpoints	Safety factors	Probabilistic approaches. "Mode of Action" biomarkers
Methods:		
limitations in:	selection of	number of test species
- test species	"target" taxa	number of exposure routes Cost/effective tests
- endpoints	multispecies tests	Multivariate analysis Biomarker-based endpoints
Techniques:		
problems of:	Standardization	Biotechnol. applications
- reproducibility	GLP, intercalibration	Statistical improvement
- data analysis	EC_{50}, NOEC, LOEC	Time-concentration-effect modelling

Variations in test quality have been considered here as the second source for variability (Table 1). These aspects should be regarded as a source for errors, which should only affect variability and uncertainty when not detected. Good Laboratory Practice (GLP) and quality assurance (QA) and quality control (QC) programmes are appropriate techniques for maintaining, evaluating and enhancing performance. However, the situation is different for the HRA of existing substances. In these cases, preexisting (sometimes very old) data must be considered, and the assessment requires a reliability and quality assessment of each single one of these data. Therefore, data quality becomes an additional source of uncertainty.

Finally, we must recognize that most hazard and risk assessments transform the information produced by each experiment into a single data point or figure; i.e. a standard acute toxicity test on fish produces replicate information for five different concentrations at daily intervals, but all this information is transformed into a single number, the 96h LC_{50}, mostly used without consideration of experimental confidence limits. Different statistical approaches have been suggested to resolve this problem. A full description of these issues is outside the scope of this symposium, but the importance of the subject can be justified mentioning two recent meetings on this topic; the 1995 conference „Asking the Right Questions: Ecotoxicology and Statistics" supported by the SETAC and published as vol. 5, number 3 of Ecotoxicology, and the OECD workshop on „Analysis of Data from Aquatic Ecotoxicity Tests" to be held in October 1996. Table 1 summarizes the uncertainty problems/solutions of HRA.

Biological thresholds in control and monitoring programs.

Traditionally, in the EU, the emission of pollutants has been controlled by chemical analysis. However, biological methods are current practice in North American countries; and are under consideration in some European countries (Whitehouse et al., 1996). Chemical methods use the thresholds developed by HRA, i.e. Water Quality Objectives (Bro-Rasmussen et al., 1994). They have two main controversial aspects:

a-The chemical composition of the sample must be fully known, only those substances identified and analyzed will be assessed.

b-The assessment focuses on the individual threshold for each substance, the evaluation of synergistic effects is problematic.

The advantage of toxicity-test-based methods is to avoid both problems. Biological methods require the development of specific thresholds: i.e. Toxicity-based Acceptability Criteria. These criteria should consider at the same time the uncertainties regarding the ecological relevance of the selected endpoints and the variability of the test results. Two alternatives for effluent control can be used as examples. In the *off-line* systems, a test battery substitutes the chemical analysis, and the dilution factor (sewage flow *versus* receiving water flow) substitutes the exposure assessment. The threshold or Toxicity-based Consent (TBC) is expressed in terms of a minimum effluent dilution (% v/v) below which the $L(E)C_{50}$s should not fall (Whitehouse et al., 1996). In the *on-line* systems, the test organisms are directly caged into the effluent or downstream the dumping point using appropriate equipment (Mulliss et al., 1996), the threshold or Acceptability Level is expressed in terms of a certain level of response (i.e. less than X% of mortality).

The application of biological methods in monitoring programs has experienced an exponential growth regarding the use of biomarkers. The information produced by biomarkers mostly focuses on the assessment of exposure conditions; however, additional possibilities have been described elsewhere (Tarazona et al., 1995). In addition, toxicity tests can be also incorporated into monitoring programs. In this case, the difference between the endpoint value and the acceptability threshold (this difference represents the safety factors applied by HRA) must be achieved using concentration procedures. An example can illustrate this point. The EU-HRA guideline considers as ecologically-safe (PNEC) value 1/1000 of the lowest acute $L(E)C_{50}$ of either fish, daphnids, or algae. The same argument, used in the opposite direction, indicates that if a sample concentrated by a factor of 1000 does not reach the 50% effect-level when tested on fish, daphnids and algae, then, it can be considered as ecologically safe. The level of uncertainty should be the same in both cases; when applied to chronic NOECs, this factor becomes 10. Obviously, concentration procedures are specific for each chemical group. Nevertheless, recent works suggest that for certain groups, i.e., highly lipophilic chemicals, Toxicity Identification Evaluation procedures (Castaño et al., 1994) or semipermeable membrane devices (SPMDs) (de la Torre et al., 1995) can become potential tools for the development of toxicity-based monitoring thresholds.

References

Bartell SM, Gardner RH, O'Neill RV (1992) Ecological risk estimations. Lewis Pub, Boca Raton, 250 pp

Bro-Rasmussen F, Calow P, Canton JH, Chambers PH, Silva Fernandes A, Hoffmann L, Jouany JM, Klein W, Persoone G, Scoullos M, Tarazona JV, Vighi M (1994) EEC water quality objectives for chemicals dangerous to aquatic environment (List 1). Rev Environ Contam Toxicol 137:83-110

Castaño A, Cantarino MJ, Castillo P, Tarazona JV (1996) Correlations between the RTG-2 cytotoxicity test EC50 and in vivo LC50 rainbow trout bioassay. Chemosphere (in press)

Castaño A, Vega M, Blazquez T, Tarazona JV (1994) Biological alternatives to chemical identification for the ecotoxicological assessment of industrial effluents: the RTG-2 in vitro cytotoxicity test. Environ Toxicol Chem 13:1607-1611 de la Torre AI, Fernandez C, Tarazona JV, Muñoz MJ (1995) Detection of arochlor, DDT, malathion, and HCB using semipermeable membranes as concentration method. Chemosphere 31:2727-2737

ECB (1996) Technical guidance document for the hazard and risk assessment of new and existing substances. European Chemical Bureau, Ispra.

Harras MC (1996) Regulatory use of ecotoxicity statistics: a US perspective. Ecotoxicol. 5:145-154

Mulliss RM, Revitt DM, Shutes RBE (1996) The determination of the toxic influences to *Gammarus pulex* (amphipoda) caged in urban receiving waters. Ecotoxicol 5:209-215

Muñoz MJ, Ramos C, Tarazona JV (1996) Bioaccumulation and toxicity of hexachlorobenzene in *Chlorella vulgaris* and *Daphnia magna*. Aquat Toxicol (in press)

Renner R (1996) Ecological risk assessment struggles to define itself. Environ Sci Tech 33:172A-174A

Scheuplein RJ (1995) Information needed to support hazard identification and risk assessment of toxic substances. Toxicol Lett 79:23-28

Tarazona JV, Carballo M, Castaño A, Muñoz MJ. (1995) The role of cell biology on the application of toxicology to environmental sciences: biological assessment of pollution and biomarkers. In Cajaraville MP (ed) Cell Biology in Environmental Toxicology, UPV, Bilbao, pp 15-28

USEPA (1992) Framework for Ecological risk assessment. EPA/630/R92/001, Risk Assessment Forum, Washington DC

van Asselt MBA, Beusen AHW, Hilderink HBM (1996) Uncertainty in integrated assessment: A social scientific perspective. Environ Model Assess 1:71-90

Vandenberg JJ (1995) Risk assessment and research: an essential link. Toxicol Lett 79:17-22

Whitehouse P, Crane M, Redshaw CJ, Turner C (1996) Aquatic toxicity tests for the control of effluent discharges in the UK - the influence of test precision. Ecotoxicol 5:155-168

Wilson JD, Cibulas W, DERosa CT, Mumtaz MM, Murray E. (eds) (1995) Decision support methodologies for human health risk assessment of toxic substances. Toxicol Lett Special Issue 79:1-316

The Precautionary Principle and Science-based Limits in Regulatory Toxicology: The Human Experience, Individual Protection

D.M. Pugh
Veterinary College, Ballsbridge, Dublin 4, Ireland.

The When and How of European Chemical Safety Law

Responding to various concerns over the safety of chemicals, the Commission of the EU, since its foundation, has created, extended and continually revised the legislation which gives effect to those concerns. The many legal instruments issued by the appropriate Directorates General have seen the introduction of new controls in some member states as well as the overall harmonization of controls amongst those member states which had national legislation already in place. In keeping with the major purpose of the EU, to provide for unrestricted intra-community trade, chemical safety legislation applies to specified commodity classes such as medicines, cosmetics, food, drinking water, pesticides, animal feed additives, industrial chemicals and so on. Also apparent, however, are non-sectoral motivations, notably, a wish to improve worker safety in the case of exposure to chemicals during their manufacture, sale, use or disposal. That last concern, the safe disposal of chemicals, hints at that other great motivation in the EU, a drive to protect the environment.

These few words of introduction will have made very plain that the body of EU law having a chemical safety content is enormous. However, mindful of an undertaking neither to bore with an exhaustive catalogue of EU legislation nor with the painful detail of its content even when that is toxicological, what follows is a when and how scheme which exemplifies at the simplest level the triggers which lead to legislation and the tools which have been used to provide the effect mechanism in that legislation. This scheme pinpoints danger and exposure as the triggers for regulation, and in that way underscores what is for us the self-evident need to use evaluative toxicology as the best way to identify, in the biological sense, the potentials of chemicals to be harmful. It is, of course, in the development of the means which make thresholding possible that toxicology makes its biggest contribution to the legislative tools exemplified. Also implied is a role for the exposure/effect relationship as a tool central both to the risk assessment element of a toxicological evaluation and to the style of the provisions in the resulting legislation. In a nutshell, the purpose of chemical safety legislation is the modification of human behaviour towards chemicals, be those humans researchers, manufacturers, traders or users and its best effector mechanism is the limiting of human exposure to chemicals.

WHEN AND HOW TO LEGISLATE

Essentials	1)	when substance is harmful	
	2)	if exposure of man to substance is likely	
Refinements	1)	consider size of risk	
	2)	consider consequences of regulating	
Tools	1)	Banning = no risk	e.g. Delaney Clause
	2)	Thresholding = negligible risk	e.g. $1x10^{-6}$ in a lifetime
	3)	Balancing = risk vs. benefit	e.g. worker safety
	4)	Labelling = informed of risk	e.g. user safety
	5)	Packaging = accidents less likely	e.g. distributor, user and child safety

Important though such legislation is, it is plain that the EU Commission has not relied on legislation alone but has created other structures which contribute to the same aim of chemical safety, deriving directly from the legislation in the case of Scientific Advisory Committees, sometimes contributing more generally as with its Member State Institutions and sometimes set up specifically to administer the legislation as with the Medicines Agency.

Having dealt with the motivations to provide the regulations and the mechanisms which the regulations bring to bear, next I would like to describe some aspects of what I believe to have been a very important milestone in the road towards chemical safety in the EU, both for the Commission and for toxicology. The Classification, Packaging, and Labelling of Dangerous Substances Directive of 1967 exploited two of the previously mentioned strategies used in chemical safety legislation. These were the requiring of packaging to be appropriate to the properties of the contained substance and the requiring of the presentation to the user of information appropriate to the safe use and disposal of the contents. This foundation legislation has since been "amended" eight times and "adapted to technical progress" a further twenty-one times to-date. Most significant among all of these was Directive 79/831. Commonly known as the 6th Amendment, it added to the 1967 Directive the requirement for the pre-marketing notification, via a dossier, of those safety studies whose results justified the proposed classification and labelling of the substance. By this indirect means the 6th Amendment required a toxicological evaluation not just of dangerous substances, as its full title would suggest, but of all substances to be placed on the community market after 1981. Indeed, its scope was to be so all-embracing that this was best defined by excluding those few classes of chemicals to which it would not apply. One major significance of the 6th Amendment to the theme of this symposium is that it has been responsible solely or in part for the production of many follow-up legal instruments some of which have come to bear on all other sectors of regulated chemicals, thus increasing even further its already very broad impact. Of these 6th Amendment derivative measures, the need to use the toxicological (and other) test methods specified in its Annex V in the safety dossiers submitted to other sectors, e.g. for human medicines, veterinary medicines and animal feed additives is worth specific mention. In this way it has had a huge effect on the standardisation of toxicological testing methods across the broad spectrum of legal instruments which govern chemical safety in EU member States.

Zero Exposure

What these legal instruments also have in common is a reliance on what is known as the command and control format. That format, simply put, requires the observation of a specified limit to what is permissible and, in the case of chemical safety, is expressed in the specification of exposure conditions which must not be exceeded e.g. as in contamination of air at the workplace. At an extreme far removed from the 6th Amendment, the command and control approach to legislation achieves zero exposure by the banning of named dangerous chemicals e.g. as in the Marketing and Use Directive, 76/769 EEC. Continuing that approach to safety, many of the sectoral pieces of community law contain a provision which allow Member States temporarily to withdraw products from their markets when new information has given rise to concern over their safety, pending an evaluation by the Commission of the grounds for withdrawal. Some readers will know of the year long involvement of the Commission with the Danish and German governments' decisions temporarily to ban the sale and use of the EU approved animal feed additive Avoparcin in their markets.

Tolerable Exposure

By and large, however, chemical and chemical-containing products both gain access to the Community market and remain there by virtue of the tolerable exposure thresholds defined by their safety studies. These in turn depend on the continuing belief among toxicologists in the inverse of the dictum of Paracelsus, that is that a small enough dose or exposure, even if repeated, is usually without detectable adverse affect on the recipient, animal or man.

In fulfillment of the dictum of Paracelsus, the chemical presence or biological effect routes discussed in the two preceding papers lead to the definition of those thresholds of exposure which are judged to be safe, acceptable or tolerable. Toxicology, ably assisted by analytical chemistry, provides in this way the starting point which leads to the all-important definition of those conditions of use or exposure whose observance is the key to achieving acceptable safety in use from all regulated chemicals, substances or products.

Benefits to Toxicology

It can freely be acknowledged, that much of the growth in toxicology in the last fifty years can be attributed to the ever-increasing requirement of the public for higher levels of assurance as to the safety of chemicals. This requirement, expressed in the many revisions and extensions of Community law over the years, has often provided toxicologists with new challenges, e.g., at the simplest levels in devising test procedures capable of identifying and quantifying the potential of a chemical to induce a particular adverse affect. Later, such tests have been assessed for consistency of performance in inter-laboratory ring-tests.

This consumer-driven wish, expectation even, for assured safety was evident in the originally empty test-method-defining annexes of the 6th Amendment. At that point the aspirations of those who framed the text of the directive exceeded in some cases the ability of toxicologists to provide from then existing practice, the necessary validated, standard test methods. Toxicology of course responded by proposing methods to the Commission, which in turn, funded and organized their validation. The contemporaneous contribution of the OECD in securing international agreement for the adoption of a series of toxicological testing methods was quickly recognized and supported by the Commission which then adopted those procedures in the interests of global harmonization and with the prospect of the mutual recognition of safety study results. A second benefit enjoyed by toxicology, again funded lavishly by the Commission and responding to a consumer-led preference, has been the dramatic growth of in vitro techniques. In brief, toxicology has not only benefited from but has contributed vitally both to the philosophy underpinning EU chemical safety legislation and to the detail by which that law delivers its intents.

The Precautionary Principle

However, the precautionary principle element of the title of this paper and I will speak makes plain, and perhaps to the disappointment of toxicologists, that there is sometimes more than the discipline of toxicology involved in defining the exposure limits which underpin chemical safety legislation.

By way of providing some definitions for this paper, it could be said that the precautionary principle means the taking of an action against a particular substance or preparation in the absence of a range of well-conducted studies sufficient to justify that action. The action may be taken on the basis of a suspicion or concern which is so great (and hopefully so well reasoned) that the delay, which would be inevitable were investigations to be completed, is judged to be unacceptable when weighed against the risk attaching to further exposure. The banning of the organochlorine pesticides in the late 1950's occurred as much because their presence in corpse fat was shown to be increasing year on year as because of damning, new toxicological findings in man or other mammals. In the case of chemicals not yet marketed, the early removal of those shown to be microbial mutagens from further testing and development is another type of example of the precautionary principle in practice. Similarly, the maximum acceptable concentration for any pesticide in drinking water, set at 0.1 µg/L in the 1980 Drinking Water Quality Directive reflected a not unreasonable consumer preference for pesticide free water. As a limit, it was in fact an analytical surrogate for zero, consumer driven, with little or no input from toxicology and exemplifies well a triumph for the precautionary principle over the scientific method-based defences of the pesticide industry.

Science-based Limits

By and large, science-based limits are, by contrast with those set in a precautionary way, derived from studies which are designed to reveal an adverse effect or effects and reflect also a subsequent process of extrapolation down to an exposure limit which would be expected to be free of adverse effect for man. Such limits are frequently higher than those identified via precautionary approaches. For example, in the case of the widely used herbicide Atrazine, a contaminant level of 0.1 µg/L in drinking water is permitted by a precautionary approach, while a science-based limit has permitted up to 2 µg/L, i.e. 20 times more to be present in drinking water in Northern Italy for a specified period of time.

Food Additives and Consumer Safety

Having described the precautionary principle and science-based limits, for the rest of this paper discussion will be confined to the safety of materials added to food and in particular, to those substances which occur in human food as residues of agents which had been used as in - feed enhancers of animal production. Against a background of the various chemicals which might be added to food, these residues are placed among the unintentional additions to foods of natural origin.

Animal feed additive residues were chosen for the discussion of the use of toxicology in EU chemical safety legislation because for many it will probably be an unfamiliar group of regulated, potential human toxicants with a rather individual range of attributes. However, despite this unusual array of attributes, the strategy common to all other science-based limits expressed in Community chemical safety law remains the one which is applied to animal feed additives. That strategy entails just three steps, namely

a) identifying biological effects, usually in a surrogate species
b) establishing the concentration/ time conditions which allow those effects to become evident,
c) regulating so that the substance when in use will not achieve exposure conditions likely to result in adverse effects in man.

This simple strategy has long been applied to the exposure of people to all of the various chemicals added to food intentionally or otherwise It will be familiar in the system which lead to the definition of Acceptable Daily Intakes (ADIs) in the case of food additives and is used in various other expressions of exposure conditions which would be acceptably safe in other circumstances e.g., the Threshold Limit Values (TLVs) or Short Term Exposure Limits (STELs) applied to industrial chemicals at the workplace.

All such systems depend on identifying, usually in long-term animal studies, exposure conditions at which an effect ceases to be apparent, the so-called No Effect Level (NEL) or one of its variants, the No Observable Effect Level (NOEL) or the more specific No Observable Adverse Effect Level (NOAEL).

This animal-derived threshold is then extrapolated to man in a simple process which allows for rat/ man weight differences, rat/ man sensitivity differences and inter-individual variation in man. This process, by means of a safety-factor and a body weight factor, converts the experimentally derived rat NEL to an amount which an average sized person could consume daily for life, the familiar ADI.

As pointed out already, however, command and control type legal instruments are used in the consumer safety area and these have a further requirement, that of inspectability and enforceability. In the area of food safety this requires for simplicity of administration the definition of the maximum acceptable concentrations of the target chemical in named food items all of which could then be subject to monitoring by analysis. The simple exceeding of the defined concentration, known as the Maximum Residue Level (MRL), is usually enough to trigger the imposition on the manufacturer or supplier of the foodstuff that penalty defined by law. The concentration limits in foodstuffs are set by distributing the ADI amongst a generous overestimate of the mean daily consumption of a range of appropriate foodstuffs. In the particular case of animal feed additive substances the market-basket overestimate used in the EU allows for a daily per capita consumption of 300g muscle, 100g liver, 50g each of fat and kidney, 1.5l of milk and 100g of egg. Distribution of the ADI to achieve MRLs throughout these food amounts plainly builds in an added margin of safety for all but the most over-indulgent of consumers. It is also a device which attempts to allow for a range of inter-individual and international variation in dietary habits while using a single, internationally-adopted concentration value for each MRL.

To have defined target MRLs, however, provides not only the action threshold for an enforcement agency but also the target to be met by the animal producer in his marketed produce. In the case of an animal feed additive substance, the item so far missing from my account of the safety management scenario is the device by which the proper use of the feed additive in growing animals is shown not to result in concentrations which will exceed the permitted MRLs in the finished carcass. This desirable state is achieved by requiring the feed-additive manufacturer to supply in his registration dossier the details of studies which characterise the depletion of residual additive from the appropriate tissues of each animal species in which the additive is to be used. Ideally, an additive would leave no residues. In practice this is not always so and an interval which must be allowed to elapse between the last additive-containing feed and slaughter is defined from the results of the depletion study. This interval, usually called the withdrawal period, is built into the tightly-specified, permitted conditions of use of the additive. Adherence to a single fixed interval is valid because the permitted conditions of use for each feed additive define also the concentration at which the additive may be present in animal feed and, hence, define also the daily dose for the animal species in question.

The Future

It is generally accepted that the derivation of ADIs their subsequent adoption and observation by food and feed manufacturers, as well as their market-basket based monitoring and, for some chemical classes at least in recent times, their enforcement by law, has, for more than four decades, delivered acceptable safety to consumers, i.e., protection of the individual has been achieved.

That, of course, is not an entirely satisfactory way to have to conclude on the effectiveness of chemical safety legislation in achieving those goals stated in the "whereases" which introduce Council Directives. Unfortunately, there is very little published material which touches on the topic of the performance assessment of regulations. That found does at least make the gap understandable, however. To quote from the book "Regulating Industrial Risks - science, hazards and public protection", "Improvement of performance and results can be achieved only if there are reliable and respected techniques for evaluating both the results of regulating and the effectiveness of the regulatory process itself. Except in relatively straightforward cases however, implementation usually eludes numerical or objective measurement ... there are few established procedures to guide evaluations". (Baram, 1985). That book was sourced from the library of the EU Information Service in Dublin, was its most modern contribution in the area and was the only item of relevance found in an extensive series of Commission's-based enquiries.

To conclude it is obvious that evaluation of the effect of safety regulations is an area awaiting exploration, at least at the level of the Commission. That it seems not yet to have been done is an omission many toxicologists will no doubt find surprising when set against the enthusiasm of the Commission for requiring the validation of toxicological practices.

Toxicologists working at the coal-face of regulatory toxicology have found no difficulty in proposing ways of fine tuning their crude but seemingly effective device of an NEL-based thresholding of exposure of the individual. On both fronts then, there remains ample scope for future improvements, both in principle and in practice.

References

Baram M (1985) Implementation and evaluation of regulations. In: Otway H and Peltu M (eds) Regulating industrial risks - science, hazard and public protection, Butterworths, London, pp 57-75

Faure M (1994) The EC directive on drinking water: Institutional aspects.In: Bergman L and Pugh DM (eds) Environmental Toxicology, Economics and Institutions - the Atrazine case study, Kluwer, Dordrecht, pp 39-88

The Environmental Experience: Ecosystem Protection.

Finn Bro-Rasmussen,
Technical University of Denmark, Ecology & Environmental Science, DK-2800-Lyngby, Denmark

In the words of Genesis in the Holy Book, man is given to rule over all the earth, over the fish of the sea and the birds of the air, over livestock and all creatures that move along the ground. Reading these words and contemplating his own technological achievements, man feels continuously re-assured that this interpretation is both correct and justified.

Less fortunate is man's environment and all other living species that were not created in Gods own image and to his likeness. These receive the protection, only, which springs from man-given laws - administered by man-made agencies and established to protect 'man as part of the environment'.

Introduction

Human toxicology is the science dealing with poisons and poisonous effects. Man (or the mammal) is the target organism and the object of interest, including individual functions and optimal health of man. Primarily, therefore, human toxicology is studying processes and life functions, and accordingly, series of guidelines and methodologies for the testing of chemicals (cf. fig. 1) have been developed in order to accommodate such studies (OECD, 1993).

Ecotoxicology on the other hand, is devoid of similar positive goals. Almost as a godchild's gift, the ecotoxicology has been given the task to evaluate and predict the potential impact of chemicals on species and/or populations of the environment in order to protect these against extinction or lasting reductions. Repeatedly it is claimed that protection of biota at the level of individuals is not part of ecotoxicological objectives (Bro-Rasmussen et al., 1984).

TOXICOLOGY - MAN & ENVIRONMENT

Experimental and/or epidemiological studies of adverse effects from chemical substances upon biological systems

Human toxicology

- Uptake
- transport
- metabolism
- effects
of chemical substances

Protection of man
- individuel functions and optimal health

Ecotoxicology

- Distribution
- fate
- effects
of chemical substances

Protection of species
- ecological function and structure

Toxicological tests for effects - mammals

Acute toxicity - *oral, dermal, inhalat.*
Acute irritat./corros. - *dermal, ocular*
Skin sensitisation
Toxicity/repeated dose - *oral, dermal, inhalation*
Sub-chronic toxicity - *oral, dermal, inhalation*
Teratogenicity
Reproduction-toxicity
Reproduction-toxicity - *2 generations*
Toxico-kinetics
Delayed neurotox. - *acute, subchronic*
Carcinogenicity
Teratogenicity
Chronic toxicity tests
Combined chronic tox./Carcinogenicity

Possibly supplemented with:
Immunotoxicity tests
Genetic toxicity tests

Ecotoxicological tests for effects

Algae - test	*Growth inhibition*
Daphnia *spp*	*Acute toxicity*
	Reproduction
Fish tests	*Acute toxicity*
	Prolonged toxic.
	Early life-stage
Act. Sludge	*Respir. inhibition*

Possibly supplemented with:
Tests for toxicity on *birds, earthworms, bees and plants*

Figure 1. **Objectives and methods**

In practice, this leads to the well-known situation that ecotoxicology in its routine activities hardly reaches beyond the counting of lethalities, and it seldom demonstrates more than a statistical evaluation of the development of harmful effects. We are reminded that mortality is the cessation of all biological activity - leaving nothing further to examine (Woin, 1995). In recent years, only, it slowly emerges that the objects and primary targets for ecotoxicological studies ought to be all impacted constituents of the environment, including flora and fauna, and the need for our understanding of the coherent ecosystem. It is the individual as well as the integrated functions (and malfunctions) that matters, including also structural diversities of natural communities and ecosystems (cf. fig. 2) - in order for us to develop adequate means and measures in our protection of environmental qualities.

The absence of (eco)toxicological data

and the grossly inadequate procurement of data, incl. the shortage of research on secondary and other effects is a dominating problem.

In our present-day industrialised world, the total production in volumes of synthetic chemicals is approaching that of primary productivity in nature. The increasing 'chemicalisation' of daily life leads to an increasing demand for more knowledge on the significance of the chemical impact on man and on environment. As a consequence we experience an urgent need for the testing and approval of individual chemicals. Measured against these demands the two toxicological disciplines both suffer from a shortage of fundamental data and lack of basic knowledge for identification and assessment of hazards connected to production, use and disposal of chemicals. During recent years, this has become evident in connection with the urgent need for development of programmes and models for the prioritisation (van d. Zandt and Leeuwen, 1992) and selection of existing chemical substances for risk assessments (EU, 1995).

Less noticed, but essential nonetheless, is the increasing shortage of extended studies on modes and mechanisms of effects of industrial chemicals, including also secondary and other subsequent - cascading - effects (see fig. 3). This is a subject area in which the human toxicology has already focused on effects, such as reduction in male reproduction rates (*cf.* increasing research activities on estrogenic effects (Miljøstyrelsen, 1995)), or the indirect influence on morbidity, life expectancy and life quality and probably also on increased cost of public health care. Admittedly, however, it is also an area in which the causal relations and the pathways - direct or indirect - of chemical impact to human health are still insufficiently clarified.

As for the ecotoxicology the way to go is even longer and the long-term perspectives are more vague. The research efforts which are required are much greater, if scientific proofs have to be established for all cause-effect relationships, and before regulatory consequences generally can be made

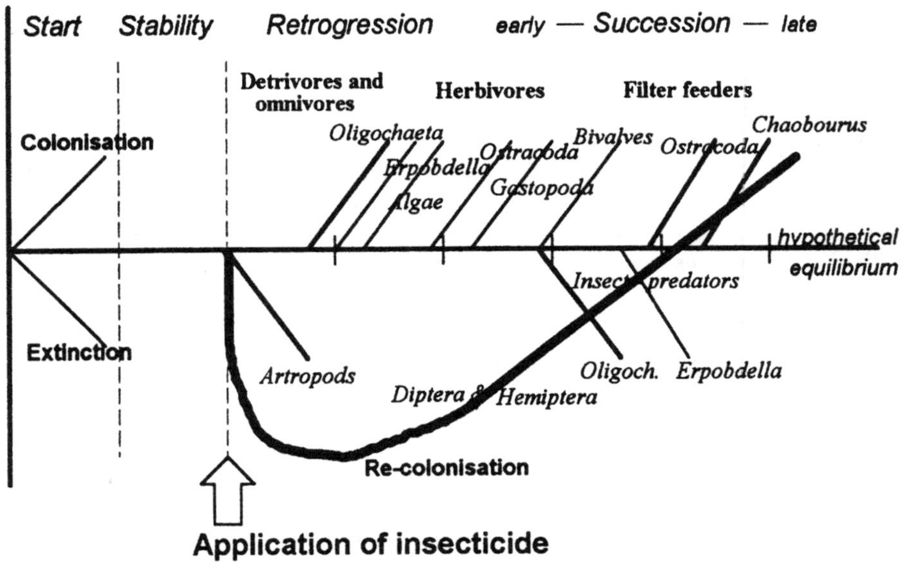

Figure 2 **SUMMARY OF EFFECTS INCLUDING INVERTEBRATE COMMUNITY DEVELOPMENT IN AN ISOLATED EUTROPHIC POND AFTER LETHAL INSECTICVIDE EXPOSURE** *(after P. Woin, 1995)*

---oooOOOooo---

PRIMARY EFFECTS	SECONDARY EFFECTS (Example)	DERIVED EFFECTS
Human toxicity	Reduced life time Increased morbidity Reduced life quality	Anxiety Increased health costs Areas inhabitable for man
Ecotoxicity	Changed interaction between species (predator - prey- symbionts) Extinction of vulnerable species	Changes in composition Loss of biodiversity Population extinction

Figure 3. CASCADING EFFECTS

acceptable. The question, therefore, must necessarily be raised whether we can afford to wait until evidence is ultimately established on the causal connections between on the one hand the impact of chemicals on the environment, and on the other hand observations on disruption of community patterns, reductions in biodiversities, and irreversible changes in ecosystems functions (Woin, 1995).

Establishment of quality objectives or limit values

for the protection of either man or of the environment differs in techniques, in safety requirements and - not the least - in levels of protection.

Limit values are frequently interpreted as a means of regulating chemical uses, whereas in practice they are much rather serving the purpose of being 'indicators of permissable pollution'. Realistically, therefore, it would be more appropriate to consider limit values to be standards or norms (VROM, 1994) (cf. fig. 4), which may then be:

- *effect-related*, such as environmental quality standards, e.g. WQO or quality objectives/standards for receiving waters,
- *exposure-limiting*, such as emission or discharge limit values,
- *process-regulating*, such as product norms or specifications,
- *use-regulating*, such as behavioural or handling norms and standards.

Among those, only the last mentioned norms and/or standards can be said to serve use-regulating purposes, whereas effect-related and exposure-related limit values both are insufficient means of protection, *unless they are specifically supported* by direct use regulating rules, including banning or phasing out procedures. This latter situation has been experienced during recent years by withdrawal of earlier approved pesticides as a necessity for the enforcement of drinking water limit values for pesticide residues.

It is the protection of human health – not the reduction of the environmental impact – which is normally advocated, when a higher degree of safety is required and, thus, when the possibility of regulating chemical use patterns are considered. As for the environment, it is generally found acceptable, from the side of authorities or from the public including politicians, that more lenient regulations and less restrictive requirements are established and enforced. This is clearly demonstrated, when environmental quality standards or discharge limit values are legally accepted at levels which are higher than the values otherwise corresponding to the requirements of a Zero-Effect (Miljøstyrelsen, 1983)

Differences are also found between the areas of human health and environmental protection, when we are dealing with the evaluation of uncertainties and the application of precautionary safety factors for use in the establishment of maximum exposure limits (see fig. 5). This is most conspicuous in two different subject areas, namely

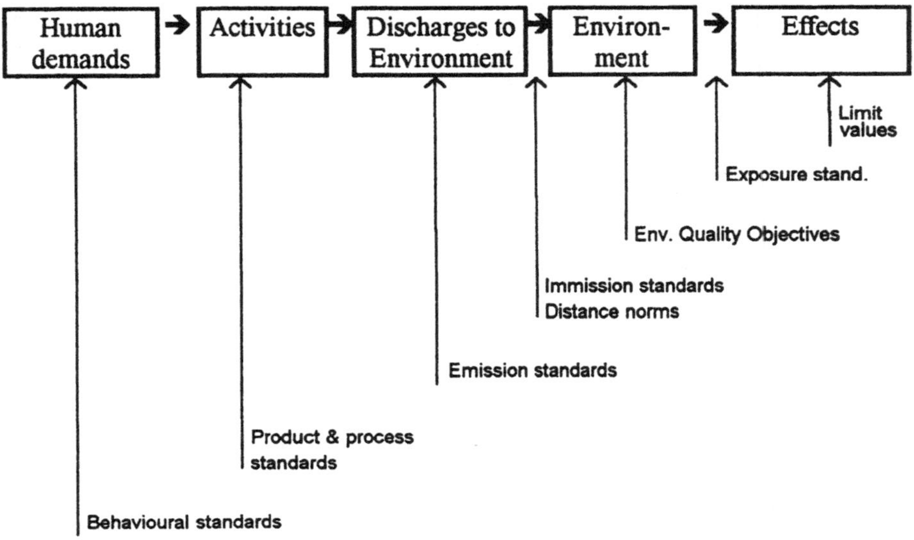

Figure 4 **Environmental norms, standards & limit values**

---oooOOOooo---

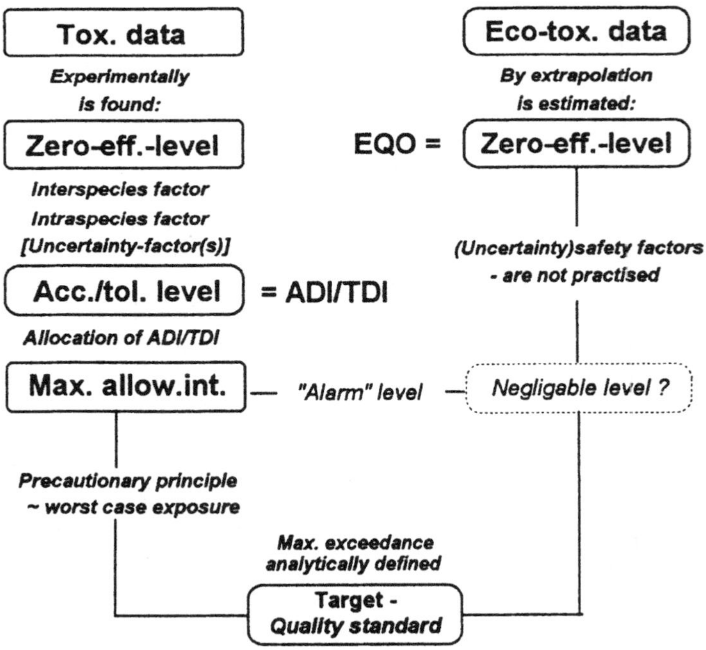

Figure 5. **Where is the limit?**

- in relation to the use of experimental toxicity data, which are considerably more demanding for the establishment e.g. of human health tolerance levels for chemicals in food or drinking water, than is the case for the same chemicals e.g. for the setting of environmental quality standards, and
- when considering the possibility of applying specific safety/uncertainty factors in the setting of limit values. Such practice is normal for the establishment of human health standards, while it is rarely the case in the setting of environmental quality standards. On the contrary, cost-benefit analyses or technological reasoning are often found to be preferable, and gradually, therefore, it is becoming more and more difficult to give assurance of environmental protection at the Zero-effect level.

Uncertainties in risk assessments

have achieved increased interest in recent years, primarily in the field of human toxicology, but gradually also in ecotoxicology.

From the early days of human toxicology, safety requirements and precautions were formulated. They served the purpose of trustworthy transfer of data and experiences from animal experimentation into human health protection measures. This was - and still is - especially important when ADI/TDI-values had to be established from experimental Zero-Effect-levels by application of so-called Safety-factors (SF-values).

Such factors were first introduced in order to account for variability (Dourson and Stara, 1983) as a source of error when dealing with

- differences in individual sensitivity, i.e. by applying an *intra-species factor*, and
- the transfer of data from animals to man, i.e. by means of an *inter-species factor*.

There is a sound and causal reasoning and also statistically based experiences, behind the use of such factors, although undoubtedly, at the time of their introduction they were merely suggested as sufficiently 'wide' to satisfy any precautionary requirements (US-EPA, 1993). To-day, however, there is a trend to underline the vulnerability of biological systems towards chemical exposure on the one hand, and an increasing understanding of toxico-kinetics and -dynamics on the other, which has raised a requirement for a critical evaluation of the safety requirements, and an attention towards the continued use of safety factors (Renwick, 1993; ECETOC, 1995), now called *uncertainty factors* (UF).

In this situation focus is on the one side on the need for transparency of uncertainty factors in order to avoid 'overprotection', and on the other on increasing public demand for safety as well. It is hereby stressed, that arbitrary application factors per se do not stimulate any research for unknown impacts or motivate further search for non-assessed risks which they intend to substitute.

Limit values are to-day established and assessed individually on the basis of different schemes of safety/uncertainty considerations (IPCS and WHO, 1994; Bro-Rasmussen, 1995) (see fig. 6). This, of course, opens the way for the introduction of 'politically' formulated, precautionary principles such as specifically phrased quality requirements. This is e.g. seen for drinking water and food standards, when these are based on 'analytical detection limits' which define a restrictive 'legal no-contamination' level instead of a toxicologically set 'no-effect level'. Similarly, of course, settings of 'liberal protection levels' can be politically negotiated and accepted, as it is frequently practised in the field of setting protection standards for the working environment.

It is a characteristic of the situation, that exactly this discussion on safety/uncertainty in connection with the establishment of limit values, quality standards and risk assessments, dominantly is tied to the fields of human toxicology and environmental medicine. A corresponding analysis of safety factors and of the elements of uncertainty in the area of eco-toxicology has actually been initiated in recent years (Calabrese and Baldwin, 1993) (see fig. 7), but it has hardly attracted attention outside the academic discussions. This is exemplified by the fact, that environmental quality objectives (or standards) in the OECD as well as in the EU continuously are established on the basis of extrapolations from experimental effects data to narrowly defined, and statistically assessed Zero-Effect levels, theoretically including a 5% risk element, and without any use of further safety (uncertainty) factors.

The assessment of health risks vs. environmental risks

includes prediction and uncertainty evaluations in exposure assessment as well as in effects assessments.

As part of the most recent development of EU risk assessment directives (EU, 1993; EU, 1994), it has become apparent, that the concepts of hazard and risks - expressed as effect potential and exposure probability, measured in PNEC- and PEC-values respectively - shall become integrating elements of any future chemical evaluation. Test methods and assessment principles which have been developed and laid down in the combined Technical Guidance document (EU, 1995) are introduced, and the results will be related to human health as well as to environmental protection needs.

This is a development, which is most challenging and requiring as far as:

- *Quality and credibility* of assessment schemes and practices,

and

- *Quantitative data* requirements are concerned.

It is a situation in which present as well as future test systems and chemical evaluation procedures are bound to deal with the chemical impact on man as well as on environment, *and* in which harmony must be established between protection of human health and the safeguarding of environmental values.

	WHO *)	Renwick	ECETOC	US-EPA
Interspecies *kinetic* *dynamic* *oral intake* *inhalation*	**10** *4* *2,5*	**(10)** *4* *2,5*	 *4* *1*	**10** *(= factor* *10A)*
Intraspecies *kinetic* *dynamic* *general populat.* *Work place*	**10** 3,2 3,2	**(10)** *3,2* *3,2*	 *3* *2*	**10** *(= factor* *10H)*
Extrapolation: sub-chronic ⇒ chronic			3 2-3	10 *(= factor* *10S)*
Extrapolation: LOAEL ⇒ NOAEL	(1 -) 10	> 1	3	10 *(= factor* *10L)*
Spec. considera-tion (*e.g.* cancer)	1 - 10	1 - 10		1 - 10
Database insufficient	1 - 10	1 - 10		> 1 - 10
Residual uncertainty				< 1 - 10 *(= MF)*

*) WHO will normally not accept total-factor > 10^4 (*cf.* ref. 14, IPS & Who, 1994

Fig. 6 **Uncertainty/safety factors used in human risk assessment**

ECOLOGICAL RISKS

1. Interspecies variation	10 - 1000
2. Intraspecies variation	
Non-endangered species	10
Endangered species	10
3. LL (= less-than-lifetime) exposure	
≥ 15 % of lifetime	1
< 15 % of lifetime	10
< 25 % of lifetime	10
4. Acute toxicity ⇒ Chronic NOAEL	
Lethality (LD - LD_{50} range)	50 (95%)
	100 (99%)
5. LOAEL ⇒ NOAEL	10
6. Residual uncertainty (Modifying factor)	1 - 10

---oooOOOooo---

Tests according to OECD, EU/CSTE etc. recommend
min. 3 single species representing 3 (sub)phylae:
fish (*vertebrate*)
daphnia (*invertebrate*)
algae (*phyton*)

Figure 7 **Uncertainty factors in Ecological risk assessment**
(after Calabrese, 1993)

Furthemorer, this has to be done by referring to the origin of impacts, i.e. by aiming not only at risk reductions, but also by direct regulation up-stream at the stages of production and application.

It is in order to meet such demands, that

1. the egocentric acceptance of higher non-human than human risks must be reduced,
2. human health protection should not erroneously be considered sufficient to cover also the needs of environmental protection,
3. the ultimate consequences of chemicals impacting natural ecosystems and their biota must be fully studied and the significance of environmental disturbances fully appreciated,
4. the environmental risk evaluation connected to the impact of chemicals must be extended to achieve a full coverage of environmental needs, and last, but not least,
5. a better understanding must be developed and supported concerned with the understanding of ecological systems, and separate quality objectives must be developed and enforced in order to protect natural flora and fauna communities – independently and in their own rights.

References

Bro-Rasmussen F. Christiansen K. and Folke J. (1984) From Ecotoxicology to Assessment of Environmental Hazards of Chemicals. In: Hommage au Professeur René Truhaut, Paris, 19. October. pp 156-163.

Bro-Rasmussen F. (1995). Where is the limit? In: Report (in Danish) from a Consensus conference arranged by the Danish Board of Technology in Denmark, Copenhagen.

Calabrese E.J. and Baldwin L.A. (1993) In: Performing Ecological Risk Assessments. Lewis Publishers, Chelsea, MI, USA. pages 257

Dourson M.L. and Stara J.F. (1983) Regulatory History and Experimental Support of Uncertainty (Safety) Factors. Reg. Toxicol. Pharmacol. 3:224-238

ECETOC (1995) Technical Report no. 68; Assessment Factors in Human Health Risk Assessment, B-1160 Brussels, Belgium

EU (1993) Commission directive 93/67/EEC on Risk Assessment for New Substances

EU (1994) Commission regulation No. 1488/94 on Risk Assessment for Existing Substances

EU (1995) Technical Guidance documents, November 1995 in support of directives on Risk Assessment for New and Existing Chemicals; cf. Comm. Directive 93/67/EEC and Council Regulation 1488/94

IPCS and WHO (1994) Environmental Health Criteria 170. Assessing human health risks of chemicals: Derivation of Guidance values for health-based exposure limits, Geneva.

Miljøstyrelsen (1983) (Danish National EPA) Guidelines (in the Danish) for Receiving Water Quality Planning, Volume I and II; No. 1/1983 and 2/1983.

Miljøstyrelsen (1995) (Danish National EPA) Male Reproductive Health and Environmental Chemicals with Estrogen Effects. Project report no. 290

OECD (1993) Guidelines for the Testing of Chemicals, Volume 1 and 2; OECD, Paris

Renwick A.G. (1993). Data-derived Safety Factors for the Evaluation of Food Additives and Environmental Contaminants. Food Add. and Contam. 10(3): 275-305

US-EPA (1993) A Guidebook to Comparing Risks and Setting Environmental Priorities. Doc. EPA 230-B-93-003, Washington DC, USA (cit. ECETOC (1995) Technical Report no. 68)

van d. Zandt P. and van Leeuwen C.J. (1992) A Proposal for Priority Setting of Existing Chemical Substances. (so-called 'EPS-report' commissioned by the EU/DG XI).

VROM (1994) Environmental Quality Objectives in the Netherlands. Ministry of Housing, Planning & Environment, The Hague, The Netherlands. (ISBN 90 6092 783 4)

Woin P. (1995) Xenobiotics in Aquatic Ecosystems. Thesis from the Department of Ecology, Lund University, Sweden

Is Sustainable Development a Practical Possibility Given the Continued Use of Plant Protection Products ? - The Scientific View -

Monika Herrchen and Werner Klein
Fraunhofer-Institut for Environmental Chemistry and Ecotoxicology,
D-57392 Schmallenberg, Germany

Sustainability: An Often Discussed Paradigm

Prior to discussing the possibilities and limitations in sustainable use of plant protection products a comprehensive definition of sustainability should be given. Recently, the American Society of Agronomy adopted the following definition (Abrol and Katyal, 1990):

"A sustainable agriculture is one that, over the long term, (i) enhances environmental quality and the resource base on which the agriculture depends, (ii) provides for human fibre and food needs, (iii) is economically viable, and (iv) enhances the quality of life for farmers and the society as a whole".

At the agroecosystem level, sustainabiliy is defined as "the ability of an agroecosystem to maintain productivity when subject to a major disturbing force" (Cornway, 1987) or "sustainability concerns whether a given level of productivity can be maintained over time" (Marten, 1988). A somewhat broader definition of sustainability in an industrial society has been given recently by an Enquiry Commission of the German Government (Anonymous, 1994). Four major basic rules should be obeyed for a sensible use of resources and environmental protection:

(1) Adequate reduction of use of renewable resources and oriented at regeneration rates of these resources
(2) Sensible use of non-renewable resources
(3) Consideration of impact capacities of the environmental compartments ("critical load" concepts, partly controversially discussed)
(4) Consideration of inherent time scales for the ecological systems (reaction and regeneration times).

Before judging on different agricultural management practices and technologies in the light of these sustainability criteria a set of *environmental impact endpoints* or *categories* should be defined. These categories represent environmental items, which might be affected by the agricultural technologies including plant protection products. The categories can also be allocated easily to the above sustainability criteria and thus form a link in between these criteria and the concrete management practices. So, the frame for a multicriteria assessment of agricultural management practices based on the concept of sustainability is given.

Table 1: A relation between agricultural management practices, environmental impact categories and sustainability criteria

Management practices (examples)		Environmental impact categories (examples, multiple allocations usual, but not given)		Sustainability criteria (examples, multiple allocations not always given)
Fertilization, manure, sludge application	⇒	Eutrophication, nitrification,heavy metal burden, acidification, use of internal ("farm derived") resources, renewable resources	⇒	Sensible use of non-renewable resources, adequate reduction of use of renewable resources, critical load
Tillaging / non-tillaging	⇒	Erosion, soil depletion energy consumption, use of external (fossil fuel) resources	⇒	Sensible use of non-renewable resources
Pesticide application	⇒	Impact on ecological balance, filtering function, groundwater contamination, habitat function, agricultural worker protection	⇒	Consideration of impact capacities (critical load), enhancement of quality of life for farmers and society, economical viability
Crop rotation	⇒	Soil productivity, fertility	⇒	Provision of renewable resources, provision for human fibre and food need
Genetic engineering	⇒	Human health, environmental safety, consumer protection	⇒	Enhancement of quality of life for society

The recently discussed non-tillaging practices do have a positive influence on the corresponding impact category by reducing the soil erosion and depletion, by reducing the energy consumption, and also by increasing the use of internal renewable resources instead of externally produced fertilizers. Thus, the sustainability criterium of sensible use of non-renewable resources is fulfilled. On the other side, an increased use of herbicides might be needed resulting in a negative influence on the impact category of impact on ecological balance and the groundwater contamination. So, the sustainability criterium to consider the impact capacities must carefully be observed; the criteria to enhance the quality of life for farmers and society, as well as economical viability should be fulfilled. This example clearly demonstrates, that different agricultural practices should be weighted against each other by comparing their influences on all corresponding impact categories. The final evaluation using the different sustainability criteria should be made after giving weight to these criteria on the basis of socio-economic aspects. A modern instrument to perform such a comparative multicriteria evaluation on different technologies is that of ecobalances of technologies. It will be presented at the end of the manuscript.

Agricultural Technologies and Their Competing Influences on Environmental Items with Special Emphasis on Plant Protection Products

Soil management technologies: Soil erosion by water is a serious, worldwide problem. Mitigation efforts are predominantly focused on conservation tillage practices. In fact, sediment concentrations are significantly reduced with long-term no-till management practices (i. e. fully undisturbed soil, undisturbed residue removed prior to precipitation). Kling et al. (1990) reported a 51- 67% reduction compared to tillaging practices. Besides consolidation, weathering, drying and wetting sequences as well as residues are factors influencing the erosion, and thus they also contribute to variabilities between different individual events. A broad overview on different observations on consequences caused by long-term no-till management practices is given in table 2. The examples (Aase and Pikul, 1995; Buhler, 1995; Gish et al., 1995; Kampmann et al., 1994) show various advantages for a maintenance and improvement, respectively, of soil functions. Also, the energy consumption and use of non-renewable resources might be reduced. On the other side, a major disadvantage is the need to use higher application rates of plant protection products. This might result in an increased impact on the non-target areas depending on the pesticide design and application techniques (see below). Under the light of sustainability the table can be interpreted in a way that *"positive influences of the non-till practices on environmental items"* directly contribute to a sustainable agriculture. For an even better interpretation, which also allows for concrete measures, a further combination with differently weighted sustainability criteria can be made: an interpretation with special emphasis on socio-economic aspects might result in other measures compared to an interpretation focussing on the long-term enhancement of environmental qualities.

Crop Rotation: An important aim of sustainable development is the sensible use of renewable resources, i.e. atmospheric nitrogen, sun energy, water from precipitation and soil organic matter. The inclusion of polyannual legumes is a powerful instrument because of its implications to maintain soil fertility. It was demonstrated by Caporali and Onnis (1992) that the nitrogen residual effect of a preeceding 4-year lucerne ley was 130 kg ha^{-1} of inorganic nitrogen. Both, organic and inorganic N-sources produced similar grain and oil yields. The long-term crop rotation - on the other hand - might result in a diversity increase of species living on and in the arable soil. As regards pests, the communities might be balanced for several species (e.g. nematodes) but also imbalanced for others (e.g. weeds) and thus result in an increased use of selected plant protection products. The increased use of plant protection products might lead to a certain amount of non-extractable residues and to an impact on the adjacent non-target ecosystems.

Fertilization: Consequences of the use of fertilizers such as eutrophication and groundwater contamination are obvious; only one interesting aspect will briefly be focused on: "Farming soils, not fields" is a strategy to increase fertilizer profitability in the case of such fields that are composed of heterogenous soil sub-units. Modern machines do have installed sensors triggering individually the fertilizer application rate. Such an optimum soil treatment might result in additional returns over variable costs of up to $ 23/acre (Carr, 1991).

Pest Control by Chemical Plant Protection Products: Considering the fact that plant protection products are active ingredients with biocidal properties they have to have effects at least on the target area. Sustainability requires that the arable soils keep their productivity, their filtering capacity and - at least reversibly- their habitat function. The latter means that following the end of land utilization the original, site-specific population diversity will again succeed. Effects outside the target area should not cause long-term damage to the ecological balance. The term "*damage to the ecological balance*" is in our opinion a partly social and political one, since it is connected to the societies' acceptance of changes in the ecosystem functions. As long as no mutual agreements on the acceptance exist, any changes should be avoided at all or at least they have to be reversible.

This requires for a continued use of plant protection products that application technologies have to be optimized to minimize the release into ecotons and to exclude properties resulting in long-term impacts. With respect to the latter, the Council Directive 91/414/EEC and the Annex VI (Uniform Principles) try to ensure an adequate degradability and accumulation potential. They aim at negligible amounts of the parent compound at the beginning of the next vegetation period and in food to be consumed as well as of non-extractable residues (Klein, 1993).

Table 2: Non-tillaging/conservation tillage and its influences on selected environmental items

Observed consequences (as compared to tillaging)	Influence on the following impact categories is Positive	Negative
increased crop yield as compared to fallow-crop treatment	productivity, fertility	
increase of soil organic C	productivity, fertility, reduced groundwater contamination by sorption of pesticides	
reduced sediment runoff	reduced soil erosion and pesticide export	
increased water use efficiency	reduced use of water resources	
increased population density of earthworms	habitat function	
increased microbial acitivities	filtering function	
formation of macropores	reduced groundwater contamination by increased pesticide degradation in macropores (discussed controversially)	groundwater contamination by pesticides (preferential flow)
appearance of general trends in population dynamics of weeds (mostly increase of populations)		impact on ecological balance outside the target area because of increased pesticide use
increased use of different kinds of pesticides and higher application rates		impact on ecclocigal balance outside the target area

For an evaluation of damages in non-target ecosystems a risk analysis is usually undertaken, based on the probabilities of transport from the target to the non-target area (Klein, 1993; Koerdel, 1996). Though in principle a complete matrix should be constructed, in which the transfer risks between target and non-target areas are defined for all major combinations, the present experimental data are far from comprehensive and so recourse must be made to the small number of data sets and suitable models. In particular, the relative importance of alternative movement paths to the non-target areas have to be quantified, so that relevant management practices can concentrate on controlling the major routes. In each case it is essential to build predictive models of the transfers, so that the matrix can be expanded to consider a wide range of site, pesticides and agricultural conditions. These models can range from fully mechanistic numerical models, to evaluative models based on expert judgement. It is also important that results are based on experimental data of movement. For the paths of spraydrift, surface runoff and runoff in drainage systems the experimental data basis is such that a first approach can be made to construct the matrix: Spraydrift amounts have been extensively studied. A summary of 119 experimental measurements (Ganzelmeier, 1995) gives the measured deposition amount at a sequence of distances from the application edge. These experimental data were generalized by the use of a mathematical model (Klein, 1994). Also, transport of pesticides by surface runoff has been investigated in numerous experiments. Main factors determining that movement are pesticide properties (solubility, sorption and degradability), characteristics of the soil surface, and meteorological conditions. However, it is difficult to generalize experimental results into a predictive mathematical model, so that an evaluative model using expert judgement is the actual method of choice. The runoff in drainage systems also is difficult to predict using simulation models because of the importance of preferential flow. Research models are difficult to parameterise and cannot actually be generalized to a wide variety of situations. For fairly soluble plant protection products in general the lowest risk for the non-target area is from spraydrift, and the relative importance of surface runoff and drainflow is higher and approximately equal. The relative importance of the two runoff components will depend on the hydrological setting. Consequently, an optimal pesticide design with respect to solubility, degradability and sorption is an instrument to allow for sustainable use of these products. Again, a summary of optimization potentials and their influences on different environmental items is given in table 3. In summary, an optimization within any of the given technologies and pesticide properties always means an improvement of the continued use of plant protection products under the aspect of sustainability.

Genetic Engineering: An approach to reduce the amount of chemical plant protection products are strategies of biological control. The numerous approaches (Cook, 1988) can be subdivided into the categories *"regulation of the pest population"* (use of biological control agents, e.g. insects, plant pathogens), *"exclusionary systems of protection"* (use of beneficial organisms as a living barrier) and *"systems of self defense"* (genetic engineering of crops to become resistant either against the pest itself or against a plant protection product). Recently, the latter approach is discussed controversially because of safety and ethic reasons. Genetic engineering resulting in resistance against a pesticide allows for an optimal dosaging and application technique of the chemical and thus results in reduced application rates and export from the target area. Genetic engineering resulting in pest resistant crops even goes a step further.

Table 3: **Optimized application technologies and pesticidal properties and their influences on selected environmental items**

Optimized technologies and pesticide properties	Influence on the following impact categories Positive	Negative
Spraying equipment, modern nozzles resulting in optimal particle size distribution, incorporation into soil, controlled release	Impact on the ecological balance outside the target area (terrestrial and aquatic systems)	
Application rate and time related to environmental conditions	Impact on the ecological balance outside the target area	
Biodegradable (DT_{90} < 100 d)	Soil productivity, filtering function, groundwater contamination	Long-term functionality of the pesticide
Low accumulation (BCF < 100)	Consumer protection, impact on ecosystems (also after long-range transport)	
Low formation of non-extractable residues (< 70 % in 100 d)	Soil productivity, filtering function, long-term groundwater contamination (after remobilization)	
Solubility, sorption constant	Impact on the ecological balance outside the target area	

Using life-cycle assessment (LCA) methodologies to evaluate different optimization strategies, which aim at sustainable agriculture

The above tables offer alternative and/or complementary strategies to reach sustainability in agriculture. However, a selected strategy mostly results in both positive and negative influences on different environmental items. Thus, the optimum for sustainable agriculture is difficult to identify. The life-cycle assessment methodology presents a possibility to identify the environmentally best out of alternative products, technical processes or technologies. It also facilitates a final evaluation using weighting criteria (i.e. in our case differently weighted sustainability criteria).

The life-cycle assessment (LCA) methodology performs an input/output analysis (inventory), classifies all emissions and resource and energy consumption along the life-cycle of a technical product into environmental impact categories, characterises them by referring to reference compounds and normalises by relating to the entire impact of a reference area.

A similar approach can be made for comparing the environmental burden by different agricultural technologies including pesticide application (Ceuterik, 1996): the functional unit could be - e.g. - the cultivation of 1 ha of land including the succession of crops, crop rotation and the effects of possible set-aside of land. An inventory analysis of the resource and energy consumption, of all material and chemical input and output fluxes including those from pesticide production processes, construction of agricultural maschines etc. and also economic fluxes (investments) has to be made. A first idea is given on the scheme on the following page. In a second step these fluxes are classified into impact categories identical to those given in the tables (e.g. impact on the ecological balance, groundwater contamination and soil functions). The characterization and normalization steps have to some extent to be modified and/or extended compared to the "classical" LCA approaches: more site-specific aspects have to be taken into account, and the fate of the emissions has to be considered as well as new references (for not so far used impact categories) have to be defined. After having done so, the contributions of the alternative technologies to the different impact categories are quantified or at least qualitatively described. In a last step, the different impact categories can be given weight by first allocating them to the above defined sustainability criteria (multiple allocation usual) and secondly weighting the sustainability criteria themselves: if - e.g. - the consideration of impact capacities (critical load concept) is given most weight, the corresponding impact categories "impacts on the ecological balance", "eutrophication", and "heavy metal burden" are also given most weight. In this case, the agricultural technology(ies) contributing less to these categories are the most sustainable ones. If the socio-economic aspects (enhancement of quality of life for farmers and society) are given most weight, the impact categories "agricultural worker protection" and "consumer protection" are the most relevant ones. These technologies which give most improvement to these categories are also aiming at most sustainability.

Conclusions

The question "is sustainable development a practical possibility given the continued use of plant protection products?" can clearly answered with "Yes". Necessary prerequisites to make the term "sustainability" operable are well defined sustainability criteria, environmental items which do reflect these criteria and can also be used to picture the impacts caused by different agricultural technologies including pesticide application. The use of plant protection products should be judged upon in the light of sustainability preferably in combination with other agricultural technologies. The adoption of life-cycle assessment methodologies to solve such a multicriterial problem seems to be a promising way.

Scheme: Adopting the LCA methodology: Input/output analysis for a functional unit of cultivating 1 ha of land

material and ecological input: * energy use including sun energy * renewable/non-renewable resources including their extraction * fertilizers, pesticides, genetically modified pesticides including their production * seeds * agricultural maschines including their construction * constructions on the field, eg. drainage systems * water (irrigation, precipitation)	➔	cultivation of 1 ha land using different technologies (non-tillaging, crop rotation, pesticide application	➔	*material and ecological output:* * crop yield (kg/ha) * sediment and water fluxes (leaching, surface runoff, sedimentation, drainage runoff...) * P/N/K-emissions to soil and water * pesticide emissions to non-target ecosystems including groundwater * emissions by fuel use
economic input: * services * investments	➔		➔	*economic output* * crop yield ($ / ha) * return of investment

References

Aase JK and Pikul JL Jr (1995) Crop and soil responses to long-term tillage practices in the northern Great Plains. Agronomy Journal 87(4): 652-656

Abrol IP, Katyal JC (1990) Managing soil fertility for sustained agriculture. In: Singh RP (ed.) Sustainable agriculture - issues, perspectives and prospects in semi arid tropic. Vol. 2, Indian Soc. Agronomy, Proc Symp, February 6-10, New Delhi, pp 235-265

Anonymous (1994) Die Industriegesellschaft gestalten - Perspektiven für einen nachhaltigen Umgang mit Stoff- und Materialstroemen, Enquete-Kommission "Schutz des Menschen und der Umwelt" des Deutschen Bundestages (ed.), Economica Verlag, Bonn

Buhler DD (1995) Influence of tillage systems on weed population dynamics and management in corn and soybean in the central USA. Crop Science 35(5): 1247-1258

Caporali F and Onnis A (1992) Validity of rotation as an effective agroecological principle for a sustainable agriculture. Agriculture, ecosystems and environment 41: 101-113

Carr PM, Carlson GR, Jacobsen JS, Nielsen, GA, Slogley EO (1991) Farming Soils, not fields: A strategy for increasing fertilizer profitability. J Prod Agric 4: 57 - 61.

Ceuterik D (ed.) (1996) International conference on application of life-cycle assessment in agriculture, food and non-food agro-industry and forestry: achievements and prospects. Preprints, VITO, Brussels

Conway GR (1987) The properties of agroecosystems. Agric Syst 24: 95-117

Cook RJ (1988) Biological control and holistic plant-health care in agriculture. American Journal of Alternative Agriculture 3(2): 51-62

Ganzelmeier H, Rautmann D, Spangenberg R, Streloke M, Herrmann M, Wenzelburger HJ, Walter HF(1995) Studies on the spray drift of plant protection products. In: Mitteilungen aus der Biologischen Bundesanstalt für Land- und Forstwirtschaft, Heft 305, Blackwell Wissenschaftsverlag GmbH, Berlin/Wien

Gish TJ, Shirmohammadi A, Vyravipillai R, Wienhold BJ (1995) Herbicide leaching under tilled and no-tillage fields. Soil Sci Soc Am J 59(3): 895-901

Kampmann T, Gottesbüren B, Knuesting E, Malkomes H-P, Bartels G (1994) Criterions of stress in agricultural ecosystems, its connection and assessment of data and conclusions for agriculture. Effects of a long-term application of plant protection products used in different intensities. In: Mitteilungen aus der Biologischen Bundesanstalt für Land- und Forstwirtschaft Berlin-Dahlem No. 295, pp 353-384

Klein AW, Goedicke J, Klein W, Herrchen M, Koerdel W (1993) Environmental assessment of pesticides under Directive 91/414/EEC. Chemosphere 26(5): 979-1001

Klein M (1994) MOPED Manual (Model for Pesticide Drift), Schmallenberg

Kling KW, Flanagan DC, Norton LD, Laflen JM (1990) Rill erodibility parameters influenced by long-term management practices. Transactions of the ASAE 38 (1): 159-164

Koerdel W, Herrchen M, Armstrong A (1996) Impact of pesticides on non-target ecosystem. submitted for publication in Pesticide Outlook

Marten GG (1988) Productivity, stability, sustainability, equitability and autonomy as properties for agroecosystem assessment. Agric Syst 26: 291-316

Regulations and Risk Assessments on the Ecotoxicological Impact from the Use of Plant Protection Products in the European Union – an Industry Viewpoint

Nazim Punja
Agricultural Research & Development, Merck Sharp & Dohme (Europe) Inc, Brussels

Introduction

In order to prevent the loss or deterioration of food, the use of plant protection products will continue to be needed in agriculture to control pests which damage or destroy important food crops, pre- or post-harvest. With the onset of pest resistance, old products need in time to be phased out and replaced by new ones with different biochemical modes of action. Much of the research and development (R&D) required to discover new products is carried out by the crop protection industry.

A myriad of regulations have emerged over the last 25 years to evaluate, approve, review and control the use of these plant protection products considered essential in modern agriculture to sustain or enhance our standards of living. Much of the data required to satisfy regulatory requirements, including ecotoxicological studies, are also generated by industry.

Plant protection industry operates within the constraints of commercial, legal, socio-economic and political pressures. A mutually acceptable balance between these apparently countervailing forces is a necessary pre-requisite for achieving the common goal of 'sustainable development'. For balancing the ecological equation in favour of nature conservation, high intensity agricultural production reduces the rate of conversion of forest, jungle, wetlands and other ecologically diverse habitats into tillable land, and judicious use of plant protection products plays an important role in sustainable agricultural production.

This paper outlines the status of current regulatory data requirements on ecotoxicology pertaining to approval for the use of plant protection products in the European Union (EU) and comments on various important regulatory decisions and decision-making processes which influence further investment in R&D.

Ecotoxicological Data Requirements in the EU

The Council Directive 91/414/EEC

Council Directive 91/414/EEC concerning the authorization of plant protection products in the EU market includes chemical pesticides as well as micro-organisms and viruses. The directive has now been incorporated in the national legislation of most of the EU member states. The directive and subsequent amendments to the directive, *inter alia*, give guidelines on data requirements for active substances (Annex II) and formulated products (Annex III) (EU 1991-1996).

The guidelines are still incomplete and are gradually being developed and adopted. In the meantime, applicants for the approval of new active substances and authorization of plant protection products have to work with incomplete guidelines and rely, where appropriate, on national guidelines.

Annex VI ('Uniform Principles') of the 91/414 directive (as amended by the 94/43/EC directive) was annulled in June 1996 by the European Court of Justice. The Court's decision was based on Specific Principles under Section 2.5.1.2 of Annex VI which allowed member states the option to provisionally authorize products, the use of which may result in residues in groundwater in excess of the 0.1 mg/litre limit set by the Council Directive 80/778/EEC. Uniform Principles are currently being redrafted by the EU Commission.

Conflicting Regulations: National versus EU Data Requirements

The 'subsidiarity clause' in the Maastricht Treaty allows implementation of national legislative provisions in conflict with Community provisions provided they are essential for the protection of the national interest. The legality of the application of this clause to restrict the use of plant protection products is doubtful.

Denmark and Sweden have legislated the so-called 'cut-off criteria' for certain inherent properties of pesticides, the breach of which makes the offending pesticide products unregistrable in these EU member states. The environmental cut-off criteria are similar to those of Annex VI to the 91/414/EEC directive (as amended by the 94/43/EC directive), but they are more rigid, as they do not contain the exception provisions of the EU (Tables 1 and 2). Although the decision to annul was made for a different reason, the annulment of Annex VI by the European Court of Justice now provides another opportunity for these countries to renegotiate the insertion of more severe criteria, including lower acceptable Toxicity to Exposure Ratios (TERs), *vide infra*, in the revised directive. In addition, national legislations to reduce the number of pesticides over specified times have resulted in de-registration of pesticide products based on the implementation of the cut-off criteria.

Table 1 Decision-making criteria (Annex VI to directive 94/43/EC[1])

'No authorization, if'[2]	unless'
Soil	
DT_{50} > 3 months and DT_{90} > 1 year or non-extractable residues >70% of initial dose & mineralisation rate <5% after 100 days (3 soil types at 20°C and 1 soil type at 10°C)	No accumulation in soil leading to phytotoxicity or unacceptable residues in succeeding crops or unacceptable environmental impact
Water	
residue of a.s. > 0.1 mg/l (combined residues of a.s. and metabolites >0.5 mg/l) in ground/drinking water	MS may conditionally authorise for 5 years (renewable only once) while field monitoring studies are being conducted
Non-target species	
Birds & other terrestrial vertebrates $TER_{a;st}$ <10 and TER_{lt} <5; BCF <1	No unacceptable impact in field use
Aquatic organisms Fish/Daphnia TER_a <100, TER_{lt} <10 Algal growth inhibition TER <10 BCF >1000 (readily biodegradable a.s.) BCF >100 (not readily biodegradable a.s.)	No unacceptable impact on viability of exposed species (predators) occurs, directly or indirectly, in use
Honey bees Q_{HO} or Q_{HC} >50	No unacceptable effects on bee larvae, behaviour or colony survival in field use
Beneficial arthropods >30% of test organisms affected in lethal or sub-lethal tests	No unacceptable impact in field use
Earthworms TER_a <10; TER_{lt} <5	Earthworm populations not at risk in use
Soil micro-organisms N or C mineralisation affected by >25% after 100 days	No impact on microbial activity

[1] Now annulled

[2] Results from maximum application rate

a Acute st Short-term lt Long-term BCF bio-concentration factor
Q_{HO} Hazard quotient (oral) Q_{HC} Hazard quotient (contact)

Table 2 **Environmental fate and ecotoxicological cut-off criteria**

Properties	EU	Denmark	Sweden	Netherlands
DT_{50} in soil	>3 months	>3 months (except 8 months for herbicides which require persistence)	>4 months at 25°C; >6 months at 18°C; >9 months at 15°C; >1 year at 10°C	>2 months at 20°C unless there is no environmental toxicity or mobility problem and no unacceptable residue in ploughed field 2 years after the last application
DT_{90}	>1 year			
Mobility in soil		Aged soil column leaching (% in): - soil clay carbon >2 >10 ~2 >3 5-10 ~2 >5 <5 ~1 → lysimeter or field study; reject if >5g/ha/year	Koc ≥ 50 and DT_{50} at 20°C ≥ 1 month	
Bio-accumulation	100-1000; >1000 and >95% (14 day) de-puration	K_{ow} >1000; DT_{50} >20 days; aqueous hydrolysis >28 days; BCF >100	BCF 500; in fish >2000 and DT_{50} at 20°C ≥ 1 month	
Toxicity to aquatic organisms				1/TER:- PEC (96-hour) ≥0.1 LC_{50} fish PEC (96-hour) ≥ 1 LC_{50} crustaceae PEC (96-hour) ≥ 1 EC_{50} algae

Risk Assessment

Study guidelines, albeit incomplete, have recently been issued in the EU after lengthy consultations with representatives of competent national regulatory authorities, industry, non-governmental organizations and academia. The guidelines are based on a tier approach starting with laboratory studies, the results of which may trigger further longer term laboratory or field studies. The guidelines are stated to be flexible, reliant upon expert judgement and amenable to adjustment if scientifically justified (EU, 1991-1996). Regulatory decisions are based on the EPPO risk assessment schemes (EPPO, 1993/94). It is too early to judge the success of these guidelines. Many of the study protocols are not yet developed or finalized.

In the risk assessment scheme currently used in the EU, regulatory decisions are based primarily on consideration of the toxicity of the single most sensitive ecological species tested in the laboratory. In the case of aquatic organisms, the indicative test species for an insecticide is often *Daphnia magna*. Test organisms are chosen because of the ease of culturing and testing them in the laboratory. *Daphnia* is a convenient organism for both acute and reproductive toxicity tests because it reaches reproductive age in only a few days and reproduces rapidly. It is ironic that an organism with these attributes is least likely to be permanently damaged, e.g. from chemical spillage, because any survivors will reproduce very rapidly. Even if all free swimming *Daphnia* were to be killed, the daphnid population can rapidly re-establish itself from hatching of the generally much more tolerant resting eggs (*ephippia*). Daphnia, nevertheless, is a useful indicator of the toxicity of chemicals to a wide range of organisms.

Exposure of the test organisms is estimated for the highest potential concentration of the active substance in the relevant environmental compartment or habitat (e.g. soil, water, seed or grass), giving, e.g., Predicted Environmental Concentration (PEC) for earthworms in soil, or for fish in water, or Estimated Theoretical Exposure (ETE) for birds.

Toxicity end-points in the test organism, e.g. LC_{50} (acute) or NOEC (long term), is set against the PEC or ETE to give the Toxicity Exposure Ratio [e.g. TER (short-term) = LC_{50} / PEC ; TER (long-term) = NOEC/ PEC]. The TER value thus obtained is then compared against the appropriate 'safety factor' (e.g. 5, 10 or 100) for regulatory decision (Table 1). TER values act as useful screens to eliminate the obviously innocuous substances from further consideration. High toxicity of the active substance also triggers the need to test its formulated product. Breach of any of the TER values, however, should not automatically lead to rejection of the proposed use but to the need for further information to evaluate what happens under more realistic and practical conditions.

The following examples illustrate some differences between ecotoxicological effects observed in the laboratory and under natural conditions.

Beneficial Arthropods

The 30% toxicity trigger set for beneficial arthropods (Table 1) is well within the natural fluctuations of population density. For instance, official field data from southern EU countries show that normal fluctuations in population density are often greater than 40% for *Chrysoperla carnea* and up to 60% for *Aphidius* species. It is well known that arthropod populations normally recover rapidly after being temporarily affected by pesticide applications.

Aquatic Organisms

End-points of toxicity (LC_{50} and NOEC values) are obtained in the laboratory in clean water and at constantly renewed concentrations, conditions more severe than encountered in the natural environment. For instance, the 72-hour EC_{50} values for permethrin and cypermethrin, synthetic pyrethroid insecticides, on *Daphnia magna*, *Cleon dipterum* and *Asellus aquaticus* are increased (i.e. acute toxicity decreased) 200-400 fold in the presence of sediment as a result of lowered bioavailability, under conditions more akin to those prevailing in nature (Riley, 1990; Hill, 1985).

Factors Influencing Ecotoxicological Risk Assessment

Factors which influence concentrations of pesticides in surface waters include agronomic practices (rate and mode of application, target crop and season), climatic conditions (temperature, humidity, wind and rainfall), physical-chemical properties of the pesticide (volatility and water solubility), the body of water (area, depth, and flow) and surrounding field (soil, grassland, slope and distance from crop to the body of water). The major route of contamination from short-term exposures is from spray drift, run-off or over-spray, and for long-term exposures from drainage and discharge. The major routes of dissipation of pesticides from the environment include volatilization, sedimentation, hydrolysis, photodegradation, metabolism and biodegradation.

Exposure concentrations of pesticides in different surface waters can be estimated using various environmental fate models using relevant data inputs. These inputs include agronomic conditions, properties of the pesticide, prevailing environmental and climatic conditions, type of receiving waters, size, location and geographical information, and rates of contamination and dissipation. Inputs also include susceptibility of the organisms, bioavailability as well as time of sensitive stages of the organisms against the use regime, to help decide which type of ecosystems are at risk, during what periods, and which of the organisms are to be protected. Ecology of the target populations should be considered in relation to behaviour, habitats, feeding habits, life-cycles, recovery potential and population dynamics, including natural fluctuations.

The models must be fully validated using realistic and practical field conditions because worst-case compounding of inputs could give totally unrealistic predictions. FOCUS (Forum for the Co-ordination of pesticide fate models and their Use), a group consisting of experts from regulatory authorities, industry and research institutes has recently been set up to deal with simulation modelling of surface water. Refinement and validation of these models may be the most pressing need for better ecotoxicological risk management decisions.

Refinement of Ecotoxicological Risk Assessment Models

The likelihood of a substance to cause an adverse effect in the environment is dependent upon a combination of factors, including toxicity, exposure and probability of exposure. Toxicity is the inherent potential of the substance to cause injury. Hazard is a function of toxicity and exposure. The lower the exposure, the lower the hazard. Risk is the probability of the adverse effect occurring. Therefore risk must include the assessment of the probability that exposure will occur.

Ecotoxicological risk assessment differs from human risk assessment in two fundamental ways. Firstly, end-points for risk assessment in humans deal with the protection of the individual organism. In ecotoxicological risk assessment, on the other hand, individual organisms in the environment are part of an extensive food chain, and are either devouring or being devoured. Such an individual organism may be regarded as inexpendable only if it is a rare or endangered species. Ecotoxicological risk asessment is aimed at protection of the functions of populations, communities and ecosystems. Populations are less sensitive than their most sensitive members, and communities and ecosystems are less sensitive than their most sensitive components. As minor effects on the diversity of species may be tolerated, assessment of the impact on the survival or reproduction and growth of organisms is more useful than measurements of physiological effects on individual organisms. Secondly, in ecotoxicological risk assessment, unlike human risk assessment, toxicity can be measured in a number of different organisms to obtain a range of toxicities. Ecotoxicological risk assessment should consider the role of these organisms in the ecosystem structure and function, e.g., in nature reserves, fishing sport, open drain or ditch. Furthermore, it must be recognised that ecosystems have resiliency and redundancy, and that functions are conserved even when the ecosystem is under considerable stress. This recognition of the difference between human and ecotoxicological risk assessments should be acknowledged and implemented by appropriate regulatory decision-making schemes.

TERs from single exposures do not give the quantitative likelihood of the occurrence of such exposures until they are refined by the use of probabilistic models which consider the variable nature of the actual environmental exposures, taking into account natural variations by giving distributions of multiple exposures. Such models can be used to estimate how frequently exposures to a particular substance will exceed any given threshold of toxicity effects on particular organisms in the environment. A range of susceptibility of species to substances must also be taken into account; data from sensitive as well as from less sensitive groups of organisms must be used.

Probabilistic risk assessments can be focused on certain scenarios with different ranges of exposures or with different combinations or sensitivities of species, and therefore can provide more realistic information than assessments which rely only on point-source measurements or worst-case estimates. Such assessments also allow the rational identification and ranking of different risk scenarios for exposure reduction and site-specific risk mitigation.

Industry and the Incentive to Invest

It takes on average 8-9 years between discovery and registration of a plant protection product for first sale in the EU, and another 6-7 years to break-even the development expenditure incurred, often leaving less than 9 years before patent expiry and decline of the market (Figure 1, Series 1). A two-year earlier registration can result in a significant improvement in financial return (Figure 1, Series 1).

The ratio of the number of new chemicals biologically screened to each active pesticide discovered which is finally commercialised is currently >10,000 : 1. Many potential plant protection products are thus rejected by the industry during the R&D phase for a variety of reasons, not least of which being the ecotoxicological acceptability. The cost of research for the discovery of a new product, usually treated as 'sunk' cost, is some $50-100 Million, and the cost of developing the product for registration of major uses amounts to another $50 Million. Thus $100-150 Million have to be invested in R&D for each new pesticide. This is a high-risk business, affordable by relatively few large multi-national companies which currently invest up to 12% of their revenue in R&D.

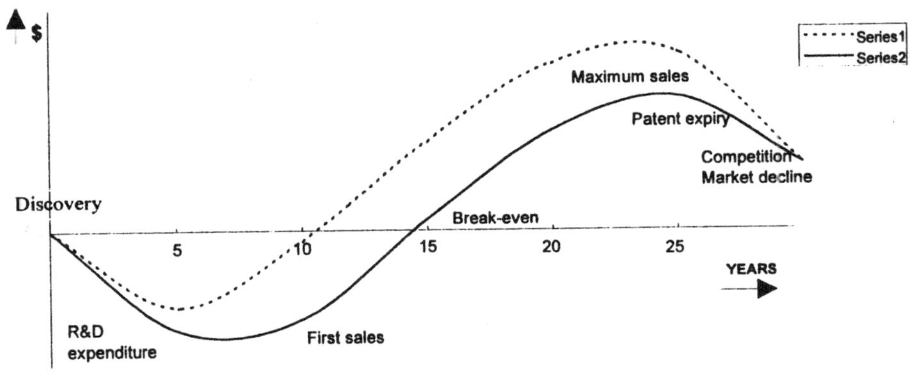

Figure 1. R&D break-even curve

Plant protection industry welcomes sensible regulations, restrictions and data requirements which can be justified by good science. Effective labelling of information on the proper use of the product in practice would also result in lower exposures to non-target organisms. The cost of ecotoxicological testing of an active substance or plant protection product in accordance with current guidelines accelerates dramatically as one moves from the lower to the higher test tiers. For instance, some short-term acute toxicity tests cost $20,000, while field trials can cost over $2 Million. Concern for 'expenditure of resources' should also include cost to industry, ultimately leading to cost to the end-user, and should ensure that resources, including organisms, are not wasted in pursuance of unnecessary or excessive regulatory demands.

Sustainable agricultural development and the preservation of the environment are the common goals of all the 'partners' – industry, regulators, farmers and the general public – involved in the discovery, manufacture, regulation and use of plant protection products, and benefiting from their use. In order to attain these common goals, i.e., to produce plant protection products which have acceptable quality, efficacy and safety, industry must receive a fair return on its investment and adequate patent and data protection as an incentive to invest in R&D, vital for further investigation and new discoveries.

Conclusion

Global harmonization of data requirements, based on good science, benefit-risk evaluation and use of the appropriate risk assessment procedures, would ensure that optimum data are available for making timely regulatory decisions on the ecotoxicologically acceptable plant protection products. The continued use of such products in agriculture is necessary to maintain and improve our quality of life without endangering that of future generations.

References

EPPO (1993/94) European and Mediterranean Plant Protection Organisation and the Council of Europe (EPPO/CoE) Environmental Risk Assessment Schemes, EPPO Bulletin, Volume 23 (1), 1993 and Volume 24 (1) 1994.

EU (1991 - 1996) Amendments to the Council Directive 91/414/EEC which give guidelines on data requirements include: Commission Directives 94/37/EC (physical-chemical properties), 94/43/EC ('Uniform Principles'; efficacy and decision-making guidelines), 94/79/EC (mammalian toxicology and metabolism), 95/35/EC (derogation from GLP), 95/36/EC (environmental fate), and 96/12/EC (ecotoxicology).

Hill, IR (1985) Effects on non-target organisms in terrestrial and aquatic environments. In: The Pyrethroid Insecticides (Leahey, JP, ed.) Taylor & Francis, pp 151-262.

Riley, D (1990) Current testing in the sequence of development of a pesticide. In: Pesticide effects on Terrestrial Wildlife (Sommerville, L and Walker, CH, eds.) Taylor & Francis, pp. 11-24.

Challenges in Analytical Toxicology

(Chair: J. Segura, Spain)

Analytical Development for Low Molecular Weight Xenobiotic Compounds[1]

Hans H. Maurer, Joachim W. Arlt, Thomas Kraemer, Christian J. Schmitt and Armin A. Weber
Institute of Pharmacology and Toxicology, Department of Toxicology, University of Saarland, D 66421 Homburg (Saar), Germany

[1] Dedicated to Prof. Dr. R. Klaus Müller, Leipzig, on the occasion of his 60th birthday

Abstract

Specific and sensitive detection or precise quantification of xenobiotics in biosamples (e.g. blood, urine, saliva, sweat, hair) are great challenges in analytical toxicology. GC-MS is the most sensitive, specific and universal analytical method for low mass xenobiotics. Precise quantification can be performed using the selected ion mode (SIM) and stable isotopes as internal standards. Negative chemical ionization (NCI) can improve severalfold the sensitivity for the determination of compounds with electronegative sites (e.g. halogens). For screening and identification of most of the basic and neutral drugs (e.g. drugs of abuse, psychotropics, hypnotics, analgesics, cardiacs) in urine, a systematic toxicological analysis procedure (STA) was developed using GC-MS after acid hydrolysis, extraction and acetylation. For detection of acidic xenobiotics (e.g. anticoagulants, ACE inhibitors, diuretics, antirheumatics) in urine, a further GC-MS procedure was developed using extractive alkylation. For the detection of non-volatile xenobiotics (e.g. toxic peptides like α- and β-amanitin or phase II metabolites) electrospray LC-MS procedures were developed. The procedures and examples show that in analytical toxicology GC-MS is the method of choice for low mass xenobiotics while LC-MS is that for non-volatiles.

Introduction

In analytical toxicology xenobiotics have to be specifically and sensitively identified and quantified in different biosamples like blood, urine, saliva, sweat or hair. Thin-layer chromatography (TLC), gas chromatography (GC), high-performance liquid chromatography (HPLC), capillary electrophoresis (CE) or immunoassays (IA) can be used to solve particular analytical problems (De Zeeuw 1989; Tracqui et al. 1995; Tagliaro et al. 1996). However, GC-MS is the most sensitive, specific and universal analytical method for low mass toxicants (Maurer 1992a-b; Solans et al. 1995; Gonzalez et al. 1996; Maurer et al. 1996). Quantification can be performed in the selected ion mode (SIM) using stable isotopes as internal standards (Maurer 1992a-b; Moeller et al. 1992; De la Torre et al. 1995). Negative chemical ionization (NCI) can improve severalfold the sensitivity for determination of compounds with electronegative sites (e.g. halogen atoms in the xenobiotic or in the derivatizing agent) compared with conventional positive CI or electron impact ionization (EI). However, before quantification, especially in clinical and forensic toxicology, the usually unknown toxicants must first be screened and identified preferably in urine providing higher concentrations than other body fluids (Maurer 1992a; Pfleger et al. 1992; Solans et al 1995; Maurer and Arlt 1996; Maurer et al. 1996). However, enantiomers e.g. of amphetamines cannot be differentiated using such achiral GC-MS procedures. If necessary, the enantiomers can be separated after conversion to diastereomers using chiral reagents (Maurer and Kraemer 1992) or directly using chiral GC columns.

In the following, GC-MS procedures are presented for precise and sensitive quantification of drugs in plasma and for screening and identification of most of the drugs in urine. Finally, electrospray LC-MS procedures are presented for the detection of low mass xenobiotics non-volatile in GC.

Materials and methods

Chemicals: All chemicals used were obtained from E. Merck (Darmstadt, Germany) or Sigma-Aldrich (Deisenhofen, Germany) and were of analytical grade.

Apparatus for GC-MS procedures: A Hewlett Packard (HP, Waldbronn, Germany) 5890 II gas chromatograph combined with an HP 5989A MS Engine mass spectrometer and an HP MS ChemStation (DOS series) with HP G1034C software were used. The GC conditions were as follows: splitless injection mode; column, HP capillary (12 m x 0.2 mm I.D.), cross-linked methylsilicone, 330 nm film thickness; injection port temperature, 280°C, carrier gas, helium; flow-rate, 1.9 ml/min; column temperature, programmed from 100-310°C at 30°/min, initial time 3 min, final time 8 min. The MS conditions were as follows: for identification EI full scan mode, ionization energy, 70 eV; for quantification NCI (methane) full scan mode, ionization energy, 230 eV.

Apparatus for LC-MS procedures: An HP 1050 HPLC connected with an HP 59987A API-ES electrospray interface and an HP 5989B MS Engine were used. The MS conditions were as follows: SIM mode with the ions *m/z* 919 and 920 for α- and β-amanitin. (Details on the tuning conditions are described in Maurer et al., 1996). The isocratic separation was achieved on a Chromcart CC 125/2 Kromasil RP-18 narrowbore column with 5 µm particle size and a Kromasil guard column from Macherey-Nagel (DŸren, Germany). The HPLC mobile phase was a 22:78 mixture (v/v) of methanol-ammonium acetate (0.02 mol/l, adjusted to pH 5). The flow-rate for the LC-MS was 75 µl/min.

Plasma and urine samples: Blank plasma and urine samples were collected from healthy volunteers. Authentic samples were available from clinical or forensic cases.

Sample preparation of plasma for GC-MS quantification of morphine, codeine, 6-monoacetylmorphine, cocaine and benzoylecgonine: One ml of plasma was mixed with 2 ml of phosphate buffer (pH 7.6) and 50 µl deuterated standards resulting in a plasma concentration of 100 ng/ml each. The drugs were isolated by solid phase extraction using 3 ml-Chromabond C-18 columns (Macherey-Nagel). The drugs were eluted with 2ml of acetone-dichloromethane (3:1, v/v). The eluent was evaporated to dryness under a stream of nitrogen at 60°C. The residue was derivatized with a mixture of 100 µl of pentafluoropropionic anhydride (PFPA) and 75 µl of pentafluoropropanol for 30 min at 60°C. The reagents were evaporated under a stream of nitrogen at 60°C. The residue was dissolved in 30 µl ethyl acetate and 1-2 µl were injected into the GC-MS (Maurer 1992b; Moeller et al. 1992)

Sample preparation of urine for GC-MS screening of basic and neutral drugs (systematic toxicological analysis, STA): A 5 ml volume of urine was refluxed with 1.5 ml of 30% hydrochloric acid for 15 min. Following hydrolysis, the sample was basified with 2 ml of 10 mol/l aqueous sodium hydroxide and the resulting solution was mixed with 5 ml of 30% aqueous ammonium sulphate to obtain a pH between 8 and 9. This solution was extracted with 5 ml of a dichloromethane-isopropanol-ethyl acetate mixture (1:1:3; v/v/v). After phase separation by centrifugation (5 min, 1500g), the organic layer was evaporated to dryness. The residue was derivatized by acetylation with 50 µl of acetic anhydride-pyridine (3:2; v/v) for 30 min at 60°C. After evaporation of the derivatization mixture, the residue was dissolved in a volume of 50 µl of methanol and 0.5-1 µl were injected into the GC-MS (Maurer et al. 1996).

Sample preparation of urine for GC-MS screening of acidic drugs using extractive alkylation: A 2 ml portion of urine was alkalized with 2 ml of buffer (1 mol/l Na_2HPO_4/ 2 mol/l NaOH) containing 0.02 mol/l tetrahexylammonium (THA) hydrogensulfate to obtain a pH of 11.5 - 12.0. After addition of 6 ml of a solution of 1 mol/l methyl iodide in toluene, the sample was shaken in a water bath at 50°C for 30 min with a multifix shaker S 300 (Schwinherr, Schwäbisch Gmünd, Germany). The two phases were separated by centrifugation at 1500 g for 3 min. SPE columns (diol phase, 10 ml/ 500 mg) conditioned with 5 ml of methanol and 5 ml of toluene were used to remove the coextracted THA salts from the organic phase. The THA salts were adsorbed on the diol sorbent. Also adsorbed analytes were selectively eluted with 3 ml of diethylether. The combined eluate was carefully evaporated at 60°C to dryness (vaccum 300-500 mbar). The residue was dissolved in 50 µl of ethylacetate and 1 µl was injected into the GC-MS (Maurer and Arlt 1996).

Sample preparation of urine for LC-MS detection of amanitins: A 5 ml volume of urine was first diluted 1:1 with distilled water. LiChrolut RP-18 (500 mg, Merck) columns were washed with ten reservoir volumes of aqueous acetic acid (1%), activated with 3 ml of methanol and rinsed with 3 ml of water. The sample was loaded onto the column, washed with 3 ml of aqueous acetic acid (1%) and then eluted with 6 ml of methanol. The eluate was evaporated to dryness at 65°C. The residue was redissolved in 100 µl of mobile phase and 20 µl were injected into the HPLC system with complete loop filling mode (Maurer et al. 1996).

Results and discussion

GC-MS quantification of drugs of abuse in plasma: The described procedure allowed the sensitive and precise quantification of morphine, codeine, 6-monoacetylmorphine, cocaine and benzoylecgonine in plasma. In Fig. 1 NCI mass chromatograms of ions *m/z* 411, 537 and 557 (morphine PFP) and ions *m/z* 414, 540 and 560 (d3-morphine PFP) of a plasma extract containing 20 ng/ml of morphine are shown. The detection limit in the NCI (0.5 ng/ml) mode was severalfold lower than in the EI mode. Such procedures are very precise because of the use of deuterated standards and very sensitive because of the use of the NCI mode. Therefore, they are state-of-the-art today in clinical toxicology, legal medicine and doping control.

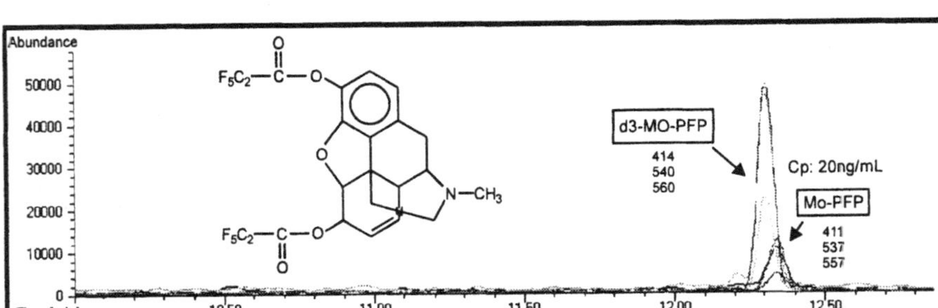

Fig. 1. NCI mass chromatograms of ions *m/z* 411, 537, 557 (morphine PFP) and ions *m/z* 414, 540, 560 (d3-morphine PFP) of a plasma extract containing 20 ng/ml of morphine.

GC-MS screening for basic and neutral drugs in urine (systematic toxicological analysis, STA): Rapid acid hydrolysis of urine was performed, because most of these drugs are excreted in urine as conjugated metabolites. Derivatization was essential for sensitive detection of the polar metabolites. Acetylation has been approved for a comprehensive screening.

For detection of the different drug classes, mass chromatography was used with typical ions selected from the mass spectra of the drugs and their metabolites. Generation of the mass chromatograms could be started by clicking the corresponding pull down menu which executes our macros. The identity of the peaks in the mass chromatograms was confirmed by library search of the underlying mass spectra with that recorded during the corresponding studies.

This STA procedure is exemplified for the detection of designer drugs. As shown in Fig. 2, designer drugs like MDMA, MBDB and their metabolites can be detected using mass chromatography with the ions *m/z* 58, 72, 86, 150, 162, 164, 176 and 178 (Maurer 1996). The mass spectra underlying the peaks could be identified by computerized comparison with reference spectra (Pfleger et al. 1997). However, the basis for developing such procedures was the study of the metabolism of these drugs (Maurer 1996; Ensslin et al. 1996a-b).

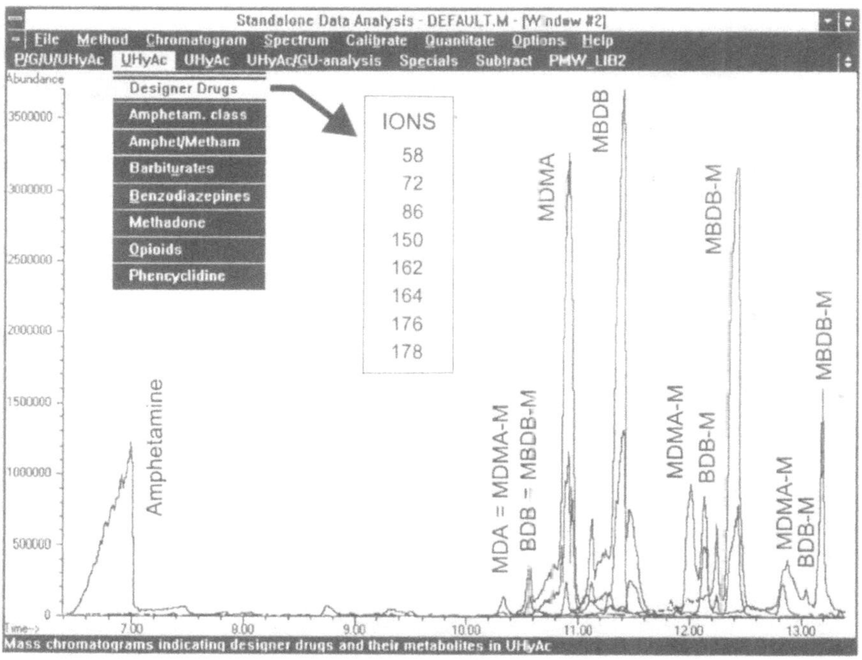

Fig. 2. Mass chromatograms with typical ions indicating amphetamine, the designer drugs MDMA, MBDB and their metabolites in urine. The merged mass chromatograms can be differentiated by their colours on a colour screen.

Interferences by biomolecules or further drugs could be excluded because of the mass spectral identification. The mass spectra of all the endogenous biomolecules, detectable using the described procedure, are included in our library and handbook (Pfleger et al. 1992, 1997). The limit of detection in urine ranged between 1-100 ng/ml (S/N 3) of the different drugs so that most of the barbiturates and other sedative-hypnotics, anticonvulsants, benzodiazepines, antidepressants, neuroleptics, stimulants, designer drugs, phencyclidine, opioids, non-opioid analgesics, antihistamines, antiparkinsonians, beta-blockers, antiarrhythmics, laxatives and their metabolites can be detected in urine at least within several hours after ingestion of a single therapeutic dose (Maurer 1992a).

GC-MS screening for acidic drugs in urine using extractive alkylation: Acidic xenobiotics and their metabolites were isolated and methylated in one step by extractive alkylation. Removal of the coextracted THA salts from the organic phase was essential to avoid pyrolysis of the THA in the injection port of the GC leading to a decrease in the efficiency of the capillary column. The diol phase yielded the best reproducibility and recovery under the described conditions.

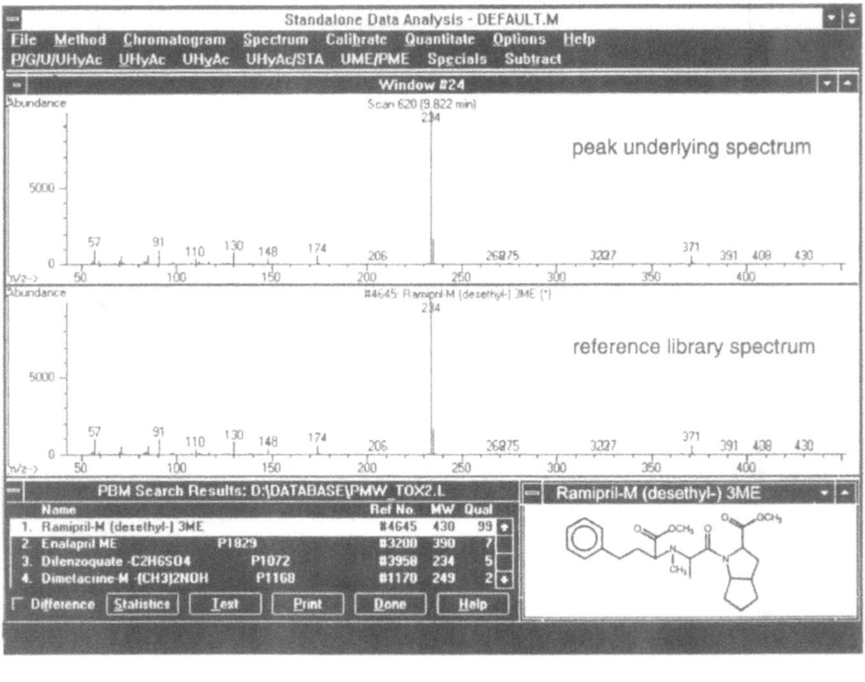

Fig. 3. Mass spectrum underlying a peak in the mass chromatograms indicating ACE inhibitors and their metabolites in urine, the reference spectrum and structure of the methylated desethyl metabolite of ramipril and the hit list found by library search.

This procedure is exemplified for the detection of ACE inhibitors and their metabolites in urine. Again, the presence of these compounds can be indicated by mass chromatography with the ions *m/z* 157, 169, 172, 215, 220, 234, 248, 262 and 365. The mass spectra underlying the peaks can be identified by computerized comparison with reference spectra (Pfleger et al. 1997) as shown in Fig. 3. The limit of detection ranged between 10-50 ng/ml (S/N 3) of the different drugs in the full scan mode (Arlt and Maurer 1996).

LC-MS detection of amanitins in urine: The applied electrospray LC-MS technique allowed the separation, ionization and detection of the amatoxins. In Fig. 4 the electrospray mass spectra, structures, empirical formulas and molecular masses of α- and β-amanitin are shown. Since both spectra contain the ions *m/z* 919 and 920 in different abundances, both ions were selected as diagnostic ions for detection of the amatoxins in the SIM mode. The limit of detection for both toxins was 10 ng/ml (S/N 3).

Symptoms of an intoxication with amanita mushrooms do not appear before a lag time of about 8 to 24 h. In our experience, urine concentrations of α- and β-amanitin at this time ranged between 50 to 500 ng/ml so that the described LC-MS procedure is suitable to diagnose an intoxication.

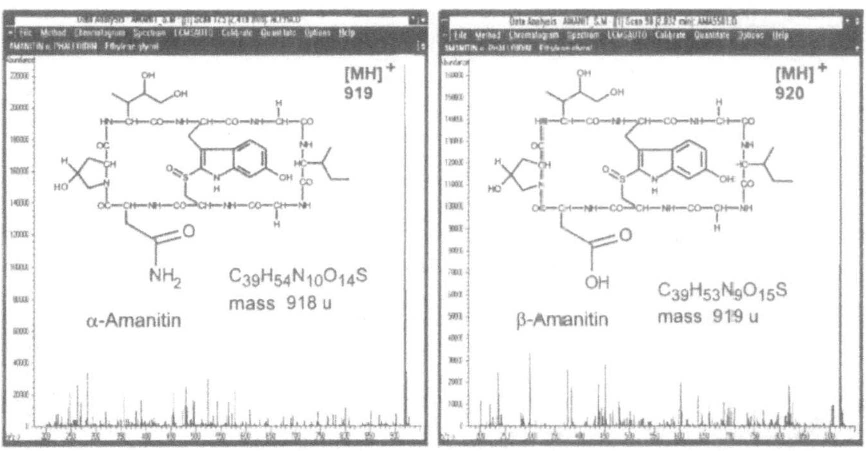

Fig. 4. Electrospray mass spectra, structures, empirical formulas and molecular masses of α- and β-amanitin.

Conclusions

The GC-MS procedure for quantification of morphine after PFP derivatization was very precise using the stable isotopes as internal standard and very sensitive using chemical ionization in the negative mode (NCI). Such procedures are state-of-the-art today in clinical toxicology, legal medicine and doping control.

The GC-MS procedure for detection of basic and neutral drugs (STA) allows the precise and sensitive detection of most of the toxicologically relevant drugs like barbiturates and other sedative-hypnotics, anticonvulsants, benzodiazepines, antidepressants, phenothiazine and butyrophenone neuroleptics, central stimulants, designer drugs, phencyclidine, opioids, non-opioid analgesics, antihistamines, antiparkinsonians, beta-blockers, antiarrhythmics, laxatives and their metabolites.

The GC-MS procedure for detection of acidic drugs using extractive alkylation has proved to be suitable for simultaneous screening of acidic drugs like coumarin anticoagulants, ACE inhibitors, diuretics, non-steroidal antirheumatics and oral antidiabetics in urine as well as in plasma and gastric contents.

The LC-MS procedure allows the precise and sensitive detection of α- and β-amanitin in urine, thus allowing the specific diagnosis of an intoxication with amanita mushrooms.

These examples show that GC-MS is the method of choice for precise and sensitive identification and quantification of low mass xenobiotics volatile in GC while ESI-LC-MS is that for the detection of non-volatile xenobiotics in biosamples.

References

Arlt JW, Maurer HH (1996) On the toxicological detection of 4-hydroxycoumarin anticoagulants and their metabolites in urine by GC-MS after extractive alkylation. Proceedings of the 32nd Intern. Meeting of TIAFT. Interlaken, August 1996, in press

De la Torre R, Ortuno J, Gonzalez ML, Farre M, Cami J, Segura J (1995) Determination of cocaine and its metabolites in human urine by GC/MS after simultaneous use of cocaine and ethanol. J Pharm Biomed Anal 13:305-312

De Zeeuw RA (1989) Modern chromatographic procedures in systematic toxicological analysis. J Chromatogr 488:199-213

Ensslin HK, Kovar K-A, Maurer HH (1996a) Toxicological detection of the designer drug 3,4-methylenedioxyethylamphetamine (MDE, "Eve") and its metabolites in urine by FPIA and GC-MS. J Chromatogr B 683:189-197

Ensslin HK, Maurer HH, Gouzoulis E, Hermle L, Kovar K-A (1996b) Metabolism of racemic MDE in humans. Drug Metabol Dispos 24, in press

Gonzalez G, Ventura R, Smith AK, De la Torre R, Segura J (1996) Detection of anti-inflammatory drugs in plasma and urine by GC-MS. J Chromatogr A 719:251-264

Maurer HH (1992a) Review: Systematic toxicological analysis of drugs and their metabolites by gas chromatography-mass spectrometry. J Chromatogr 580:3-41

Maurer HH (1992b) Methods for GC-MS. In: Pfleger K, Maurer HH, Weber A (eds), Mass spectral and GC data of drugs, poisons, pesticides, pollutants and their metabolites, 2nd ed, part 1. VCH publishers, Weinheim, pp 3-32

Maurer HH (1996) On the metabolism and the toxicological analysis of designer drugs. Ther Drug Monit 18:465-470

Maurer HH, Kraemer T (1992) Toxicological detection of selegiline and its metabolites in urine using fluorescence polarisation immunoassay (FPIA) and GC-MS and differentiation by enantioselective GC-MS of the intake of selegiline from an abuse of methamphetamine or amphetamine. Arch Toxicol 66:675-678

Maurer HH, Kraemer T, Ledvinka O, Schmitt CJ, Weber AA (1996) Gas chromatography- mass spectrometry (GC-MS) and liquid chromatography-mass spectrometry (LC-MS) toxicological analysis. Studies on the detection of clobenzorex and its metabolites within a systematic toxicological analysis procedure by GC-MS and by immunoassay, and studies on the detection of α- and β-amanitin by atmospheric pressure ionization electrospray LC-MS. J Chromatogr. B, in press

Moeller MR, Fey P, Rimbach S (1992) Identification and quantification of cocaine and its metabolites, benzoylecgonine and ecgonine methyl ester, in hair of Bolivian coca chewers by GC/MS. J Anal Toxicol 16:291-296

Pfleger K, Maurer HH, Weber A (1992) Mass spectral and GC data of drugs, poisons, pesticides, pollutants and their metabolites, 2nd ed. VCH publishers, Weinheim

Pfleger K, Maurer HH, Weber A (1997) Mass spectral library of drugs, poisons, pesticides, pollutants and their metabolites, 3rd ed. Hewlett Packard, Palo Alto, in prep.

Solans A, Carnicero M, De la Torre R, Segura J (1995) Comprehensive screening procedure for detection of stimulants, narcotics, adrenergic drugs, and their metabolites in human urine. J Anal Toxicol 19:104-114

Tagliaro F, Smith FP, Turrina S, Equisetto V, Marigo M (1996) Complementary use of capillary zone electrophoresis and micellar electrokinetic capillary chromatography for mutual confirmation of results in forensic drug analysis. J Chromatogr A, in press

Tracqui A, Kintz P, Mangin P (1995) Systematic toxicological analysis using HPLC/DAD. J Forensic Sci 40:254-262

Air Pollution and Allergy

(Chair: M. Lövik, Norway)

Pollution and the Development of Allergy: The East and West Germany Story

Thomas Nicolai and Erika von Mutius
Universitätskinderklinik, Lindwurmstr.4, D-80337 München, Germany

Summary

Allergic diseases are partly genetically determined, but environmental factors have a strong influence on the expression of allergic symptoms in genetically predisposed subjects. In particular, outdoor air pollution has received widespread attention as a potential manifestation factor. The unification of Germany provided a unique opportunity to study the impact of radically different environmental and social conditions on the development of allergies in two genetically homogenous populations. A high car density and NO_2 exposure were typical for many West German cities. Severe pollution due to heavy industrialization and private coal burning for heating purposes were the main sources of air pollution in East German cities.

We assessed the prevalence of asthma and allergic disorders in 9-11 year old children in in East Germany (Leipzig and Halle) and in West Germany (Munich). All fourth grade pupils in Munich (n=7,445) were compared with those in Leipzig and Halle 1991 (n=3,105). Hay fever, skin test reactivity to common aeroallergens and asthma were considerably more prevalent in West Germany as compared to East Germany. When atopy was taken into account, there was no longer a significant difference in the prevalence of asthma between the two parts of the country.

Introduction

Allergic diseases cluster in families (Dold 1992), and this indicates a genetic background. However, allergies are very variable in their symptomatology and in the severity of their manifestations even within families. Studies in twins (Edfors-Lubs 1971) have shown that concordance among monozygotic twins is relatively low when compared to that of other genetically determined illnesses. Therefore, environmental factors must influence the expression of allergic diseases in genetically predisposed subjects.

Exposure to air pollutants (SO_2, particulate matter, vehicle exhausts, ozone) has been suspected as risk factor for allergic respiratory disorders (Dockery 1989, Ware 1986). Not all studies have confirmed these suspected effects of outdoor pollution. The concomitant presence of different pollutants in western urban areas has made it difficult to attribute specific health effects to specific substances. Sometimes it is impossible to differentiate the effects of environmental exposure from genetic differences between populations exposed or unexposed to the pollutants studied.

The studies reported here were initiated on the assumption that the unification of Germany provided a unique opportunity to study the impact of different environmental conditions on the expression of allergies in a genetically homogenous population. For the 45 years between the end of the war and the fall of the Berlin wall, there was almost no significant demographic exchange between both sides of Germany. Private car ownership and NO_2 exposure were characteristic for West German cities, heavily polluting industrialization, few private cars, and use of lignite for heating was typical of East Germany.

The objective of this study was to compare the prevalence of allergic sensitization and diseases between children in East and West Germany. If the differences in environmental conditions contribute to expression of allergic disorders, significant differences in the prevalence of allergic sensitization and allergic diseases between the two sides of the country should be found.

Methods

All fourth grade pupils (n=7,445, response rate 87.0%) of all primary schools in Munich were included in a first study in 1989/90. All schoolchildren (n=3,105, response rate 84.8%) attending classes of the fourth grade of a random sample of 39 schools in Leipzig and of 23 schools in Halle were studied in East Germany in 1991/92. To keep the study populations as similar as possible with regard to ethnic background, the analysis was restricted to German children.

An identical self-administered questionnaire was distributed to the parents of the children in all study areas. The questionnaire included questions concerning sociodemographic characteristics, physician's diagnoses and typical symptoms of respiratory and allergic disorders and their possible etiologic factors. Atopic sensitization was assessed by skin prick tests to 6 common aeroallergens using an identical multitest device in all 3 study areas (Berbig 1991). Bronchial hyperresponsiveness (BHR) was assessed by cold air hyperventilation challenge (Nicolai 1993, v. Mutius 1994).

Results

The lifetime prevalence of asthma was higher in West Germany (9.3%) than in East Germany (7.2%, $p<0.05$). When only current asthma with symptoms in the past 12 months was considered, the difference between both study areas (5.9% versus 3.9%, $p<0.0005$) was even larger. However, the lifetime prevalence of recurrent wheezing (26.8% versus 17.0%, $p<0.0005$), nocturnal cough (6.1% versus 3.4%, $p<0.0005$) and frequent cough with exercise or during foggy or cold weather (14.2% versus 11.7%, $p<0.05$) was higher in East Germany when compared with West Germany. Attacks of shortness of breath were reported with similar frequency in both study areas. Lifetime prevalence of physician diagnosed bronchitis in East Germany was increased when compared with West Germany (33.7% versus 15.9%, $p<0.0005$). Hay fever was reported more frequently in West Germany than in the eastern part of the country (8.6% versus 2.7%, $p<0.0005$).

Atopic sensitization to mites (10.3% versus 4.2%, $p<0.0005$), grass pollen (21.3% versus 7.9%, $p<0.0005$) and cats (7.3% versus 2.9%, $p<0.0005$) as assessed by skin prick tests were all significantly more frequent in children in West Germany. No significant difference in baseline measurements of pulmonary function could be demonstrated in West and East German children. The prevalence of BHR was significantly higher in West Germany when compared to the eastern part of the country (v. Mutius 1994).

The relationship of both, asthma and bronchitis to skin test reactivity, BHR, and study area was further assessed in a multivariate logistic regression analysis (v. Mutius 1994). In a first step, age, sex and the study area were entered in the model. The male sex as well as the West German study area were significant risk factors for the development of asthma, whereas only the East German study area was significantly associated with a diagnosis of bronchitis. After introducing sensitization to mites, pollen, cats or dogs in the model, the West German study area was no longer significantly associated with the higher prevalence of asthma. The East German study area, however, remained a highly significant risk factor for the development of bronchitis. BHR was a strong risk factor for asthma but not for bronchitis in either area.

In each study area, the prevalence of atopic sensitization decreased significantly with increasing numbers of siblings (v. Mutius 1994). This relation remained significant when controlling for a family history of asthma or atopy, sex, education, smoking, the presence of pets at home, the age of the subject, the month of testing and the study area.

High rates of road traffic impaired pulmonary function and increased respiratory symptoms in children (Wjst 1993) in Munich. However, the prevalence of allergic rhinitis and asthma was not associated with increasing car traffic. The decrease seen in peak flow rates was comparable with the effect of passive smoke exposure in the same population (Dold 1992).

Discussion

The prevalence of hay fever and skin test reactivity to common aeroallergens is considerably higher in West Germany as compared to East Germany. Furthermore, the prevalence of asthma was also higher in the West German study area (v. Mutius 1992, 1994) as has been confirmed by other studies (Krämer 1992). However, when atopy was taken into account, there was no longer a significant difference in the prevalence of asthma between the two parts of the country. Increasing exposure to car traffic was associated with diminished lung function and an increased prevalence of unspecific respiratory symptoms, but not of asthma and allergic rhinitis (Wjst 1993) or skin test positivity.

In West Germany the prevalence of atopic sensitization to most aeroallergens was 2.5 - 3 times higher when compared to the eastern part of the country. Another study (Krämer 1991) found less six year old children in the East to have positive RAST results for house dust mite or pollen compared to the same age group studied in the West.

Our findings suggest that outdoor air pollution by SO_2, particulate matter and vehicle exhausts (v. Mutius 1992, 1994, Wjst 1993) is not a strong determinant for the development of atopy. Measurements of house dust mite allergen in East and West German homes, however, point towards a higher mite exposure in the East (Leupold 1994). However, East German children reacted less frequently to house dust mites than West German children. Similar findings are reported from other surveys in cities (Kahle 1992, Krämer 1991): 6.8% of the children in Halle compared to 13.7% in Cologne and 13.2% in Düsseldorf were RAST-positive for house dust mite.

The prevalence of bronchitis, wheezing and cough was higher in East Germany than in the western part of the country. This is in agreement with another study (Krämer 1992) comparing six year old children in Halle and Leipzig in East Germany with Duisburg and Düsseldorf in West Germany. Atopy and BHR were not related to a diagnosis of bronchitis in East Germany. Leipzig and Halle are heavily polluted by SO_2 and particulate matter. An increased prevalence rate of cough has been observed among children intermittently exposed to very high concentrations of SO_2 at moderate levels of particulate SO_4 (Dodge 1985). The American Six Cities Study reported a positive correlation of the prevalence of bronchitis and chronic cough with exposure to particulate matter (Dockery 1989). However, no association between the prevalence of persistent wheeze or asthma and exposure to particulate matter, NO_2 or SO_2 was observed. Our findings of an increased prevalence of bronchitis and frequent cough in Leipzig, with higher levels of SO_2 and particulate matter than in Munich, thus confirm earlier reports of this assocation, although it does not allow to disentangle the effect of both factors. In turn, the higher prevalence of asthma and BHR to cold air challenge in West Germany than in the eastern part of the country, provides additional evidence that longterm exposure to SO_2 , even at the observed very high levels, and particulate matter does not affect BHR and asthma.

Our results suggest that the difference in the prevalence of asthma between West and East Germany is attributable to the difference in the prevalence of atopic sensitization between the two parts of the country. Some reports (Burr 1989, Ninan 1992, Robertson 1991) have shown an increase in the prevalence of childhood asthma in the last decades. Similar increases have been observed for the prevalence of hay fever (Ninan 1992, Wüthrich 1986) and eczema (Ninan 1992, Taylor 1984), and a causal link has been suggested. Our findings support the hypothesis that increased atopic sensitization is a strong determinant for the increase in the prevalence of asthma observed in western countries in recent years.

All study populations comprised German children sharing a homogenous genetic background. Therefore, differences in the prevalence of diseases between both parts of Germany reflect environmental influences. Exposure to outdoor air pollution by SO_2, particulate matter and vehicle exhausts was associated with bronchitis, unspecific respiratory symptoms and decrements in pulmonary function, but not with asthma and atopy. Conversely, factors associated with West German living conditions may induce the development of allergic sensitization. Atopy was found to be a major determinant for the inception of asthma.

References

Berbig B, von Mutius E, Nicolai T, Stiepel E, Adam D (1991) Comparing the Multitest to a conventional skin prick test (in German). Allergologie 14:51-57

Burr ML, Butland BK, King S, Vaughan-Williams E (1989) Changes in asthma prevalence: two surveys 15 years apart. Arch Dis Child 64:1452-1456

Dockery DW, Speizer FE, Stram DO, et al (1989). Effects of inhalable particles on respiratory health of children. Am Rev Respir Dis 139:587-594

Dodge R, Solomon P, Moyers J, Hayes C (1985) A longitudinal study of children exposed to sulfur oxides. Am J Epidemiol 121: 720-736

Dold S, Wjst M, von Mutius E, Reitmeir P, Stiepel E (1992) Genetic risk for asthma, allergic rhinitis, and atopic dermatitis. Arch Dis Child 67:1018-1022

Dold S, Reitmeir P, Wjst M, von Mutius E (1992) Effects of passive smoke exposure on the respiratory tract in children (in German). Monatsschr Kinderheilkd 140:763-768

Edfors-Lubs ML (1971) Allergy in 7,000 twin pairs. Acta Allergol 26:249-285

Kahle S, Behrendt H, Ring J, Vieluf D, Hinrichs J, Krämer U, Dolgner R, Schlipköter HW (1992) Epidemiologische Untersuchungen zur Prävalenz allergischer Erkrankungen bei 6jährigen Kindern in Halle (Sachsen-Anhalt). Allergologie 15: 58

Krämer U, Altus C, Behrendt H, Dolgner R, Gutsmuths FJ, Hille J, Hinrichs J, Mangold M, Paetz B, Ranft U, Röpke H, Teichmann S, Willer HJ, Schlipköter HW (1992) Epidemiologische Untersuchungen zur Auswirkung von Luftverschmutzung auf die Gesundheit von Schulanfängern. Forum Städte Hygiene 43: 82-87

Krämer U, Behrendt H, Dolgner R,Kainka-Stänicke E, Oberbarnscheidt J, Sidaoui H, Schlipköter HW (1991) Auswirkungen der Umweltbelastung auf allergologische Parameter bei 6jährigen Kindern. In: Ring J (ed) Epidemiologie allergischer Erkrankungen. MMV-Verlag München, pp 165-178

Leupold W, Walther K-U, Wahn U (1994) Vorkommen von HDM-Antigen (Der pI/Der fI) in Wohnungen in Relation zu Umgebungs- und Krankheits-Faktoren. Abstract at the "16. Jahrestagung der Gesellschaft für Pädiatrische Pneumologie", München
von Mutius E, Fritzsch C, Weiland SK, Roell G, Magnussen H (1992) Prevalence of asthma and allergic disorders among children in united Germany: a descriptive comparison. Br Med J 305:1395-1399
von Mutius E, Martinez FD, Fritzsch C, Nicolai T, Roell G, Thiemann HH (1994) Prevalence of asthma and atopy in two areas of West and East Germany. Am J Respir Crit Care Med 149:358-364
von Mutius E, Martinez FD, Fritzsch C, Nicolai T, Reitmeir P, Thiemann HH (1994) Skin test reactivity and number of siblings. Br med J 308:692-695
Nicolai T, von Mutius E, Reitmeir P, Wjst M (1993) Reactivity to cold air hyperventilation in normal and asthmatic children in a survey of 5697 school children in Southern Bavaria. Am Rev Resp Dis 147:565-572
Ninan T, Russell G (1992) Respiratory symptoms and atopy in Aberdeen schoolchildren: evidence from two surveys 25 years apart. Br Med J 304:873- 875
Robertson CF, Heycock E, Bishop J, Nolan T, Olinsky A, Phelan PD (1991) Prevalence of asthma in Melbourne schoolchildren: changes over 26 years. Br Med J 302:1116-1118
Taylor B, Wadsworth J, Wadsworth M, Peckham C (1984) Changes in the reported prevalence of childhood eczema since the 1939-45 war. Lancet i:1255- 1257
Ware JH, Ferris BG Jr., Dockery DW, Spengler JD, Stram DO, Speizer FE (1986) Effects of ambient sulfur oxides and suspended particles on respiratory health of pre-adolescent children. Am Rev Respir Dis 133:834-842
Wjst M, Reitmeir P, Dold S, Wulff A, Nicolai T, von Loeffelholz-Colberg E, von Mutius E (1993) Road traffic and adverse effects on respiratory health in children. Br Med J 307:596-600
Wüthrich B, Schneyder UW, Henaue SA, Heller A (1986) Incidence of pollinosis in Switzerland. Results of a representative demographic survey with consideration of other allergic disorders (in German). Schweiz Med Wochenschr 116:909

Nasal Lavage Biomarkers in Air Pollution Epidemiology

PA Steerenberg[1], PH Fischer[2], L van Bree[3] and H van Loveren[1]
[1] Laboratory for Pathology and Immunobiology,
[2] Department of Chronic Diseases and Environmental Epidemiology,
[3] Laboratory of Health Effects, National Institute of Public Health and the Environment, P.O. Box 1, 3720 BA Bilthoven, The Netherlands.

Introduction

Epidemiological studies have shown associations between air pollution and respiratory health effects (reviewed in Bascom et al., 1996). Adverse health effects at current levels of exposures have been found for ozone, acid aerosols and particulate matter smaller than 10 µm (PM10). Effects of ozone include pulmonary function decrements, increased non-specific bronchial responsiveness and inflammatory responses (Graham and Koren, 1990; Jorens et al., 1992; Koren et al., 1989). Data based on health effects associated with acute exposure to urban particulate air pollution is rapidly growing by consistent data from epidemiological studies (Bascom et al., 1996). These data indicate that increases in daily mortality, hospitalisation, asthma exacerbation, respiratory symptoms and lung function decline are associated with ambient low levels of urban particulate air pollution measured as PM10 and lower. Possibly, air pollution induced inflammatory reactions and related inflammatory mediators are involved in altered lung function and may have implications for respiratory allergy. It is possible that ozone induces mild to moderate immune disorders by shifting vulnerable balances in the immune system, such as T help-1 and T help-2 subset ratios, resulting in altered resistance to respiratory tract infections or altering the expression of respiratory allergy. Epidemiological studies focused on inflammatory responses include the use of biomarkers.

In clinical studies in which humans are exposed to ozone, the lower airway has been investigated by studying biopsies and bronchial lavage fluid (BAL) (Koren et al., 1989; Aris et al., 1993). These techniques cannot be applied in epidemiological studies with large human populations, because the techniques are invasive and very time consuming. Therefore, studies have been undertaken to develop non-invasive methods to detect immunological disorders in airways to be applied in epidemiological investigations. For this reason, biomarkers in nasal lavages (NAL) were used to correlate levels of ozone and inflammatory responses in the upper airways

The nose is the prime port of entry for inspired air, and therefore, the first region of the respiratory tract that comes into contact with airborne pollutants. It has been shown that in experimental studies in which volunteers were exposed to ozone and in epidemiological studies at elevated concentrations of photochemical air pollution the number of polymorphonuclear leukocytes (PMN's), like neutrophils, and to a lesser extent eosinophils has been increased in NAL (Frischer et al., 1993; Graham and Koren 1990; Seltzer et al., 1986), and in BAL (Graham and Koren, 1990). For this reason, an increase of the number of PMN's may be considered as a biomarker of effect for exposure to ozone.

Soluble biomarkers for ozone exposure present in the NAL have also been studied. So far, only albumin and uric acid concentrations after exposure to ozone were shown to be significantly increased in NAL and in BAL (Graham and Koren, 1990; Koren et al., 1989). An increased albumin or uric acid concentration is an indication for an increased permeability, while uric acid secretion is also a measure for scavenger activity. Histamine, eosinophilic cationic protein (ECP) and myeloperoxidase (MPO) can be used as markers of increased ozone exposure as well (Frischer et al., 1993). No studies have been presented yet in which a comparison is made between NAL and BAL, with respect to these mediators. In general, cytokines have so far not been measured in epidemiological studies of air pollution in NAL, although they play an important role during inflammatory reactions. This may change since studies with human epithelial cells isolated from the nasal mucosa show increased production of interleukins (e.g. IL-6 and IL-8) upon exposure to ozone (Devlin et al., 1994).

The goal of this review is a summing-up of results obtained with nasal lavage in experimental and epidemiological studies to determine the effects of air pollution. Compared to studies on ozone, studies on urban particulate air pollution are limited and therefore most of the reviewed data is based on ozone studies. As the number of studies on particular matter will be growing rapidly in the next years, the feasibility of nasal lavages will investigated also for this goal. Directions will be given in which this line of investigation may develop.

Methods of nasal lavage

Depending on the scope of the study for nasal lavages a number of techniques is available. In experimental settings in which small groups of volunteers or patients are subject of the study, intensive lavages can be carried out. However, when large groups are monitored on the same day, simple and reproducible methods are required. From 1870 on, nasal lavages have been used. At that time this technique was used for cleansing the nose, whereby the fluid was introduced through one nostril and was run out through the other nostril. For the purpose of analyst markers of inflammatory responses, modifications have been made. One of the most simple and reproducible technique was first described by

Graham and Koren (1990). With this method the handling of the fluid and the apparatus is in hands of the field worker. In our epidemiological studies in schoolchildren we found this technique very reliable and easy to handle under the field conditions in which 40-50 children have been lavaged (Fisher et al., submitted) . In short: Subjects need only a small training to become acquainted with the NAL-procedure. Sitting volunteers bend the neck 45° backwards and elevate the palate to close their nasopharynx. Then 10 ml of sterile phosphate-buffered saline without Ca^{2+} and Mg^{2+} , heated to 37°C, is instilled (for schoolchildren at the age of 7-10 years, 8 ml), 5 ml in each nostril with a 10 ml polystyrene pipette. The saline is held in the nasopharyngeal region for 10 sec and then the volunteer flexes the head and lets the washing solution flow into a funnel attached to a centrifuge tube (15 ml). To avoid mucus in the lavage fluid, the funnel outlet is taped with a polyamide gauze (100 mesh). The samples are centrifuged (250 g/10 min at 4°C) and the supernatant frozen at -80°C for later analysis. Part of the cells are formaldehyde (4%) fixed and cytospin preparations are made of the remaining cells, which are stained later on with the May-Grünwald-Giemsa method for cell differentiation. For differential cell counting only cytospin preparations are included when a minimum of 50 cells are found in five microscopic fields (x400). Viable cells, determined by trypan blue (0.5%) exclusion, are counted and calculated for the number of total cells per lavage (excluding large epithelial cells) as well as the number of cells /ml of recovered fluid. In this way the average amount of fluid recovered is 6.6 ± 1.3 ml (Steerenberg et al., 1996). One drawback of this technique is the low yield of leukocytes and cytokines in the lavage fluid.

A more advanced method is that of the group of Greiff et al. (1990), in which a compressible plastic container "nasal pool" (NP) is used. This device has a nasal adapter, or nozzle, attached to its upper ending. The nasal adapter is inserted into the nasal opening of the subject sitting in a forward-flexed neck position (60°) from the upright position. The NP device , filled with 14 ml fluid is fully compressed by the subject. Approximately 12 ml fluid is thus introduced into one nostril. The instillate can be maintained in the nasal cavity as long as the container is compressed and the subject feels comfortable, which may be for 15 min or even longer.

A similar but more labour-intensive method is introduced by the group of Peden (Noah et al., 1995), in which the nasal lavage is performed using a disposable metered-dose nasal inhaler filled with sterile, room-temperature normal saline solution. This device is actuated 5 times in one nostril while occluding the other nostril, then the subject is instructed to exhale through the lavaged side into a specimen cup. Total input of saline solution is approximately 7 ml, and median return is 37% for normal subjects.

A very elegant method is the use of a urethral catheter provided with an inflatable balloon. The balloon is inserted into the nostril, and gently inflated by a syringe. While the subject is sitting in writing position, a syringe is filled with 10 ml of pre-warmed Hanks' balanced salt solution and the fluid is instilled by gently compressing the syringe. After 5 min the fluid is recovered by retracting the syringe. This procedure is carried out in each nostril (Grünberg et al., in press).

Biomarkers measured in NAL

Parameters that are assayed in nasal lavage fluid include cells and a number of (immune) mediators and exudate markers.

Cells: After nasal lavage in a normal population the average proportion of neutrophils is 88.8 ± 24% of all leukocytes. The average proportion of eosinophils is 11.1 ± 24% and for lymphocytes 0.2 ± 0.6%. (Steerenberg et al., 1996). In asthmatic and allergic subjects eosinophilic and mast cells will be more prominently present (Bascom et al., 1996; Noah et al., 1995; Juliusson et al., 1991). In all studies PMN's correlated with increased ozone exposure (Graham and Koren, 1990; Koren et al., 1989; Frischer et al., 1993; Graham et al., 1988; Fischer et al., submitted), and can therefore be considered as a measure for a non- specific inflammatory reaction.

Myeloperoxidase (MPO). The granules of neutrophil contains a large number of different enzymes. One of them is MPO, which is present in the primary granules. The MPO-H_2O_2-halide system is toxic to many organisms including bacteria, fungi, viruses, mycoplasmata and parasites (Klebanoff and Clark, 1978), as well as to many cells of the host (Klebanoff et al., 1976). About 95% of the cellular MPO in the circulation is present in the neutrophils whereas the remaining MPO is contributed by the less numerous monocytes (Bos et al., 1978). It is expected that by activation of the neutrophils degranulation will cause an increased MPO concentration in NAL. Moreover, the concentration of MPO correlated with the number of leukocytes in NAL (Steerenberg et al., 1996) and increased after experimental (Graham and Koren, 1990) and ambient exposure (Frischer et al., 1993) to ozone.

Eosinophilic cationic protein (ECP). Eosinophilic granulocytes may be activated locally at the site of the inflammatory reaction. After stimulation they are capable of releasing highly toxic proteins e.g. ECP that are able to disrupt the integrity of the epithelial lining of the airways (Svensson et al., 1990). It is not clear whether ECP is produced locally or whether it is due to the leakage of plasma levels of ECP (Klementsson et al., 1990). In general, high levels of ECP are in accordance with high numbers of eosinophilic granulocytes in allergic patients (Svensson et al., 1990; Togias et al., 1988). Moreover, the ECP concentration is increased after elevated ambient ozone concentrations (Frischer et al., 1993).

Tryptase. This tetrameric neutral protease is found in basophilic granulocytes and mast cell secretory granules (Juliusson et al., 1991; Castells and Schwartz, 1988) and together with histamine it is released during degranulation (Schwarz et al., 1981) Tryptase is absent in NAL of healthy subjects (Juliusson et al., 1991; Castells and Schwartz, 1988), but it seems that tryptase is detectable immediately after ozone exposure (Koren et al., 1989).

Interleukin 8 (IL-8). This interleukin belongs to the super family of host defence molecules characterised by potent neutrophil-activating properties *in vitro* (Van Damme, 1991). In animal studies intradermal injection of picomolar amounts of IL-8 induces neutrophil and neutrophil-dependent plasma protein extravasation (Rampart et al., 1989). IL-8 can be produced in vitro by several cell types including macrophages, lymphocytes, neutrophils, fibroblasts and endothelial cells in response to a variety of stimuli such as endotoxin, viruses, other cytokines and ozone (Jorens et al., 1992; Bazzoni et al., 1991; Van Damme et al., 1988). IL-8 has been implicated in a wide variety of pathological conditions characterised by neutrophil infiltrations. For instance, IL-8, was present in samples from patients with pulmonary fibrosis (Carré et al., 1991) and adult respiratory distress syndrome (Hoek et al., 1993). Moreover, IL-8 challenge in the nasal mucosa in atopic and non-atopic subjects induces a significant increase of neutrophils and eosinophils infiltration, indicating that IL-8 is a powerful polymorphonuclear leukocyte chemoattractant in mucosa (Douglass et al., 1994).

Interleukin 6 (IL-6). This is a multifunctional cytokine produced by a variety of cells, including T-lymphocytes, macrophages, neutrophils, eosinophils, fibroblasts, endothelial cells and mast cells (Snick, 1990; Juliusson et al., 1991). IL-6 has been found to play a central role in inflammatory reactions. In an immortalised human airway epithelium cell line, IL-6 production was increased after ozone exposure (Devlin et al., 1994)

Exudate markers. Small changes in permeability of membranes by ozone may increase the concentrations of low molecular compounds (uric acid (M= 168) and urea (M = 60)). Larger changes may increase the concentration of macromolecules such as albumin (M = 65,000, Koren et al., 1989). Moreover, it is suggested that uric acid is probably the most important scavenger of ozone in human blood plasma and that uric acid secretion by the airway mucosa may serve as an important protective response to respiratory ozone exposure (Peden et al., 1993). In a number of studies increased albumin concentration correlated with ozone exposure (Graham and Koren, 1990).

Nitric oxide (NO). NO is produced by many cells within the respiratory tract and endogenous NO may play an important signalling role in the physiological control of airway function and in the pathophysiology of airway diseases (Barnes and Kharitonov, 1996). Endogenous NO is generated from the amino acid L-arginine by the enzyme NO synthase, of which three distinct isoforms exist. The concentration of exhaled NO is increased in patients with inflammatory diseases of the airways such as asthma, cystic fibrosis, and lower respiratory tract infection. Furthermore it has been hypothesised that NO regulates the T helper 1 and T helper 2 balance in favour of T helper 2 type IgE mediated allergic reactions. Besides its presence in exhaled air, NO can also be measured in nasal lavages (Barnes and Lieuw, 1995; own observations).

Experimental and epidemiological studies

In humans, studies concerning the effects of air pollutants on the immunological responses in the the respiratory mucosa have certain limitations. In controlled human exposure fibre-optic bronchoscopies with bronchoaveolar lavage (BAL) and forceps biopsies from the airway mucosa in the more central airways have been proven safe procedures. Transbronchial lung biopsies, however, are associated with certain risks of pneumothorax and bleeding. Much more limitations are present when epidemiology is the subject of investigation. The number of participants and the invasiveness and laboriousness of the BAL technique make the feasibility of this technique unlikely. The possibility of conducting epidemiological studies on the immune effects of air pollution became, however, feasible with the studies of Graham et al. (1988) when they demonstrated increases in neutrophil counts in NAL following exposure to 0.5 ppm ozone for 4 h on two consecutive days. In these experiments the increase persisted for 22 h. In a follow-up study by Graham and Koren (1990), the exposure level of ozone was 0.4 ppm for 2 h, measuring NAL immediately following exposure and again prior to the BAL 18h postexposure. The number of neutrophils increased in NAL and BAL; however, no quantitative correlation was found following ozone exposure. Furthermore, it was demonstrated that ozone caused an increase in tryptase and albumin. These first studies comparing NAL and BAL after exposure to ozone indicated that effects observed in NAL resemble partly the situation in the lower respiratory tract. It has been studied whether the number of recovered cells and the concentration of cytokines in normal, asthmatic and allergic non-asthmatic school-aged children differ (Noah et al., 1995). It was found that the total cell count in the nasal fluid did not differ significantly between the three groups. Eosinophils were present in greater numbers in lavages from allergic asthmatic children than in lavages from normal children; a similar trend for increased eosinophils in the allergic non-asthmatic group compared with the normal group was noted, but this difference was statistically not significant. No statistical differences were found for albumin and histamine. IL-8 and ECP concentrations differed significantly in asthmatics compared with both normal and allergic non-asthmatics. Overall, IL-8 in nasal fluids correlated significantly with ECP. Allergic non-asthmatics did not have significant increases in cytokines or other mediators compared to normal subjects (Noah et al., 1995).

The increased utilisation of NAL initiated studies to investigate the kinetics of cell washout (Hauser et al., 1994) and the inter- and intraindividual variability in lavage cells counts and mediators (Steerenberg et al., 1996; Hauser et al., 1994). In the study of Hauser et al., (1994) the baseline lavage was followed by a lavage 48 later and in a second experiment the baseline lavage was followed by a lavage 72 hours later. For both studies the follow-up cell counts in any given individual did not consistently decrease or increase relative to baseline. Generally, it was found that the decrease in the number of cells was more evident at 48 hour than at 72 hours after the baseline lavage. To explain these results, it has to be taken into account that the physiological process of saline washing cells out of the nasal cavity may lead not only to lower counts, but may cause a delayed increase in cell counts. On the other hand, absence of inflammatory cells, or the presence of saline induce an inflammatory response, leading to the recruitment of cells into the nasal mucosa.

In unexposed subjects a wide interindividual variability in lavage cell counts ranging from $<10^4$ till 10^6 (Graham and Koren, 1990; Steerenberg et al., 1996) was observed. It was shown that in almost 50% of the subjects only up to 10^4 cells/ml were found.

Frischer et al. (1993) conducted an epidemiological study in schoolchildren in which the effects of high ambient ozone levels ($\geq$ 180 μg/m³) on the immune mediators in the NAL were compared with the levels of immune mediators in the NAL during low ozone concentrations (daily maximum $\leq$ 140 μg/m³). A significant increase of PMN's from low ozone days (mean 20.3×10^3) to high ozone days (mean 27.4×10^3; $p < 0.01$) was observed. Moreover, a significant increase was found for ECP and MPO. Thus it was found that ozone at ambient concentrations initiates a reversible inflammatory response of the upper airway in normal children.

For neutrophil counts the intra-individual variation is high. Fischer et al., (submitted) found a coefficient of variation (CV) of 98% using minimal 5 subsequent lavages, and, transferring the leukocyte counts in the Hauser et al. (1994) study to CV-values, a percentage of 68% is found for minimal 2 subsequent lavages (see Table 1).

Table 1 Mean individual coefficient of variation (CV) of nasal lavage parameters in children with a minimum of 5 nasal lavages

Nasal lavage parameter	Geometric mean	CV (%)	N
Neutrophils (ml^{-1})	8197	98	53
MPO (μg/l)	68	69	70
Albumin (mg/l)	80	15	66
Uric acid (mmol/l)	5	50	65
ECP (μg/l)	5	55	70
IL-8 (pg/ml)	18	54	70

By using the CV value it is possible to calculate the power of a study for assessing health effects (Steerenberg et al., 1996). This was calculated for the effect of O_3 on PMN as found by Frischer et al. (1993). They found an increase of 26% PMN in nasal lavages per 100 mg O_3/m^3. When using this effect-estimate it was calculated that, with 6 repeated measurements in a sample of 35 children, CV values must be less than 70% in order to be able to detect effects with a power of about 90%. Of course this is a rough estimate since the calculation was based only on the PMN-counts and not on all the mediators. In addition, the study of Frischer et al. (1993) concerned children, while our CV values were calculated for adults. Nonetheless this result implies that, except for leukocytes and ECP (CV was above 70%), it is theoretically possible to detect ozone effects with the protocol (6 repeated measurements among 35 children and 26% increase per 100 mg O_3/m^3). To detect effects of such a magnitude on PMN's and ECP levels, additional measurements or more children are needed. In the children study of Frischer et al. (1993) it was shown that increased ambient ozone levels elevate the number of PMNs, and the concentrations of MPO, ECP, and albumin.

Preliminary results of a study in The Netherlands show also an association between NAL parameters and ozone levels of the previous day for total cells, uric acid and IL-8. These associations were not very strong, which was probably due to the low ambient ozone concentration during the study (Fischer et al., submitted). In the same study the total cell count, neutrophils, albumin and IL-8 were negatively correlated with lung function.

In conclusion:

- NAL is easy to perform and includes biomarkers to measure the inflammatory response of the upper respiratory tract.
- At least 48 hours should be waited between repeated NALs to exclude the effects of cell washout.
- In allergic and asthmatic school-aged children the number of cells in NAL does not differ from that of normal children, but IL-8 and ECP concentrations may be increased.
- It has been shown that effects of ambient photochemical air pollution can be monitored by cells and different mediators found in nasal lavages. Whether this is also the case for particulate matter is the subject of a recent study.

References

Aris RM, Christian D, Hearne PC, Kerr K, Finkbeiner WE, and Balmes JR (1993) Ozone-induced airway inflammation in human subjects as determined by airway lavage and biopsy. Am Rev Respir Dis 148:1363-1972

Barnes C Lieuw FY (1995) Nitric oxide and asthmatic inflammation. Immunol Today 16:128-30

Barnes PJ, Kharitonov SA (1996) Exhaled nitric oxide: a new lung function test. Thorax 51:233-237

Bascom R, Bromberg PhA, Costa DL, Devlin R, Dockery DW, Frampton MW, Lambert JM, Speizer FE, Utell M (1996) Health effcts of outdoor air pollution. State of the art. Am J Res Crit Care Med 153:3-50
Bazzoni F, Cassatella MA, Rossi F, Ceska M, Dewald B, Baggiolini M (1991) Phagocytosing neutrophils produce and release high amounts of the neutrophil-activating peptide 1/interleukin 8. J Exp Med 173:771-774
Bos A, Wever R and Roos D (1978) Characterization and quantification of the peroxidase in human monocytes. Biochim Biophys Acta 525:37-44
Carré PC, Mortenson RL, King TE, Noble PW, Sable CL, Riches DWH (1991) Increased expression of the interleukin-8 gene by alveolar macrophages in idiopathic pulmonary fibrosis. J Clin Invest 88:1802-1810
Castells M, Schwartz L (1988) Tryptase levels in nasal lavage fluid as an indicator of immediate allergic response. Allergy Clin Immunol 82:348-355
Devlin RB, McKinnon KP, Noah T, Becker S, Koren HS (1994) Ozone induced release of cytokines and fibronectin by alveolar macrophages and airway epithelial cells. Am J Physiol 266:612-619
Douglass JA, Dhami D, Gurr CE, Bulpitt M, Shute JK, Howarth PH, Lundley IJ, Church MK, and Holgate ST (1994) Influence of interleukin-8 challenge in nasal mucosa in atopic and nonatopic subjects. J Respir Crit Care Med 150:1108-1113
Fischer PH, Steerenberg PA, Geerse L, Mol DQ, Nell JA, Ameling C, Van Bree L, Van Loveren H and Lebret E (1996) Associations between ozone, lung function and biomarkers in nasal lavage in schoolchildren. Submitted
Frischer ThM, Kuehr J, Pullwitt A, Meinert R, Forster J, Studnicka M, Koren H (1993) Ambient ozone causes upper airways inflammation in children. Am Rev Respir Dis 148:961-964
Graham DF, Handerson F, House D (1988) Neutrophil influx measured in nasal lavages of humans exposed to ozone. Arch Environmental Health 43:228-233
Graham DE, Koren HS (1990) Biomarkers of inflammation in ozone-exposed humans. Comparison of the nasal and bronchoalveolar lavage. Am Rev Respir Dis 142:152-156
Greiff L, Pipkorn U, Alkner U, Persson CGA (1990) The nasal pool device applies controlled concentrations of solutes on human nasal airway mucosa and samples its surface exudations/secretions. Clin Exp Allergy 20:253-259
Grünberg K, Timmers MC, Smits HH, de Klerk EPA, Dick EC, Spaan WJM, Hiemstra PS, Sterk PJ (1996) Effect of experimental rhinovirus 18 colds on airway hyperresponsiveness to histamine and interleukin-8 in nasal lavage in asthmatic subjcts in vivo. Clin Exp Allergy, in press
Hauser R, Garcia-Closas M, Ketsey KT and Christiani DC (1994) Variability by nasal lavage polymorphonuclear leukocyte counts in unexposed subjects: its potential utility for epidemiology. Arch of Environ Health 49:267-272
Hoek G, Fischer P, Brunekreef B, Lebret E, Hofschreuder P, Mennen MG (1993) Acute effects of ambient ozone on pulmonary function of children in the Netherlands. Am Rev Respir Dis 147:111-117
Jorens PG, Van Damme J, Backer W de, Bossaert L, Jongh RF de, Herman AG, Rampart M (1992) Interleukin 8 (IL-8) in the bronchoalveolar lavage fluid from patients with the adult respiratory distress syndrome (ARDS) and patients at risk for ARDS. Cytokine 4:592-597
Juliusson S, Holmberg K, Baumgarten CR, Olsson M, Enander I, Pipkorn U (1991) Tryptase in nasal lavage fluid after local allergen challenge. Allergy 46:459-465
Klebanoff SJ, Clark RA and Rosen H (1976) Myeloperoxidase-mediated cytotoxicity. In: Schultz J and Ahmad F (eds) Cancer enzymology. Academic Press, New York, pp 267-288

Klebanoff SJ and Clark RA (1978) The neutrophil. Function and clinical disorders. Elsevier, North Holland Publ Co, Amsterdam, pp 35-72, 217-488

Klementsson H, Andersson M, Baumgarten CR, Venge P, Pipkorn U (1990) Changes in non-specific nasal reactivity and eosinophil influx and activation after allergen challenge. Clin Exp Allergy 20:539-547

Koren HS, Devlin RB, Graham DE, Mann R, McGee MP, Horstman DH, Kozumbo WJ, Becker S, House DE, McDonnell WF, Bromberg PA (1989) Ozone-induced inflammation in the lower airways of human subjects. Am Rev Respir Dis 139:407-415

Noah TL, Henderson FW, Henry MM, Peden DB, and Devlin RB (1995) Nasal lavage cytokines in normal, allergic, and asthmatic school-age children. Am J Resp Crit Care Med 152:1290-1296

Peden DB, Swiersz M, Ohkubo K, Hahn B, Emery B, Kaliner MA (1993) Nasal secretion of the ozone scavenger uric acid. Am Rev Respir Dis 148:455-461

Rampart M, Van Damme J, Zonnekeyn L, Herman AG (1989) Granulocyte chemotactic protein/interleukin-8 induces plasma leakage and neutrophil accumulation in rabbit skin. Am J Pathol 135:21-25

Schwarz LB, Lewis RA, Seldin D, Austen KF (1981) Acid hydrolases and tryptase from secretory granules of dispersed human lung mast cells. J Immunol 126:1290

Seltzer J, Bigby B, Stulbarg M, Holtzman M, Nadel J, Ueki I, Leikauf G, Goetzel E, Boushey H (1986) O_3-induced change in bronchial reactivity to methacholine and airway inflammation in humans. J Appl Physiol 60:1321-1326

Snick JJ van (1990) Interleukin-6: an overview. Ann Rev Immunol 8:253-278

Steerenberg PA, Fischer PH, Gmelig Meyling F, Willighagen J, Geerse E, Van der Vliet H, Ameling C, Boink ABTJ, Dormans JAMA, Van Bree L and Van Loveren H (1996) Nasal lavage as tool for health effect assessment of photochemical air pollution. Hum Exp Tox 15: 111-119

Svensson C, Andersson M, Persson C, Venge P. Alkner U, Pipkorn U (1990) Albumin, bradykinins, and eosinophil cationic protein on the nasal mucosal surface in patients with hay fever during natural allergen exposure. J Allergy Clin Immunol 85:828-833

Togias A, Naclerio RM, Proud D, Pipkorn U, Bascom R, Liopoulos O, Kagey-Sobotka A, Norman PS, Lichtenstein LM (1988) Studies on the allergic and non-allergic nasal inflammation. J Allergy Clin Immunol 81:782-790

Van Damme J. Interleukin-8 and related molecules (1991) In: Thomson AW (ed) The Cytokine Handbook: Immunology and Molecular Biology of Cytokines. London, Academic Press Ltd, pp 201-214

Prediction of Individual Susceptibility to Toxicants

(Chair: D. Ray, United Kingdom, and M. Repetto, Spain)

Physiological Factors Predisposing to Neurotoxicity

D. E. Ray
Medical Research Council Toxicology Unit, Hodgkin Building, Lancaster Road, Leicester, LE1 9HN, United Kingdom.

Abstract

Many factors determine individual susceptibility to toxic agents in addition to their primary interaction with the target site. Absorption, delivery to target tissues, bio-activation, bio-inactivation, elimination, and adaptive or protective responses all play important parts in determining the overall response of the individual. In addition changes in the physiological significance of the function which is disrupted may be crucially important.

Pulmonary absorption can be limited by ventilation or perfusion, both of which increase with work rate. Tissue uptake can be limited by local blood flow, which is strongly influenced by local functional activity. In areas with a blood-tissue barrier, such as brain and testis, tissue uptake can be strongly influenced by developmental state, protein binding or vascular damage. Metabolic transformation can show marked inter-individual variations at both hepatic and extra-hepatic sites, due to genetic or nutritional influences. The capacity for adaptation to toxicological insult can also vary markedly, depending on functional reserve capacity as well as on inherent plasticity.

Examples used to illustrate these factors include: the influence of motor activity on the toxicity of carbon monoxide; of noise on the ototoxicity of aminoglycoside antibiotics; of brain activity on the neurotoxicity of dinitrobenzene; of acid-base balance on the toxicity of nicotine; and of developmental stage on the neurotoxicity of haloperidol. In addition disease states can influence sensitivity. Thus anaemia sensitises to manganese; calcium deficiency to lead; nerve trauma to hexane; and Wilson's disease to copper overload.

Introduction

Biological variation ensures that no two individuals show exactly the same response to the same dose of a toxic agent. For the experimentalist this variability is a problem to be controlled by identification of the major factors determining susceptibility. For the physician this is a still larger problem, since all problems of ill health concern specific individuals, yet it is difficult to predict individual responses.

Most attention in this area has rightly been directed towards genetically determined individual differences in xenobiotic metabolism. These are major factors controlling vulnerability to many agents. However in this short review I will also point to some of the important physiological variables that contribute to determining individual susceptibility. Some of these variables also are genotypically determined, and have been reviewed (Festing, 1991) in relation to inbred experimental animals, each strain of which can be considered to represent an individual. Other variables are not genotypic, and relate more to the interaction of the individual with the environment. These interactions, although undoubtedly enhancing the vulnerability of particular individuals at a given time, are more likely to lead to episodes of vulnerability seen across whole groups sharing common occupations or activities, rather than in individuals sharing a similar genetic background.

Pharmacokinetics

Pharmacokinetic factors can profoundly influence toxicity by altering the effective dose at the target site, and many pharmacokinetic factors are in turn influenced by physiological factors. Rate of absorption can be strongly influenced by local blood flow. Pulmonary uptake is additionally influenced by ventilation rate, and J.B.S.Haldane made an early demonstration of this effect in showing that work rate had a profound influence on carbon monoxide poisoning, sedentary individuals showing a slower rise in carboxyhaemoglobin levels. Similar considerations apply to the uptake of xylene (Riihimaki et al., 1979) and other volatile solvents (Cohr, 1986). Skin and gastric absorption are similarly influenced by local blood flow. Gastrointestinal absorption can also be influenced by nutritional status (Goyer, 1995). Thus rats fed lead salts show greater increases in blood and brain lead if they are also on a low calcium diet. In man there is a similar inverse relationship between lead absorption and dietary calcium. Similar experimental relationships can be shown between zinc or iron and lead, and iron deficiency enhances cadmium absorption and toxicity in man (Seth and Chandra, 1988). The mechanistic basis for these interactions is presumably direct competition for uptake.

Tissue retention can also show marked individual differences, for example the rate of elimination of toluene varying five-fold between lean and obese subjects (Cohr, 1986). Co-exposure to loop diuretics can enhance the acute ototoxicity of aminoglycoside antibiotics, at least in part by enhancing local concentration in the target organ (Harpur, 1982; 1986). This example may also provide an illustration of synergy of action at two independent sites, since it is possible that the toxic action of loop diuretics at one site within the cochlea (the stria vascularis) summates with the damage produced by antibiotics at a second site (the hair cells) to give an overall functional defect that is greater than would be predicted from either cause alone.

The rate of elimination of agents can also show unexpected variability. Thus nicotine excretion in alkalosis is decreased to 25% of that seen in acidosis, due to the strong pH dependence of re-absorption from the bladder (Haag and Larson, 1942). All of these factors would act to increase inter-individual variability of response to toxic agents.

Binding to plasma and other proteins is an important factor controlling the toxic potential of many metals and organometals. Specific deficiency is seen in Wilson's disease, where there is an underproduction of the copper binding protein ceruloplasmin. This leads to accumulation of copper in brain to 4-13 times normal levels, and to progressive lenticular degeneration in genetically sensitised individuals (Scheinberg, 1988).

Metabolism

Where xenobiotic metabolism plays a part in either the inactivation or activation of toxic agents, both individual variations in metabolic capacity and agent-agent interactions can modify toxicity.

Most people are exposed to a wide variety of xenobiotics in their normal lives, and these can cause individual predisposition. Thus moderate ethanol consumption can cause a 1.5 to 2 fold increase in blood levels of toluene by competing for metabolism (Dossing, 1986), while enhancing hepatic disposal of toluene by phenobarbital induction decreased blood levels and reduced its ototoxic potential (Pryor et al. 1991). The combination of alcohol intake with solvent exposure appears to enhance the risk of neurotoxicity in man (Cherry et al., 1992).

Metabolic capacity can also vary with age or sex: neonatal rats, for example being particularly susceptible to the pyrethroid insecticide deltamethrin (Sheets et al., 1994) and both young female rats and young female humans being relatively vulnerable to the insect repellent diethyltoluamide (Verschoyle et al., 1992). There is also well described genetic variability in human populations for the metabolic disposal of many agents (Relling, 1989). Both of these effects can enhance the effective systemic dose, and hence the neuropharmacological effect, of drugs in individuals who are poor metabolisers (Nyberg et al., 1995).

There is evidence to connect genetic variations in xenobiotic metabolic capacity with neurological disease in man. Both thiolmethyltransferase activity and N-methyltransferase activity (Steventon et al. 1994) show different activities in patients with Parkinson's and other neurological diseases. It is presumed that these deficits in systemic metabolic capacity may predispose to neurotoxicity from some endogenous or environmental chemical, although the identity of the putative toxic agent that is the substrate for these enzymes has yet to be established (Mizuno et al., 1994).

In addition to systemic metabolism of xenobiotics, local metabolism can be important in determining neurotoxicity (Lowndes et al. 1995). The local metabolic capacity of the central nervous system can be considerable, since it is concentrated into only a few cell types (Ghersi-Egea et al., 1994). However the role of individual variability in CNS xenobiotic metabolism has not yet been investigated. A dramatic illustration of the potential for such variability is provided by the effects on both brain uptake and toxicity of ivermectin and other agents of elimination of the gene encoding the cerebral endothelial multidrug resistance protein MDR-1 in mice (Schinkel et al., 1996) or of inhibiting its action pharmacologically (Didier and Loor, 1995).

Target vulnerability

Perhaps the most interesting events determining individual vulnerability occur at the target site. Whilst the basic molecular interaction between toxic agent and biological target is unlikely to be influenced by physiological factors, the consequences of this interaction can be profoundly influenced by physiology or pathophysiology. Functional vulnerability is determined by a balance between three factors: the degree of disruption of the target function, the level of biological demand on that function, and the capacity of the system for repair or reorganisation. Most toxicological studies focus only on the first aspect however.

A good example of the role of biological demand is provided by the toxicity of carbon monoxide to the cochlea. Experimental exposure to moderate levels of carbon monoxide reduces the oxygen capacity of blood and produces a temporary reduction in auditory acuity, presumably by limiting oxygen supply. If however oxygen demand is increased by sound stimulation, the same level of carbon monoxide causes a marked and irreversible loss of acuity (Young et al., 1987). Possibly analogous synergistic effects are seen in the combination of noise with either experimental antibiotic exposure (Ryan and Bone, 1982; Lim, 1986), experimental toluene exposure (Johnson et al., 1988), or occupational toluene exposure (Morata et al., 1993). In the peripheral nervous system increased motor activity can sensitise individuals or even individual limbs to lead neuropathy (Jacobs and LeQuesne, 1984). The increased metabolic and synthetic demands associated with axonal regeneration after injury can also sensitise individuals to toxic agents, as in the case of rats recovering from nerve crush which are rendered particularly sensitive to 2,4-hexanedione (Simonati et al., 1983).

Within the central nervous system there is further evidence that both local and global reductions in functional activity can protect against the neurotoxic actions of dinitrobenzene or metronidazole, and also that enhanced functional activity can sensitise specific brain areas to such damage (Ray et al., 1992; 1996). More dramatic changes in brain activity produce correspondingly more dramatic changes in susceptibility: thus removing the cortical afferents to the striata of rats completely protects them from the mitochondrial toxin 3-nitropropionic acid (Beal et al. 1993). However in the former examples the functional changes which modulate vulnerability (changes in auditory input and motor activity) are well within the physiological range which might be experienced by normal individuals. Regional brain metabolic rate can show dramatic changes under such conditions (Ray et al., 1992), and hence where toxic agents act by disruption of energy supply (Cavanagh, 1993; Romero et al., 1995) such interactions might be expected, since enhanced metabolic demand could then convert a sub-threshold deficit into a catastrophic failure of energy supply. That this does not happen more frequently is probably a reflection of the large metabolic reserve capacity of brain, and possibly a mechanism whereby brain can actively down-regulate glycolytic activity in the face of some kinds of metabolic disruption (Pazdernik et al., 1994). Hence, in the absence of toxic agents, only artificially intense functional stimulation leads to brain damage (Sloviter, 1983).

In addition to these short term physiological variations in vulnerability, there are clear developmental windows of vulnerability to neurotoxic agents shown at some stage of their lives by all individuals. These age-dependent factors may not always be recognised in assessment of individual susceptibility however. In addition to well recognised vulnerability to relatively generalised effects such as the foetal alcohol syndrome (Guerri, 1996), very specific time windows of vulnerability can also be seen. Thus prenatal or postnatal exposure of rats to haloperidol produced opposite effects on brain dopamine receptor density (Rosengarten and Friedhoff, 1979), and prenatal diazepam increased neuronal numbers in the male rat locus coerulus while postnatal exposure had no such effect (Rodriguez-Zafra et al., 1994). In the latter case this increase was probably related to a feminising effect only possible at an early stage of brain development. Vulnerability can also change in later life; thus the topography of the rat brain lesions induced by 6-aminonicotinamide alters with postnatal age from 3 to 25 months (Horita et al., 1980), an effect which may be related to linkage between vulnerability to this anti-metabolite and reduced tolerance of the brain to metabolic insult with ageing. This age-dependent vulnerability shows some parallels with the changing topography of the lesions associated with the genetically determined condition, Leigh's disease, as the motor system in the human neonate matures (Cavanagh and Harding, 1994).

Repair capacity is the final factor that may determine individual variability. This has been relatively little studied in the context of neurotoxicity, but may be one of the factors that underlie the increased susceptibility of older individuals. Although adult brain neurones cannot be replaced, re-growth of peripheral processes is possible, and astrocytes are capable of neuronal repair by removal of damaged organelles (Cavanagh et al., 1990), and may contribute to neuronal survival in a number of other ways (Eddleston & Mucke, 1993). It has been proposed that interference with repair capacity is the means by which some non-neuropathic agents may enhance (promote) a number of peripheral neuropathies, including organophosphate-induced polyneuropathy (Moretto et al. 1992).

Conclusions

The vulnerability of individuals to neurotoxic agents can be determined by factors as varied as their genetic make up, age, state of nutrition, work rate, or co-exposure to noise and to other toxic agents. This mixture of influences can lead to a wide range of individual susceptibility. The influence of most of these individual factors can however be understood at a mechanistic level and therefore, in principle at least, could be integrated into individual risk assessments. The main factors are summarised in the diagram below:

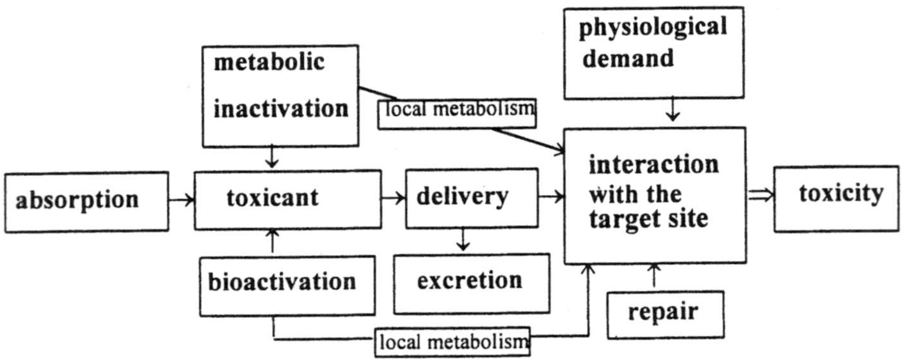

References

Beal MF, Brouillet E, Jenkins BG, Ferrante RJ, Kowall NW, Miller JM, Storey E, Srivastava R, Rosen BR, Hyman BT (1993) Neurochemical and histological characterisation of striatal excitotoxic lesions produced by the mitochondrial toxin 3-nitropropionic acid. J. Neurosci. 13:4181-4192

Cavanagh JB (1993) Selective vulnerability in acute energy deprivation syndromes. Neuropathol Appl Neurobiol 19:461-470

Cavanagh JB, Harding BN (1994) Pathogenic factors underlying the lesions in Leigh's disease. Tissue responses to cellular energy deprivation and their clinico-pathological consequences. Brain 127:1357-1376

Cavanagh JB, Nolan CC, Brown AW (1990) Glial cell intrusions actively remove detritus due to toxic chemicals from within nerve cells. Neurotoxicol. 11:1-12

Cherry NM, Labrèche FP, McDonald JC (1992) Organic brain damage and occupational solvent exposure. British Journal of Industrial Medicine 49: 776-781.

Cohr K-H (1986) Uptake and distribution of common industrial solvents. In Riihimaki V, Ulfvarson U (eds) Safety and health aspects of organic solvents. Alan R. Liss Inc., New York, pp 45-60

Didier AD, Loor F (1995) Decreased biotolerability for ivermectin and cyclosporin A in mice exposed to potent p-glycoprotein inhibitors. Int. J. Cancer 63:263-267

Dossing M (1986) Metabolic interactions between organic solvents and other chemicals. Prog. Clin. Biol. Res. 220:97-105

Eddleston M, Mucke L (1993) Molecular profile for reactive astrocytes - implications for their role in neurologic disease. Neurosci. 54:15-36

Festing MFW (1991) Genetic factors in neurotoxicology and neuropharmacology: a critical evaluation of the use of genetics as a research tool. Experientia 47:990-998

Ghersi-Egea J-F, Leninger-Muller B, Suleman G, Seist G, Minn A (1994) Localisation of drug-metabolising enzyme activities to blood-brain interfaces and circumventricular organs. J. Neurochem.62:1089-1096

Goyer RA (1995) Nutrition and metal toxicity Am. J. Clin. Nutr. 61:646S-650S.

Guerri C (1996) Teratogenic effects of alcohol: current status of animal research and in vitro models. Arch. Toxicol Suppl. 18:71-80.

Haag HB, Larson PS (1942) Studies on the fate of nicotine in the body. 1: The effect of pH on the urinary excretion of nicotine by tobacco smokers. J. Pharmacol. Exp. Ther. 76:235-239

Harpur ES (1982) The pharmacology of ototoxic drugs. Brit. J. Audiol. 16:81-93

Harpur ES (1986) The inner ear. In Cohen GM (ed) Target Organ Toxicity. CRC Press, Bocca Raton, volume II, pp 125-142

Horita N, Ishii T, Izumiyama Y (1980) Ultrastructure of 6-aminonicotinamide (6AN)-induced lesions in the central nervous system of rats. II. Alterations of the nervous susceptibility with ageing. Acta Neuropathol. (Berl.) 49:19-27

Jacobs JM, Le Quesne PM (1984) Toxic disorders of the nervous system. In Hume Addams J, Corsellis JAN, Duchen LW (eds) Greenfield's Neuropathology. Edward Arnold, pp 627-698

Johnson AC, Juntunen L, Nylen P, Borg E, Hoglund G (1988) Effect of interaction between noise and toluene on auditory function in the rat. Acta Otolaryngol. 105:56-63

Lim DJ (1986) Effects of noise and ototoxic drugs at the cellular level in the cochlea: a review. Am. J. Otolaryngol. 7:73-99

Lowndes HE, Philbert MA, Beiswanger M, Kauffman FC, Reuhl KR (1995) Xenobiotic metabolism in the brain as mechanistic bases for neurotoxicity. In Chang LW, Dyer RS (eds) Handbook of neurotoxicology. Marcel Dekker, New York, pp 1-28

Mizuno Y, Ikebe SI, Hattori N, Mochizuki H, Nakagawa-Hattori Y, Kondo T (1994) Studies on the pathogenesis of Parkinson's disease in Japan Arch. Gerentol. Geriat. 19:105-121

Morata TC, Dunn DE, Kretschmer LW, Lemasters GK, Keith RW (1993) Effects of occupational exposure to organic solvents and noise on hearing. Scand. J. Environ. Health 19:245-254

Moretto A, Betrolazzi M, Capodicassa E, Peracia M, Richardson RJ, Scapellato ML, Lotti M (1992) Phenylmethanesulphonyl fluoride elicits and intensifies the clinical expression of neuropathic insults. Arch. Toxicol. 66:67-72

Nyberg S, Dahl ML, Halldin C (1995) A PET study of D2 and 5-HT2 receptor occupancy induced by risperidone in poor metabolisers of debrisoquin and risperidone. Psychopharmacol. 119:345-348

Pazdernik T, Cross R, Nelson S, Kamijo Y, Samson F (1994) Is there an energy conservation "system" in brain that protects against the consequences of energy depletion? Neurochem. Res. 19:1393-1400

Pryor GT, Rebert C, Kassay K, Kuiper H, Gordon R (1991) The hearing loss associated with exposure to toluene is not caused by a metabolite. Brain Res. Bull. 27:109-113

Ray DE, Brown AW, Cavanagh JB, Nolan CC, Richards HK, Wylie SP (1992) Functional/metabolic modulation of the brain stem lesions caused by 1,3-dinitrobenzene in the rat. Neurotoxicology 13:379-388

Ray DE, Holton JL, Lister T & Nolan CC (1996) The glio-vascular toxicity of m-dinitrobenzene and related agents: modulation of toxicity by neuronal activation. Arch. Tox. Suppl. 18: 140-148.

Relling MV (1989) Polymorphic drug metabolism Clin. Pharmacol. 8:852-863

Riihimaki V, Pfaffli P, Savolainen K (1979) Kinetics of m-xylene in man. Influence of intermittent physical exercise and changing environmental concentrations on kinetics. Scand. J. Work Environ. Health 5:232-248

Rodriguez-Zafra M, De Bals MR, Perez-Laso C, Cales J M, Guillamon A, Segovia S (1994) Effects of perinatal diazepam exposure on the sexually dimorphic rat locus coerulus. Neurotoxicol. Teratol. 15:139-144

Romero IA, Lister T, Richards HK, Seville MP, Wylie SP, Ray, DE (1995) Early metabolic changes during m-dinitrobenzene neurotoxicity and the possible role of oxidative stress Free Radical Biol Med 18:311-319

Rosengarten H, Friedhoff AJ (1979)Enduring changes in dopamine receptor cells of ' pups from drug administration to pregnant and nursing rats. Science 203: 1133-1135

Ryan AF, Bone RC (1982) Non-simultaneous interaction of exposure to noise and kanamycin intoxication in chinchillas. Amer. J. Otolaryngol. 3:264-272

Scheinberg IH (1988) The neurotoxicity of copper. In Bondy SC, Prasad KN (eds) Metal Neurotoxicity. CRC Press, Florida, pp 56-60

Schinkel AH, Wagenaar E, Mol CAAM, Van Deemter L (1996) P-glycoprotein in the blood-brain barrier of mice influences the brain penetration and pharmacological activity of many drugs. J. Clin. Investig. 97:2517-2524

Seth PK & Chandra SV (1988) Neurotoxic effects of manganese. In Bondy SC, Prasad KN (eds) Metal Neurotoxicity. CRC Press, Florida, pp 19-33

Sheets LP, Doherty JD, Law MW, Reiter LW, Crofton KM (1994) Age-dependent differences in the susceptibility of rats to deltamethrin. Toxicol. Appl. Pharmacol. 126:186-190

Simonati A, Rizzuto N, Cavanagh JB (1983) The effects of 2,5-hexanedione on axonal regeneration after nerve crush in the rat. Acta Neuropathol. 59:216-224

Sloviter RS (1983) 'Epileptic' brain damage in rats induced by sustained electrical stimulation of the perforant path. I Acute electrophysiological and light microscopic studies. Brain Res. Bull. 10:675-697

Steventon GB, Green S, Waring RH, Williams AC (1994) D-penicillamine metabolism in neurodegenerative diseases: an in vivo / in vitro sulphydryl methylation study. Xenobiotica 24:1013-1020

Verschoyle RD, Brown AW, Nolan CC, Ray DE, Lister T (1992) A comparison of the acute toxicity, neuropathology and electrophysiology of N,N-diethyl-m-toluamide (DEET) and N,N-dimethyl-2,2-diphenylacetamide (diphenamid) in rats. Fund. & Appl. Toxicol. 18: 79-88

Young JS, Upchurch MB, Kaufman MJ, Fechter LD (1987) Carbon monoxide exposure potentiates high frequency auditory threshold shifts induced by noise. Hearing Res.26:37-43

Cellular and Molecular Mechanisms of Cutaneous Toxicity

(Chair: I. Kimber, United Kingdom)

Epidermal Cytokines and the Induction of Allergic and Non-Allergic Contact Dermatitis

Ian Kimber, Rebecca J Dearman and Marie Cumberbatch
Zeneca Central Toxicology Laboratory, Alderley Park, Macclesfield, Cheshire, UK.

Introduction

The term contact dermatitis describes the local cutaneous inflammatory reaction provoked by skin exposure to certain chemicals. Some materials have the potential directly to induce irritant reactions at the site of contact. Alternatively, the appearance of contact dermatitis may reflect an allergic reaction. Following primary exposure of susceptible individuals to a contact allergen an immune response is initiated that results in long-lived allergic sensitization. If the sensitized individual is exposed subsequently to the same chemical allergen then a more aggressive secondary immune response is elicited that results in the manifestation of allergic contact dermatitis. Clearly some chemicals are able to cause both irritant and allergic contact dermatitis; the operational distinction being that such agents are likely to induce allergic reactions in sensitized individuals at concentrations below those that are necessary for the development of irritant responses in individuals who are not sensitized. The purpose of this article is to consider the roles epidermal cytokines play in the induction of contact dermatitis, and in particular their influence on the behaviour of epidermal Langerhans cells and immunological processes.

Epidermal Cytokines

Cytokines are a large family of proteins and glycoproteins that influence markedly immune and inflammatory responses. Collectively they orchestrate immune responses through the regulation of cellular activation, division, differentiation, migration and communication. Cytokines may act in autocrine, paracrine and endocrine fashion and some are able to induce or inhibit the expression of other cytokine molecules. It is known that epidermal cells, both keratinocytes and Langerhans cells (LC), are important sources of cytokines. Some of these are expressed constitutively, whereas others are produced in response to appropriate stimuli. A list of epidermal cytokines is shown in Table 1. There is increasing evidence also that epidermal cells may produce chemotactic cytokines termed chemokines (Schroder, 1995). The importance of epidermal cytokine networks in the skin immune system has been recognized for some time and it is clear now that these molecules together influence all aspects of cutaneous responses to chemicals. One important role of epidermal cytokines is the regulation of Langerhans cell function.

Table 1 Epidermal Cytokines and their Cellular Sources

Cytokines	Keratinocytes	Langerhans Cells
Interleukin 1α (IL-1α)	+	-
Interleukin 1β (IL-1β)	-	+
Interleukin 6 (IL-6)	+	+
Interleukin 7 (IL-7)	+	-
Interleukin 8 (IL-8)	+	-
Interleukin 10 (IL-10)	+	-
Interleukin 12 (IL-12)	+	+
Interleukin 15 (IL-15)	+	+
Granulocyte/macrophage colony stimulating factor (GM-CSF)	+	-
Interferon γ (IFN-γ)	+	-
Transforming growth factor α (TGF-α)	+	-
Transforming growth factor β (TGF-β)	+	+
Tumour necrosis factor α (TNF-α)	+	-
Macrophage inflammatory protein 1α (MIP-1α)	-	+
Macrophage inflammatory protein 1γ (MIP-1γ)	-	+
Macrophage inflammatory protein 2 (MIP-2)	+	+
Monocyte chemoattractant protein 1 (MCP-1)	+	-

Data summarized from available human and/or mouse studies (Schreiber et al., 1992; Heufler et al., 1992; 1993; Matsue et al., 1992; Enk and Katz 1992a; b; Gruschwitz and Hornstein, 1992; Kimber 1994; Aragne et al., 1994; Nakamura et al, 1995; Kang et al., 1996; Howie et al., 1996; Blauvelt et al., 1996; Cumberbatch et al., 1996; Yawalkar et al., 1996; Mohamadzadeh et al., 1996).

Epidermal Langerhans Cells

Langerhans cells form a contiguous network in the epidermis where they serve as sentinels of the immune system. Following topical exposure of mice to contact allergens LC are induced to migrate away from the skin and to accumulate in draining lymph nodes. Some of the cells that reach skin-draining lymph nodes bear high levels of the chemical allergen (Kimber and Cumberbatch, 1992; Kimber and Dearman, 1996). The migration of LC is associated with their functional maturation into immunostimulatory dendritic cells (DC), such that by the time they reach the lymph nodes they are able to present antigen effectively to responsive T lymphocytes. By analogy with in vitro experiments, it is believed that the phenotypic changes that characterize LC maturation in vivo are effected by epidermal cytokines; those of greatest importance being GM-CSF and IL-1β acting individually and in concert (Witmer-Pack et al., 1987; Heufler et al., 1988; Picut et al., 1988; Ozawa et al., 1996b).

The importance of LC migration and maturation for the induction of contact sensitization is readily apparent. However, it must be recognized that stimuli other than those supplied by chemical allergens will induce the migration, and presumably the functional maturation, of LC. Thus, for instance, topical exposure of mice to sodium lauryl sulphate (SLS), a skin irritant that is believed not to cause contact sensitization, results in the accumulation of DC in draining lymph nodes (Cumberbatch et al., 1993). Moreover, irradiation with ultraviolet B (UVB) light similarly causes DC accumulation in lymph nodes (Moodycliffe et al., 1992). It is known that SLS and UVB, and also topical administration of chemical allergens, induce the rapid up-regulation of TNF-α expression by keratinocytes (Kock et al., 1990; Enk and Katz 1992a). A unifying mechanism is that TNF-α provides the signal for LC migration from the epidermis. It has been demonstrated that intradermal injection of mice with homologous recombinant TNF-α induces both a rapid reduction in the frequency of epidermal LC local to the site of exposure and an accumulation, somewhat later, of DC in the draining nodes (Cumberbatch and Kimber, 1992; Cumberbatch et al., 1994; 1995). Furthermore, systemic treatment of mice with a neutralizing anti-TNF-α antibody inhibits very significantly the stimulation by UVB irradiation or by contact allergens such as oxazolone (see Figure 2) of DC accumulation in draining lymph nodes (Moodycliffe et al., 1994; Cumberbatch and Kimber, 1995). Under such conditions, where the migration of LC is prevented or inhibited, then the induction of contact sensitization is sub-optimal (Cumberbatch and Kimber, 1995).

It now appears likely that TNF-α acts together with other epidermal cytokines, and in particular IL-1β, in the stimulation of LC migration (Kimber and Cumberbatch, 1995). As illustrated in Figure 1, intradermal injection of mice with either IL-1β or TNF-α results in an increase in the frequency of DC found within draining lymph nodes 17 hours later, although only with the latter cytokine is the accumulation of DC sufficiently rapid to be apparent within 2 hours.

The ability of exogenous IL-1β to induce LC migration might be ascribed to the fact that this cytokine has been found to stimulate TNF-α expression when injected intradermally (Enk et al., 1993a). Certainly the requirement for IL-1β to act locally on keratinocytes to induce TNF-α production would be consistent with IL-1β inducing a more delayed accumulation of DC in lymph nodes than that which characterizes injection of TNF-α. It may be that the expression of TNF-α in response to skin sensitization is largely or wholly dependent upon the stimulation by chemical allergens of increased IL-1β production by LC. Certainly it has been found that treatment of mice with a neutralizing anti-IL-1β antibody causes a partial inhibition of oxazolone-induced DC accumulation in draining lymph nodes (Figure 2).

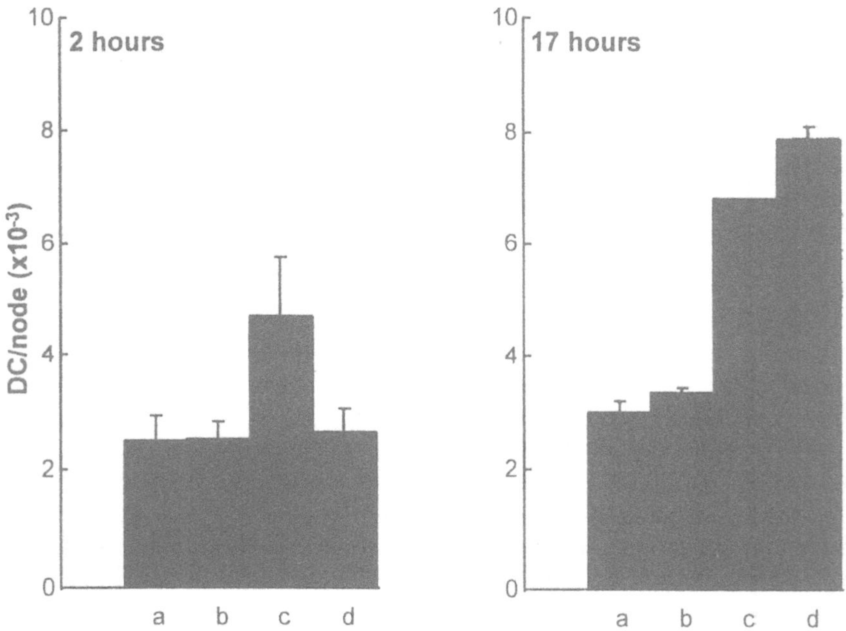

Fig. 1. Groups of BALB/c strain mice (n=10) received 30(l intradermal injections in both ear pinnae of 50ng homologous recombinant TNF-α (c) or of 50ng homologous recombinant IL-1β (d), both in 0.1% bovine serum albumin (BSA). Control mice were injected with 30μl of 0.1% BSA alone (b) or were untreated (a). The number of DC within draining lymph nodes was measured 2 hours and 17 hours following injection. The data illustrated represent the mean ±SE of 3 independent experiments (2 hours) or the mean ± range of 2 separate experiments (17 hours).

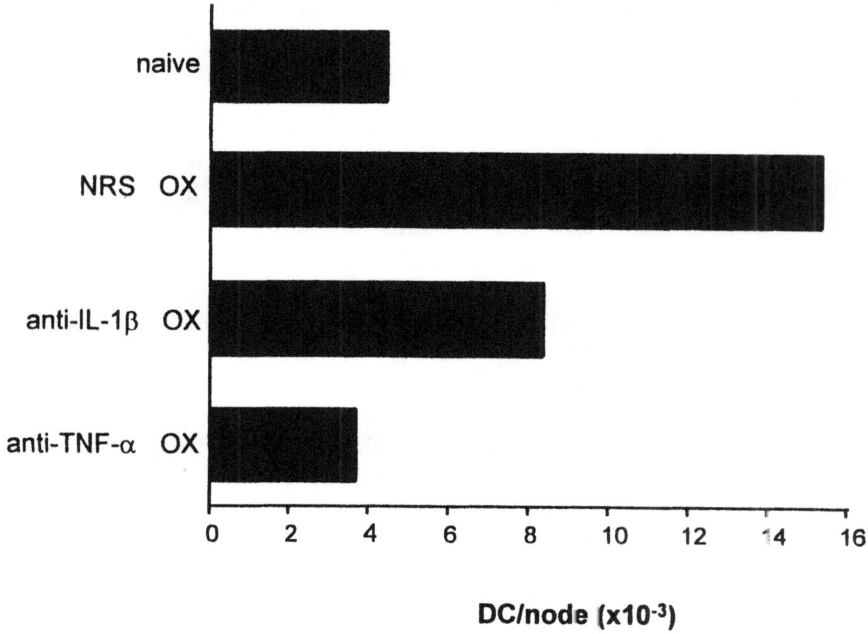

Fig. 2. Groups of BALB/c strain mice (n=10) received a single intraperitoneal injection of 100μl of either anti-IL-1β or anti-TNF-α diluted 1:5 in sterile phosphate buffered saline (PBS), or of sterile normal rabbit serum (NRS) diluted in PBS to a similar extent. Two hours later mice were treated on the dorsum of both ears with 25μl of 0.5% oxazolone (OX) in 4:1 acetone:olive oil. Control mice were untreated. Draining auricular lymph nodes were removed 18 hours later and the number of DC/node measured. A representative experiment.

The effectiveness of skin sensitization is prevented or impaired in IL-1β gene deletion transgenic mice or in mice where local IL-1β has been neutralized (Enk et al., 1993a; Shornick et al., 1996). Currently the conclusion drawn is that the induction of LC migration is dependent upon complex interactions induced by IL-1β and TNF-α with the latter providing the proximate signal for LC to leave the epidermis.

Irrespective of the mechanisms that regulate the movement and maturation of LC, the main responsibility of these cells and the DC into which they develop is the stimulation of immune responses. The vigour and quality of cutaneous immune responses are to an important extent determined by cytokines produced by LC and DC themselves and by other epidermal cytokines.

Immunoregulatory Epidermal Cytokines

The quality of induced immune responses is a function largely of the contributions made by subpopulations of $CD4^+$ T helper (Th) cells. Delayed and contact hypersensitivity responses are characteristically associated with the action of Th1-type cells (Kimber and Dearman, 1996). During the evolution of immune responses the selectivity of Th cell responses is influenced by the relative availability of cytokines, the development of Th1 cells being favoured by IL-12 and IFN-γ. Interleukin 12 appears to be particularly important in this respect and is believed to promote the differentiation of mature Th1 cells via both IFN-γ-dependent and -independent mechanisms (Manetti et al., 1993., Trinchieri, 1994; Schmitt et al., 1994). Several independent observations are consistent with important roles for IL-12 and Th1 cells in contact hypersensitivity responses. First, contact allergens induce in mice selective Th1-type immune responses (Dearman and Kimber 1991; 1992; Dearman et al., 1995), second, the majority of nickel-specific T cell clones derived from peripheral lymphocytes of nickel-sensitive donors was found to be of Th1-type (Kapsenberg et al., 1991; 1992) and third, local injection of recombinant IL-12 will enhance the acquisition of contact sensitivity in mice (Maguire, 1995).

Interleukin 12 is produced by a variety of cells, including macrophages. Of relevance to the induction of Th1-type cutaneous immune responses is evidence that both LC and keratinocytes can synthesize IL-12 (Aragne et al., 1994; Kang et al., 1996; Yawalkar et al., 1996). It has been found recently that DC produce IL-12 rapidly following interaction with T lymphocytes and are particularly effective at inducing selective development of Th1-type cells (Heufler et al., 1996). The immunostimulatory properties of the DC into which Langerhans cells mature, combined with the ability of both these populations to synthesize and secrete IL-12, will undoubtedly encourage the differentiation of Th1 cells and the induction of contact sensitization.

Other epidermal cytokines may, however, serve to antagonize Th1 cell responses or cutaneous immune responses per se. Interleukin 10, mRNA for which has been shown to be expressed by keratinocytes (Enk and Katz, 1992b), has the potential to subvert Th1 cell responses. This cytokine favours the development of Th2 rather than Th1 cells and will inhibit the immunostimulatory and antigen-presenting functions of LC and DC (Enk et al., 1993b; Macatonia et al,., 1993; Peguet-Navarro et al., 1994; Beissert et al., 1995). Importantly, it has been proposed that IL-10 may selectively inhibit Th1 cell-type responses secondary to a down-regulation of the expression by LC of CD80 (B7-1), a costimulatory molecule that has been implicated in the activation of Th1 cells (Chang et al., 1995; Ozawa et al, 1996a). In addition to regulating negatively the induction of cutaneous Th1 cell-type immune responses, IL-10 may serve as a generalized suppressor of inflammatory reactions in the skin. It has been found recently, using mice with a targeted disruption of the IL-10 gene, that in the absence of this cytokine both irritant and allergic contact hypersensitivity responses are more pronounced and more prolonged (Berg et al., 1995). The proposal is that the physiological role of IL-10 is to control the production of proinflammatory cytokines such as to constrain immunopathologic damage in the skin (Berg et al., 1995). Consistent with this is evidence that in presensitized mice IL-10 will inhibit the elicitation of delayed and contact hypersensitivity reactions (Li et al., 1994; Kondo et al., 1994). The extent to which, in practice, epidermal IL-10 regulates cutaneous immune and inflammatory responses to chemicals is not clear. In recent investigations it was found that topical exposure of mice to sensitizing concentrations of the allergen oxazolone, under conditions where the production of cutaneous IL-6 was increased, failed to stimulate the expression of IL-10 protein in skin (Holliday et al., 1996).

Concluding Comments : The Cytokine Balance

Taken together the available data indicate that cytokines play an important and in some instances possibly decisive role in the development and expression of contact dermatitis. During the induction phase of contact sensitization epidermal cytokines mediate and regulate the migration and maturation of Langerhans cells and possibly the ability of the DC into which they develop to present antigen effectively to T lymphocytes. The elicitation of allergic and irritant contact dermatitis will be favoured by the production of pro-inflammatory cytokines that effect cutaneous reactions and regulated negatively by IL-10 and possibly other epidermal cytokines that limit inflammatory processes in the skin.

References

Aragne Y, Riemann H, Bhardwaj RS, Schwarz A, Sawada Y, Yamada H, Luger T, Kubin M, Trinchieri G, Schwarz T (1994) IL-12 is expressed and released by human keratinocytes and epidermoid carcinoma cell lines. J Immunol 153:5366-5372

Beissert S, Hosoi J, Grabbe S, Asahina A, Granstein RD (1995) IL-10 inhibits tumour antigen presentation by epidermal antigen-presenting cells. J Immunol 154:1280-1286

Berg DJ, Leach MW, Kuhn R, Rajewsky K, Muller W, Davidson NJ, Rennick D (1995) Interleukin 10 but not interleukin 4 is a natural suppressant of cutaneous inflammatory responses. J Exp Med 182:99-108

Blauvelt A, Asada H, Klaus-Kovtun A, Altman DJ, Lucey DR, Katz SI (1996) Interleukin-15 mRNA is expressed by human keratinocytes, Langerhans cells, and blood-derived dendritic cells and is downregulated by ultraviolet B radiation. J Invest Dermatol 106:1047-1052

Chang C-H, Furue M, Tamaki K (1995) B7-1 expression of Langerhans cells is up-regulated by proinflammatory cytokines, and is down-regulated by interferon-γ or by interleukin-10. Eur J Immunol 25:394-398

Cumberbatch M, Dearman RJ, Kimber I (1996) Constitutive and inducible expression of interleukin-6 by Langerhans cells and lymph node dendritic cells. Immunology 87:513-518

Cumberbatch M, Fielding I, Kimber I (1994) Modulation of epidermal Langerhans cell frequency by tumour necrosis factor-α. Immunology 81:395-401

Cumberbatch M, Fielding I, Kimber I (1995) Epidermal Langerhans cell migration : signals and mechanisms. In : Banchereau J, Schmitt D (eds) Dendritic Cells in Fundamental and Clinical Immunology, Volume 2. Plenum Press, New York, pp173-175.

Cumberbatch M, Kimber I (1992) Dermal tumour necrosis factor-α induces dendritic cell migration to draining lymph nodes, and possibly provides one stimulus for Langerhans cell migration. Immunology 75:257-263

Cumberbatch M, Kimber I (1995) Tumour necrosis factor-α is required for accumulation of dendritic cells in draining lymph nodes and for optimal contact sensitization. Immunology 84:31-35

Cumberbatch M, Scott RC, Basketter DA, Scholes EW, Hilton J, Dearman RJ, Kimber I (1993) Influence of sodium lauryl sulphate on 2,4-dinitrochlorobenzene-induced lymph node activation. Toxicology 77:181-191

Dearman RJ, Basketter DA, Kimber I (1995) Differential cytokine production following chronic exposure of mice to chemical respiratory and contact allergens. Immunology 86:545-550

Dearman RJ, Kimber I (1991) Differential stimulation of immune function by respiratory and contact chemical allergens. Immunology 72:563-570

Dearman RJ, Kimber I (1992) Divergent immune responses to respiratory and contact chemical allergens : antibody elicited by phthalic anhydride and oxazolone. Clin Exp Allergy 22:241-250

Enk AH, Angeloni VL, Udey MC, Katz SI (1993a) An essential role for Langerhans cell-derived IL-1β in the initiation of primary immune responses in the skin. J Immunol 150:3698-3704

Enk AH, Angeloni VL, Udey MC, Katz SI (1993b) Inhibition of Langerhans cell antigen-presenting function by IL-10. A role for IL-10 in the induction of tolerance. J Immunol 151:2390-2398

Enk AH, Katz SI (1992a) Early molecular events in the induction phase of contact sensitivity. Proc Natl Acad Sci USA 89:1398-1402

Enk AH, Katz SI (1992b) Identification and induction of keratinocyte-derived IL-10. J Immunol 149:92-95

Gruschwitz MS, Hornstein OP (1992) Expression of transforming growth factor type beta on human epidermal dendritic cells. J Invest Dermatol 99:114-116

Heufler C, Koch F, Schuler G (1988) Granulocyte/macrophage colony-stimulating factor and interleukin 1 mediate the maturation of murine epidermal Langerhans cells into potent immunostimulatory dendritic cells. J Exp Med 167:700-705

Heufler C, Koch F, Stanzl U, Topar G, Wysocka M, Trinchieri G, Enk A, Steinman RM, Romani N, Schuler G (1996) Interleukin-12 is produced by dendritic cells and mediates T helper 1 development as well as interferon-γ production by T helper 1 cells. Eur J Immunol 26:659-668

Heufler C, Topar G, Grasseger A, Stanzl U, Koch F, Romani N, Namen AE, Schuler G (1993) Interleukin 7 is produced by murine and human keratinocytes. J Exp Med 178:1109-1114

Heufler C, Topar G, Koch F, Trockenbacher B, Kampgen E, Romani N, Schuler G (1992) Cytokine gene expression in murine epidermal cell suspensions : interleukin 1β and macrophage inflammatory protein 1α are selectively expressed in Langerhans cells but are differentially regulated in culture. J Exp Med 176:1221-1226

Holliday MR, Dearman RJ, Basketter DA, Kimber I (1996) Stimulation by oxazolone of increased IL-6, but not IL-10, in the skin of mice. Toxicology 106:237-242
Howie SEM, Aldridge RD, McVittie E, Forsey RJ, Sands C, Hunter JAA (1996) Epidermal keratinocyte production of interferon-γ immunoreactive protein and mRNA is an early event in allergic contact dermatitis. J Invest Dermatol 106:1218-1223
Kang K, Kubin M, Cooper KD, Lessin SR, Trinchieri G, Rook AH (1996) IL-12 synthesis by human Langerhans cells. J Immunol 156:1402-1407
Kapsenberg ML, Wierenga EA, Bos JD, Jansen HM (1991) Functional subsets of allergen-reactive human $CD4^+$ T cells. Immunol Today 12:392-395
Kapsenberg ML, Wierenga EA, Stiekma FEM, Tiggelman AMBC, Bos JD (1992) TH1 lymphokine production profiles of nickel-specific $CD4^-$ T lymphocyte clones from nickel contact allergic and non-allergic individuals. J Invest Dermatol 98:59-63
Kimber I (1994) Cytokines and the regulation of allergic sensitization to chemicals. Toxicology 93:1-11
Kimber I, Cumberbatch M (1992) Dendritic cells and cutaneous immune responses to chemical allergens. Toxicol Appl Pharmacol 117:137-146
Kimber I, Cumberbatch M (1995) Langerhans cell migration : initiation and regulation. In Moll H (ed) The Immune Functions of Epidermal Langerhans Cells, R G Landes Co, Austin, pp103-117
Kimber I, Dearman RJ (1996) Contact hypersensitivity : immunological mechanisms. In Kimber I, Maurer T (eds) Toxicology of Contact Hypersensitivity, Taylor and Francis, London, pp4-25
Kock A, Schwarz T, Kirnbauer R, Urbanski A, Perry P, Ansel JC, Luger TA (1990) Human keratinocytes are a source for tumor necrosis factor α : evidence for synthesis and release upon stimulation with endotoxin or ultraviolet light. J Exp Med 172:1609-1614
Kondo S, McKenzie RC, Sauder DN (1994) Interleukin 10 inhibits the elicitation phase of allergic contact hypersensitivity. J Invest Dermatol 103:811-814
Li L, Elliott JF, Mosmann TR, (1994) IL-10 inhibits cytokine production, vascular leakage and swelling during T helper 1 cell-induced delayed-type hypersensitivity. J Immunol 153:3967-3978
Macatonia SE, Doherty TM, Knight SC, O'Garra A (1993) Differential effect of IL-10 on dendritic cell-induced T cell proliferation and IFN-γ production. J Immunol 150:3755-3765
Maguire HC Jr (1995) Murine recombinant interleukin-12 increases the acquisition of allergic contact dermatitis in the mouse. Int Arch Allergy Immunol 106:166-168
Manetti R, Parronchi P, Guidizi MG, Piccinni M-P, Maggi G, Trinchieri G, Romagnani S (1993) Natural killer cell stimulatory factor (interleukin 12 (IL-12)) induces T helper type 1 (Th1)-specific immune responses and inhibits development of IL-4 producing Th cells. J Exp Med 177:1199-1204
Matsue H, Cruz PD Jr, Bergstresser PR, Takashima A (1992) Langerhans cells are the major source of mRNA for IL-1β and MIP-1α among unstimulated mouse epidermal cells. J Invest Dermatol 99:537-541
Mohamadzadeh M, Poltorak AN, Bergstresser PR, Beutler B, Takashima A (1996) Dendritic cells produce macrophage inflammatory protein-1γ, a new member of the CC chemokine family. J Immunol 156:3102-3106
Moodycliffe AM, Kimber I, Norval M (1992) The effect of ultraviolet B irradiation and urocanic acid isomers on dendritic cell migration. Immunology 77:394-399
Moodycliffe AM, Kimber I, Norval M (1994) Role of tumour necrosis factor-α in ultraviolet B light-induced dendritic cell migration and suppression of contact hypersensitivity. Immunology 81:79-84

Nakamura K, Williams IR, Kupper TS (1995) Keratinocyte-derived monocyte chemoattractant protein 1 (MCP-1) : analysis in a transgenic model demonstrates MCP-1 can recruit dendritic and Langerhans cells to skin. J Invest Dermatol 105:635-643

Ozawa H, Aiba S, Nakagawa S, Tagami H (1996a) Interferon-γ and interleukin-10 inhibit antigen presentation by Langerhans cells for T helper type 1 cells by suppressing their CD80 (B7-1) expression. Eur J Immunol 26 : 648-652

Ozawa H, Nakagawa S, Tagami H, Aiba S (1996b) Interleukin-1β and granulocyte-macrophage colony-stimulating factor mediate Langerhans cell maturation differently. J Invest Dermatol 106:441-445

Peguet-Navarro J, Moulon C, Caux C, Dalbiez-Gauthier C, Banchereau J, Schmitt D (1994) Interleukin-10 inhibits the primary allogeneic T cell response to human epidermal Langerhans cells. Eur J Immunol 24:884-891

Picut CA, Lee CS, Dougherty EP, Anderson KI, Lewis RM (1988) Immunostimulatory capabilities of highly enriched Langerhans cells in vitro. J Invest Dermatol 90:201-206

Schmitt E, Hoehn P, Huels C, Goedert S, Palm N, Rude E, Germann T (1994) T helper type 1 development of naive $CD4^+$ T cells requires the coordinate action of interleukin-12 and interferon-γ and is inhibited by transforming growth factor-β. Eur J Immunol 24:793-798

Schreiber S, Kilgus O, Payer E, Kutil R, Elbe A, Mueller C, Stingl G (1992) Cytokine pattern of Langerhans cells isolated from murine epidermal cell cultures. J Immunol 149:3525-3534

Schroder J-M (1995) Cytokine networks in the skin. J Invest Dermatol 105:20S-24S

Shornick LP, De Togni P, Mariathasan S, Goellner J, Strauss-Schoenberger J, Karr RW, Ferguson TA, Chaplin DD (1996) Mice deficient in IL-1β manifest impaired contact hypersensitivity to trinitrochlorobenzene. J Exp Med 183:1427-1436

Trinchieri G (1994) Interleukin-12 and its role in the generation of Th-1 cells. Immunol Today 14:335-337

Witmer-Pack MD, Olivier W, Valinsky J, Schuler G, Steinman RM (1987) Granulocyte/macrophage colony-stimulating factor is essential for the viability and function of cultured murine epidermal Langerhans cells. J Exp Med 166:1484-1498

Yawalkar N, Limat A, Brand CU, Braathen LR (1996) Constitutive expression of both subunits of interleukin-12 in human keratinocytes. J Invest Dermatol 106:80-83

Validation of In Vitro Methods to Single Out Photoirritants Using Mechanistically Based Tests

Wolfgang JW Pape
Beiersdorf AG, D-20245 Hamburg, Germany

Abstract

In the recent OECD draft proposal for a new guideline on acute dermal photoirritation testing, in vitro screening tests have been included as part of the sequential test strategy. These screening tests were placed directly prior to animal tests proposed. In Europe some in vitro techniques - cell culture and mechanistic tests - are under validation in a joint project of the European Center for the Validation of Alternative Methods (ECVAM) and the European Cosmetic, Toiletry, and Perfumery Association (COLIPA). Two promising cellular in vitro tests are presented and discussed as tool for the screening of photoirritancy. The first one as a general core test performed in each participating laboratory is the 3T3 mouse fibroblast Neutral Red Uptake Phototoxicity test determining the cell viability by uptake of Neutral Red as end point, whereas the second performed only by three participating laboratories was the the Red Blood Cell phototoxicity test comprising a combination of two end points, the photohaemolysis and the oxyhaemoglobin oxidation. Besides this, other mechanistic tests can be used as additional support. Identification of photoirritation is generally considered to be one area for the successful research and validation of in vitro techniques.

Introduction

Today, the most useful contributions to hazard characterisation are still provided by animal studies. Currently, acute dermal phototoxicity is tested on animals as well, using various species and different light sources. An internationally agreed test protocol does not yet exist. Recently a standard test protocol was drafted by the Organisation for Economic Cooperation and Developement (OECD) but a general agreement on a regulatory guideline could not be achieved.

According to its official statement on laboratory animals welfare, OECD in its proposal has recommended a tiered testing strategy which should include validated in vitro techniques prior to subsequent in vivo testing for confirmation or in order to determine 'no observed effect levels'. Another sequential approach was proposed by the expert group during the second workshop organized by the European Centre for the Validation of Alternative Methods (ECVAM) entitled "In Vitro Phototoxicity Testing" (Spielmann H. et al. 1994). Both strategies have been discussed recently (Koeter, 1995).

In the ECVAM workshop report 2, the authors have summarized a list of in vitro methods at different stages of developement which might be useful for phototoxicity testing in vitro and which can allow to reduce animal testing as far as possible and if photoirritant reactions are concerned. In Europe a joint validation project supported by the European Commission and the European Cosmetics Industry Association (COLIPA)was initiated in 1992 and is still ongoing.

The present paper summarizes some general aspects of some cell culture and mechanistic tests proposed to use in this validation project.

Photoirritation in vitro

Acute phototoxic, that means photo*irritant* in contrast to photo*allergic*, skin reactions may be caused by certain light-absorbing chemicals at first exposure to the sunlight. It is important that sufficient light energy is aborbed and the concentration of the chemical within the skin is high enough to provoke such adverse skin reaction. Usually, animal screening tests has been performed in order to identify such phototoxic substances. But there has always been lack of a standardized experimental routine minimizing problems of unavoidable variability. Recently, an expert committee has proposed a standard protocol for animal testing (Nilsson et al., 1993).

The legal background for the search for alternative methodologies is laid down in article 7 of the Council Directive 86/609/EEC: „An animal study shall not be performed if another scientifically satisfactory method of obtaining the results sought, not entailing the use of an animal, is reasonably and practically available.“ The scientific rationale for the use of alternatives in toxicological routine studies need to be defined as part of the validation process. Due to the public demand, and based on scientific experience, the screening of photoirritancy beside percuteanous absorption seems to be an area of successful research for alternative methodologies.

Fundamental knowledge has been developed and used in photobiological research during the last century, comprising different cellular systems and cellular components involved in phototoxic reactions. A number of basic mechanisms may be linked to photoirritation. Their knowledge should be included in a photoirritation testing strategy in vitro. Some of these essential mechanisms are the formation of light-induced radicals (type I reaction) leading to molecular changes on subcellular structures, the production of singlet oxygen by direct energy transfer (type II reaction) followed by the oxidation of biomolecules, the direct covalent binding to biomolecules, and finally the degradation of the photosensitizer, forming toxic photoproducts. All these light-induced mechanisms result in structural changes on the cellular level leading to cell damage as a prerequisite of photoirritation.

Based on these mechanistic considerations a set of different tests on mammalian cells in culture like lymphocytes, fibroblasts and/or keratinocytes, the photohaemolysis of erythrocytes and the oxidation of histidine have already been proposed (Johnson et al. 1986). In addition the photobinding to protein not only to pick up potential photoallergens (Lovell 1993) but also to study protective mechanisms is recommended, together with studies on photodegradation and the oxidation of haemoglobin (Pape et al. 1994a,b). The photodegradation may be followed by chemical analysis or by detection of increased cytotoxicity in cell culture systems. The oxidative attack evoked by photosensitizers under irradiation can also be measured by determining the peroxidation of linoleic acid (Castell et al. 1994). In principle numerous in vitro methods, if validated and accepted by the scientific community, can be used as tools to screen chemicals for the identification of acute dermal photoirritation potential. In this paper the main focus is directed versus two cellular tests i) the Neutral Red Uptake Phototoxicity Test using 3T3 mouse fibroblasts and ii) the Red Blood Cell Phototoxicity Test measuring the photohaemolytic potency and the oxidation of haemoglobin as part of the intracellular photodynamic process.

Photosensitization in vivo

The corresponding data for photosensitization used in the validation project for comparison with in vitro findings and for the choice of relevant test chemicals were based on published and historical data available in the public domain. The evaluation of the reports from the scientific literature with results obtained from animal testing and human dermatological findings was difficult. Due to different protocols used in experiments conducted for different purposes and to the qualitative description of some dermatological reports, a proper assessment was often not possible. For the validation excercise this evaluation and assessment was done by an expert subgroup, familiar with photosensitization testing.

General Technical Aspects

The validation of a toxicological in vitro method is characterized by two essential criteria which are summarized as the reliability and the relevance of the test procedure (Balls et al., 1990).

The important technical aspect of the reliability focuses on the development of an accepted high degree of standardization of the method in order to guarantee a high reproducibility and comparability of the results obtained in different laboratories. The standardisation has to be part of the test development during prevalidation phases. The relevance of a method chosen is summarized in the scientific rationale.

For the standardization of phototoxicity testing it is important to agree on a defined light source, which should be a sun simulator, and on an instrumentation for the measurement of - in particular - the UVA and UVB irradiance under routine conditions. Another essential aspect deals with solubility problems and the use of vehicles in cell culture systems. In this context the major concern focuses on hydrophobic test materials in order to minimize false-negative results in comparison to in vivo findings. This is a crucial point since cellular systems are sensitive to various organic solvents. Moreover in some cases solvents on the one hand could quench photobiological reactions, which need to be carefully considered, and on the other hand could influence the bioavailability of the test substance to a large degree. This together with specific properties of the lamp can provide false-positive results.

For the determination of the potency of a photosensitizer as laboratory light source simulated sunlight is optimal. The characteristics of photon absorption should be close to that from sunlight. Usually xenon arc lamps or metal halide doted lamps are used. Care must be taken concerning the choice of the radiation range, since most types of mammalian cells in culture are very sensitive to the UVB light. This needs to be carefully filtered off, if effects induced by UVA should preferably be studied. This holds true in particular for keratinocytes and dermal fibroblasts as possible skin target cells. Cells derived from blood like erythrocytes or lymphocytes seem to be less sensitive to UVB light and they can be exposed to the entire spectrum of the sunlight without loss of sensitivity to UVA induced photosensitzation. As a result both systems, for instance fibroblasts under UVA and red blood cells under UVB/A irradiation, can be a useful set of in vitro methods. Care must also be taken to the infrared radiation which can cause heat problems in particular when xenon arc lamps are used.

The lamp spectrum must be carefully defined and controlled during experimentation. Cut-off filter for the short-waved UV light are needed to minimize the direct cytotoxic effect of the light source. The spectral intensity of a solar simulator should be measured and controlled by spectroradiometry. For calibration purposes broadband radiometers can be used. In daily practice positive and negative controls must be used to detect irregularities in the test equipment.

The choice of the range of test concentrations is relevant for further assessments. A realistic relationship should exist between the concentrations used in vitro and the amount of test article applied in vivo. In this context, knowledge about the relevant tissue concentrations, obtained according to the application route of the test material, is needed. For that reason it was recommended to always determine concentration-response relationships instead of using single fixed doses only.

All these items have to be considered during the test development and put together in the final standard protocol. For the above mentioned tests the standard protocols are stored in the files of the INVITTOX Data Bank.

Methods Under Validation

The joint EC/COLIPA validation project on photoirritation testing in vitro was started with a phase of prevalidation. This first phase was designed as a pilot study with the aim to identify those protocols which seemed to be the most promising. Prior to the beginning of the study it was agreed that the NRU Phototoxicity test using 3T3 Balb/c mouse fibroblasts was to be performed in all laboratories as a common standard operating procedure. The other tests practiced in one or more different laboratories were as follows:

(1) the Red Blood Cell (RBC) haemolysis test and
(2) the Haemoglobin oxidation test (in combination),
(3) the Candida albicans phototoxicity test,
(4) a human lymphocyte growth inhibition test (MTT),
(5) a human keratinocyte growth inhibition test (NRU),
(6) SOLATEX-PI™, and the
(7) Skin² ™ Phototoxicity assay.

The tests are briefly described elsewhere (Liebsch et al. 1994). The human keratinocyte phototoxicity test uses the same protocol as the 3T3 mouse fibroblast NRU phototoxicity test.

The light sources used for some of these tests was the Sol 500 sun simulator (Dr. Hönle, München) supplied with a H1-Filter (50% transmission at 335 nm) for the cell culture growth inbibition tests like the 3T3 NRU and the kerytinocytes NRU phtotoxicity test. For the RBC haemolysis test combined with the Hb-oxidation a H2-Filter was used (50% transmission at 320 nm and cut-off wave-length at 305 nm) to include the so-called UVB range. This light source which allows to illuminate a larger area homogenously was also used and proposed recently for in vivo standard test protocols (Nilsson et al. 1993).

The test chemicals used during the prevalidation study are listed below. Eleven have been identified as being photoirritant (-toxic) and nine as being non-phototoxic:

1. Photoirritant Test Chemicals: Amiodarone, Bithionol,Chlorpromazine, Doxycycline, 8-Methoxypsoralene, 6-Methylcoumarin, Neutral Red, Promethazine, Rose Bengal, Tetrachlor-salicylanilide, Tetracycline

2. Non-photoirritant Chemicals: p-Aminobenzoic acid, 5-Benzoyl-4-hydroxy-2-methoxybenzenesulfonic acid, Cinnamic aldehyde, Chlorhexidine, Histidine, Laurylsulfate, Penicillin, Piroxicam, Thiourea.

Discussion of the Methods and Results

Since overviews about the experimental results have been discussed briefly in the scientific literature the reader is referred to those papers published by Spielmann et al. (1993) and Liebsch et al. (1994).

The two test protocols for the 3T3 NRU phototoxicity test and the combined RBC test including photohaemolysis and Hb-oxidation as endpoints were designed to be used in a sort of battery. The NRU test protocol is - due to the sensitivity of the fibroblasts (mouse cell line or human dermal (unpulished results)) or keratinocytes (human) to short-waved UVB light - restricted to the use of properly filtered simulated sunlight comprising only UVA and visible light. This was necessary in order to achieve sufficiently long exposure times and irradiation doses without resulting in relevant light-induced cytotoxicity. The test conditions were controlled for the light source and the cells under use by a negative control without photosensitizer and a positive control after application of a relevant range of test concentration of chlorpromazine.

To make the test sensitive the test chemicals or articles are dissolved in Earle's Balanced Salt Solution (EBSS) as a minimum medium not containing growth factors, other proteins, or chemicals sensitive to light. The chemicals are dissolved in EBSS, if necessary due to solubility problems in the presence of DMSO ot ethanol (up to 1 % final concentration) and preincubated with the cells for one hour prior to the irradiation period of 50 min (= 5 J UVA/cm^2) at room temperature. The preincubation was necessary to achieve equilibrium between solution and the cells.

The use of EBSS is limited in time and cells have to be controlled morphologically, if the incubation time is extended. After irradition, with the dark control in run in parallel, the test chemical is removed by thorough washing with normal culture medium and the cells were left in the incubator for about 20 h prior to the NRU staining procedure.

It was decided by the expert group to evaluate the concentration response curves of both the phototoxicity test and the dark control by calculating the photoirritation factor PIF. The PIF was calculated by division of the IC_{50} values of the dark reaction by that obtained from the irradiated situation. This figure was used as a first approach in deciding whether a substance is identified as a phototoxicant or not under the conditions of this test. These results can easily be compared with yes/no estimates of in vivo effects taken from the scientific literature, keeping in mind that there are nearly no two papers published in this area using similar in vivo protocols. Surprisingly, a cut-off factor of 5 was found when comparing in vivo and in vitro findings between non-phototoxic and phototoxic materials.

The Red Blood Cell test protocol, as used in our laboratory, comprises two endpoints. According to the protocol the test chemicals are dissolved in PBS, if necessary with addition of ethanol or DMSO (up to max. 1 % final concentration). The test mixture and the corresponding controls are exposed to simulated sunlight - including UVB irradiation - for 2.5 h (= 15 J UVA/cm² and 1-1.5 J UVB/cm²) at room temperature. The light exposure is followed by 30 min. dark incubation prior to further analysis. The first endpoint is the well-known and established photohaemolysis. This endpoint is generally considered to be linked to membrane- related photodynamic reactions and thus a number of photosensitzers do neither reveal photohaemolysis nor haemolysis in the dark control. But as observed during our studies in by far most cases photodynamic action is also linked to the formation of methaemoglobin. This reaction can occur under different conditions. If no haemolysis takes place its formation can be observed intracellularly. If cells are lysed in most cases methaemoglobin is formed, too. Thus photodynamic effects monitored by the oxidation of oxyhaemoglobin were detected to different degrees with all photosensitizers tested. Photohaemolysis was detected with most phototoxicants with the exception of 8-methoxypsoralene and the two tetracyclines. Due to the UVB radiation present a weak photohaemolysis was also seen with chlorhexidine.

As a result it can be stated that all photosensitizers showed methaemoglobin formation and most lead to photohaemolysis, too.

Based on these promising results the second phase of the joint EU/COLIPA project was designed as a formal validation study with the goal 'to determine if currently selected in vitro methods are capable of properly predicting the photoirritation potential to humans of chemicals applied via the systemic route or topically to the skin. The 3T3 Phototoxicity NRU Test was again performed in all participating laboratories (9 laboratories) whereas three laboratories decided to use the RBC Phototoxicity Test, too.

The major characteristics of this study were that a) the standard operating procedures (SOPs) had been finalized prior to the beginning of the study and were stored under embargo in the INVITTOX Data Bank, b) the test chemicals had been selected out of a list of substances and were coded and shipped independently by BIBRA, and c) all experimental data were finally submitted in an agreed pre-defined format to the independent statistician for evaluation according to pre-defined statistical evaluation procedures and test models.

For the NRU phototoxicity test the model was derived from phase I of the study with a defined cut-off value for the borderline between phototoxic and non-phototoxic materials. The complex combination of two end points for the RBC phototoxicity test comprised the determination of the metHb formation linked to the oxidation of oxyhaemoglobin and the resulting photohaemolysis. Both end points were screened in a range-finding experiment prior to the fine tuning experiment for the photohaemolysis to determine the 50 % grade of haemolysis induced in the light and the dark reaction.

At present final conclusions could not be drawn since the test results for the core tests are still under evaluation. But nevertheless it can be stated the the overall outcome of in particular the NRU phototoxicity test is satisfactory, whereas the combined RBC phototoxicity test protocol needs further consideration, in particular concerning the determination and evaluation of the two end points, which is more difficult but includes some insight into the mechanisms at cellular level. Detailed information will be available in the near future after finalisation of the official data evaluation.

Concerning the RBC Phototoxicity test the photooxidation of oxyhaemoglobin seems to be the more useful endpoint for a general screening.

Other interesting mechanistic tests are the determination of histidine oxidation by singlet oxygen and the photoinduced binding to protein. For this purpose human serum albumin is often used as the standard protein. Both protocols have not been part of the formal validation. But in our experience these tests can provide supportive data. Moreover, interesting results have also been obtained with a test based on skin equivalents, which can in particular be helpful if poorly soluble or disperable test materials should be studied.

Conclusions

At the present state it can be concluded from the outcome of the validation project that the 3T3 mouse fibroblast photoirritancy test reveals a good correlation with the assessment based on in vivo information. This test can be used as a tool in screening phototoxicity in order to identify phototoxic hazard.

The findings of this test can be supported by mechanistically-based tests like the combined Red Blood Cell Test to confirm a phototoxic reaction by evident oxidation of oxyhaemoglobin and/or photohaemolysis depending on the photodynamic activity of the chemical under study. Photoirritancy can further be identified by performing the Histidine oxidation test which is easily done with reliable results. Additional information can be provided by the use of tests dealing with the light-induced binding of photosensitzers to proteins and the use of reconstituted skin equivalents to overcome for instance application problems of test materials. But these tests have not been part of the formal validation study.

A general problem remains in the extrapolation between the in vitro identification of photoreactions and the transfer to an assessment of in vivo relevance, as already discussed elsewhere (Maurer, 1987). This step needs further consideration, discussion and input.

References

Balls M, Blaubooer B, Brusick D, Frazier J, Lamb D, Pemberton M, Reinhardt C, Roberfroid M, Rosenkranz H, Schmid B, Spielmann H, Stammati AL, Walum E (1990) Report and recommendations of the CAAT/ERGATT workshop on the validation of toxicity test procedures. ATLA 18: 313 - 337.

Castell JV, Gómez-Lechón MJ, Martinez LA, Miranda MA (1994) Photodynamic lipidperoxidation by suprofen and tiaprofenic acid. Photochemistry and Photobiology 59: 35-39

Johnson BE, Walker EM, Hetherington AM (1986) In vitro models for cutaneous phototxicity. In: Skin Models, Models to Study Function and Desease of Skin (eds. R Marks & G Plewig), Springer Verlag, Berlin, pp. 264-281.

Koeter, HBWM (1995) Validation: A Highly Charged Concept. Tox. in vitro 9: 851 -856 .

Liebsch M, Spielmann H, Balls M, Brandt M, Döring B, Dupuis J, Holzhuetter HG, Kalweit S, Klecak G, L'Eplattenier H, Lovell WW, Maurer T, Moldenhauer F, Moore L, Pape WJW, Pfannenbecker U, Potthast J, De Silva O, Steiling W, Willshaw A (1994) First results of the EC/COLIPA validation project: in vitro phototoxicity testing. In: Alternative Methods in Toxicology, Vol.10, In Vitro Skin Toxicology - Irritation, Phototoxicity, Sensitization (eds. A. Rougiers, AM Goldberg & HI Maibach), Mary Ann Liebert Inc., New York, pp. 243-251

Lovell WW (1993) A scheme for in vitro screening of substances for photoallergenic potential. Toxic. in vitro 7:95-102

Lovell WW (1994) Assessing phototoxicity by in vitro methods: A Review. In: Alternative Methods in Toxicology, Vol.10, In Vitro Skin Toxicology - Irritation, Phototoxicity, Sensitization (eds. A. Rougiers, AM Goldberg & HI Maibach) Mary Ann Liebert, New York, pp. 195-202..

Maier K, Schmitt-LLandgraf R, Siegemund B (1991) Devellopement of an in vitro test system with human skincells for evaluation of phototoxicity. Toxic. in vitro 5: 457-461

Maurer T (1987) Phototoxicity testingin vivo and in vitro. Fd. Chem. Toxic. 25:407-414

Nilsson R, Maurer T, Redmond R (1993) A standard protocol for phototoxicity testing. Contact Dermatitis 28:285-290

Pape WJW, Brandt M, Pfannenbecker U (1994) Photohaemolysis and Hb-oxidation combined as part of in vitro phototoxicity testing examined with different phototxic substances. Toxic. in vitro 8: 755-757

Pape WJW, U. Pfannenbecker, Diembeck W (1994) A strategic approach for in vitro phototxicity testing. In: Alternative Methods in Toxicology, Vol.10, In Vitro Skin Toxicology - Irritation, Phototoxicity, Sensitization (eds. A. Rougiers, AM Goldberg & HI Maibach), Mary Ann Liebert Inc, New York,. pp. 203-212.

Spielmann H, Balls M, Brandt M, Döring B, Holzhuetter HG, Kalweit S, Klecak G, Liebsch M, L'Eplattenier H, Lovell WW, Maurer T, Moldenhauer F, Moore L, Pape WJW, Pfannenbecker U, Potthast J, De Silva O, Steiling W, Willshaw A (1994) EC/COLIPA project on in vitro photoirritancy testing: first results obtained with a Balb/c 3T3 cell phototoxicity test. Toxic. in vitro 8:

Assessment of the Phototoxicity Risk of New Drugs

Miguel A. Miranda
Departamento de Química/Instituto de Tecnología Química UPV-CSIC, Universidad Politécnica de Valencia, Camino de Vera s/n; Apartado 22012, E-46071 Valencia (Spain)

Phototoxicity and Photoallergy

Photosensitization is the general term to define abnormal reactions of the skin, due to the exposure to UV or visible radiation and an exogenous chemical (Spikes, 1989; Harber and Bickers, 1988; Johnson and Ferguson, 1990). *Phototoxicity* and *photoallergy* are the two major types of skin photosensitization (Beijergsbergen van Henegouwen, 1981; Epstein, 1983; Epstein and Wintroub, 1985; Miranda, 1992; Spikes, 1989). The former appears more frequently and does not involve an immunological mechanism. It should occur in all individuals provided that their skin contains the photosensitizer at the appropriate concentration and is exposed to high enough light doses. Usually, the clinical picture associated with phototoxicity includes burning, itching, erythema, oedema, hyperpigmentation and desquamation. These reactions are confined to the exposed areas of the skin and appear immediately after the first exposure. Photoallergy to chemicals is much less common, although it is still important in interest and severity. It depends on the individual immunological status and is developed in some patients as a modified ability of the skin to react to the combined effect of sunlight plus a photosensitizer. It is associated with a cell-mediated immunity response (Johnson and Ferguson, 1990). Some clinical manifestations are erythema, eczema and/or vesiculobullous eruption. These cutaneous responses are usually restricted to the exposed parts of the skin, but can be extended to the unexposed areas, including previously sensitized distant sites. They do not occur after the first treatment with chemical and light; rather, there is an induction period of one or two weeks before skin reactivity can be demonstrated (Epstein, 1983; Epstein and Wintroub, 1985).

Many drugs with diverse pharmacological activities and different structural formulae are known to induce phototoxic and/or photoallergic responses either after systemic or topical application. Hence, it has been difficult to establish satisfactory structure-activity relationships (Beijersbergen van Henegouwen, 1981; Miranda, 1992; Spielmann et al., 1994). In order to advance satisfactory predictions a deeper understanding of the molecular mechanisms involved in these phenomena appears necessary.

Among photosensitizing drugs, the nonsteroidal anti-inflammatory 2-arylpropionic acids deserve special interest. Since these agents have been designed to inhibit inflammation, the fact that they are able to elicit inflammatory responses in conjunction with light is hardly acceptable (Kochevar, 1989; Miranda, 1992). Besides, the incidence of photosensitization disorders associated with 2-arylpropionic acids is higher than with other types of drugs (Kochevar, 1989). We shall make preferential use of the information available on these drugs to illustrate the general aspects of the topic.

Phototoxicity Mechanisms

The first event in the photosensitization by drugs is absorption of a photon, which promotes the drug to its excited singlet state. This is a high energy species, which can undergo deactivation following different pathways (Miranda, 1992; Moysan et al., 1994; Spikes, 1989): a) fluorescence emission, b) internal conversion to the ground state, c) intersystem crossing to an excited triplet state, d) ionization to give a radical cation plus an ejected electron, e) unimolecular reactions such as fragmentation to free radicals or transformation into photoproducts, f) intermolecular processes, such as energy transfer, electron transfer or chemical reactions. The partner reactants are usually a biomolecule or oxygen present in the medium.

After intersystem crossing (path c), the resulting *excited triplet* can undergo similar deactivation processes: g) phosphorescence emission, h) intersystem crossing to the ground state, i) ionization, j) unimolecular photochemical reactions, k) intermolecular processes (energy/electron transfer or chemical reactions). Due to the higher lifetimes of triplet states, these processes are more likely to occur than those taking place from the singlet states.

Processes d), e), j) and k) are more frequently involved in the mechanisms of photosensitization by drugs. The so-called *types I and II mechanisms* are oxygen-dependent. The former is based on initiation of a chain reaction by free radicals, while the latter is associated with singlet oxygen. By contrast, *type III* photosensitization processes do not involve oxygen. Instead, stable photoproducts of the drug or covalent photoadducts of the drug to biomolecules are in the origin of the observed photobiological effects.

A Mechanistic Approach to Phototoxicity

In view of the operating mechanisms, the following aspects should be taken into account for the systematic evaluation of a potentially phototoxic drug: a) photophysics and photochemistry of the drug (photodegradation in different media, photoproducts formed, chemical and spectroscopic evidences for the involvement of highly reactive intermediates, etc.), b) photochemical interaction of the drug with relevant biomolecules (lipids, proteins, DNA, or their simple building blocks), c) effects produced by co-irradiation of the drug with cells or other biological test systems (photosensitized damage, alteration, disfunction, etc.), d) *in vitro/in vivo* correlations, and e) structure-activity relationships (identification of the chromophores responsible for phototoxicity, quantitative correlations).

Photophysical Studies on Drugs

Relevant data on the mechanism of action of phototoxic drugs can be obtained from the study of their radiative processes. Determination of the UV-Visible absorption spectra is important, since only absorbed light can initiate a photochemical reaction. Emission spectroscopy is also very informative, and the required technology has advanced enormously over the last decades through introduction of the time-resolved mode. It is possible to calculate the singlet and triplet excited state energies from the wavelengths of the fluorescence or phosphorescence maxima, respectively. Emission quantum yields, radiative lifetimes and quenching parameters provide valuable information on the excited states, such as generation efficiency, duration and reactivity (Moysan et al., 1994).

A very useful technique is flash photolysis (Moore and Chappuis, 1988; Navaratnam et al., 1993). The drug is subjected to a very short pulse of intense radiation produced by a flashlamp or a laser. This generates an instantaneous high concentration of excited molecules, which are directly observed by emission or absorption spectroscopy. As the pulse duration is much shorter than the decay processes, the fate of excited states can be followed in real time. Hence, flash photolysis allows not only determination of excited states energies and lifetimes or triplet quantum yields, but also obtention of triplet-triplet absorption spectra. Moreover, it can be extended to detect other transient species such as free radicals, which survive for a period much longer than the radiation pulse and possess characteristic absorption features.

Photochemistry of Drugs

Most frequently, the photochemical reactivity of a drug is established by steady state photolysis. The resulting reaction mixtures are resolved by means of preparative chromatographic methods, to give the pure photoproducts. Their structures are assigned by spectral analysis. Isolation and identification of the photoproducts allows to propose photochemical mechanisms. In addition, the photoproducts may exhibit toxicity in the dark or act as photosensitizers in their own right.

The photochemical transformations of a wide number of drugs have been illustrated in a well-documented review (Greenhill and McLelland, 1990). Hence, our attention will be focused on the nonsteroidal anti-inflammatory 2-aryl-propionic acids, because of their interest as representative compounds in the field of photosensitization by drugs. It has been shown that these compounds undergo photodecarboxylation, to afford complex mixtures of photoproducts (Boscá et al., 1994 and 1995; Castell et al., 1993 and 1994; Costanzo et al., 1989; De Guidi et al., 1993 and 1994; Miranda et al., 1991; Moore and Chappuis, 1988; Reszka and Chignell, 1983). Their formation involves the intermediacy of free radicals and drug-derived peroxides. In addition, reactive oxygen species such as singlet oxygen and/or superoxide radical anion are produced. Depending on the structure of the aryl substituent, other photochemical reactions can also occur.

In certain cases, cellular damage by stable photoproducts (in the dark) has been proposed as a major mechanistic route in the photosensitization by drugs (Bakri et al., 1983; Beijersbergen van Henegouwen, 1981; Greenhill and McLelland, 1990). This does not appear to be the case with 2-arylpropionic acids. Although as a general rule the most toxic photoproducts in different *in vitro* assays are the decarboxylated 1-hydroxyethyl derivatives (Castell et al., 1993; Costanzo et al., 1989; Miranda et al., 1991), the toxic concentrations required are not likely to be reached in the tissues (Boscá et al., 1995). However, some members of the group (benoxaprofen, tiaprofenic acid, suprofen and ketoprofen) give mainly photoproducts with ethyl side chain (Boscá et al., 1994; Castell et al., 1993, 1994 and 1994 a; Reszka and Chignell, 1983). These lipophylic photoproducts accumulate in cellular membranes, where they produce cell lysis upon light exposure (Kochevar et al., 1984; Castell et al., 1994 a). Thus, the photobiological activity might be associated in these cases with the formation of photoproducts with high photosensitizing ability.

In Vivo Phototoxicity Testing

Animal models have been employed for predictive purposes with some success (Ljunggren and Lundberg, 1985; Maurer, 1987; Spielmann et al., 1994). Phototoxic reactions have been induced in laboratory animals after epidermal, intraperitoneal or oral administration of phototoxic compounds followed by irradiation with different light sources (xenon or mercury arc lamps, fluorescent tubes, monochromatic light sources, etc.). Most frequently used species are mice and guinea-pigs, although rabbits, hamsters and several other animals have been employed as well. Hairless rats, mice and guinea-pigs are now available; their use avoids problems of hair removal. Evaluation of the phototoxic reaction is dependent on the species used. In mice, it involves the quantitative assessment of oedema, by measuring either the weight increase of an exposed organ (tail, skin, etc.) or its increase in thickness (skin, ear). By contrast, in guinea-pigs erythema reactions are more pronounced.

Although the above *in vivo* methodology has been found to be suitable for several phototoxic drug groups, the use of animals is expensive, time-consuming and requires a considerable degree of technical ability. Moreover, testing in animals is difficult to standardize, due to the variability of a number of factors such as species, skin properties, hair charecteristics or metabolic patterns. Moreover, there is an ethical barrier to the unjustified or excessive use of experimental studies with laboratory animals, which becomes increasingly reduced and restricted by recent regulatory guidelines (Spielmann et al., 1994).

In Vitro Phototoxicity Assays

The phototoxic potential of new drugs should be detected at an early stage of their development by means of a simple, rapid, inexpensive and sensitive assay requiring readily available materials. Such an assay should be easily adaptable to a routine procedure, in order to screen a large number of molecules for photosensitizing ability. All these requirements can in principle be fulfilled by *in vitro* methods (Miranda, 1992; Moysan et al., 1994; Spielmann et al., 1994). A considerable number of methods have been developed both for general screening purposes and for elucidation of specific mechanisms. They typically consist in the exposure of a biological target to the photosensitizer plus ultraviolet radiation, followed by evaluation of the produced effects through measurement of an adequate endpoint. Photohemolysis and photoinduced yeast growth inhibition are old and well established *in vitro* assays supported by large databases. Other tests systems involve co-irradiation of the photosensitizer either with cells (fibroblasts, keratinocytes, hepatocytes, lymphocytes, etc.) or with isolated biomolecules (polyunsaturated fatty acids, simple aminoacids, proteins, DNA).

As stated above, we shall focus our attention on 2-arylpropionic acids as representative photosensitizing agents, trying to describe some interesting applications of *in vitro* methodologies to this family of nonsteroidal anti-inflammatory drugs.

Photohemolysis: Photohemolysis reflects the impairment of the function and integrity of cellular membranes after exposure to light plus a photosensitizing chemical (Barth et al., 1977). Hence, false negative predictions must be expected for compounds acting *via* photodamage to DNA or other biological targets different from the membrane components. With antiinflammatory 2-arylpropionic acids, photohemolysis has been quantified in two ways: 1) as the release of hemoglobin, through the absorbance of the supernatants (after centrifugation) at 540 nm and 2) by the decreasing number of intact red blood cells, which is proportional to the optical density at 650 nm. Application of this assay has confirmed the phototoxicity of benoxaprofen, naproxen, ketoprofen, carprofen, suprofen and tiaprofenic acid (Boscá et al, 1994 and 1995; Costanzo et al., 1989; De Guidi et al., 1993 and 1994, Ljunggren, 1985, Miranda et al., 1993). The variations produced by withdrawal of oxygen or incorporation of different additives (radical scavengers, singlet oxygen quenchers, etc.) are related to the involved photosensitization pathways, which allows to employ this model system for mechanistic investigations.

Photodynamic lipid peroxidation: The photodynamic action of a photosensitizer is defined as the ability to mediate light-induced oxidation of biological substrates in the presence of oxygen. It is known to produce detrimental effects on cell membranes, so it must play an important role in skin phototoxicity (Girotti, 1990).

Lipids are good model systems to investigate photodynamic properties *in vitro*. Pure linoleic acid (Boscá et al., 1994 and 1995; Castell et al., 1994; Miranda et al., 1993) has been employed to assess the phototoxicity of 2-arylpropionic acids. The linoleic acid method is simple and expedient and involves the use of a well-defined reaction medium. Through the addition of mechanistic probes, it is possible to determine the photosensitization pathways involved (Boscá et al., 1994 and 1995; Castell et al., 1994). However, it detects photosensitization at the membrane level; thus, damage to DNA or other cell constituents may pass unnoticed. In aqueous solution, ketoprofen, tiaprofenic acid and suprofen are able to photosensitize linoleic acid peroxidation, as determined spectrophotometrically, through the appearance and subsequent increase of a band corresponding to the conjugated dienic hydroperoxides at 233 nm (Boscá et al., 1994 and 1995, Castell et al., 1994; Miranda et al., 1993). In this case, a type I mechanism appears to be responsible for the observed photoperoxidation, since it is dramatically inhibited by butylated hydroxyanisole and reduced glutathion but it is not affected by diazabicyclooctane.

Photodamage to fibroblasts: A different type of approach to phototoxicity testing involves exposure of selected target cells to the simultaneous effect of potential photosensitizers and ultraviolet radiation. This is followed by evaluation of the produced cellular damage by means of an adequate endpoint. Several cell systems have been tried (fibroblasts, keratinocytes, lymphocytes, hepatocytes, etc.), but the most promising results have been obtained with fibroblasts, either as cultured cells or as continuous cell lines (3T3 or A431). On the other hand, two types of toxicity endpoints have been used: a) cell viability, determined either through the mitochondrial dehydrogenase conversion of a tetrazolium salt (MTT) to a coloured formazan product or by means of the neutral red uptake (NRU) (Duffy et al. 1989; Spielmann et al., 1994) and b) membrane damage, by measuring the decrease of intracellular lactate dehydrogenase (LDH) activity as indicator of enzyme leakage (Boscá et al., 1995; Castell et al., 1994 a).

Concerning the application of these phototoxicity tests to 2-arylpropionic acids, co-irradiation of tiaprofenic acid or ketoprofen with human fibroblasts produces photodamage to the cellular membranes, as indicated by the LDH method (Boscá et al., 1995; Castell et al., 1994 a). In an independent study conducted with 3T3 cell lines, a significant decrease of cell viability (MTT test) has been observed upon irradiation in the presence of benoxaprofen (Duffy et al., 1989).

In Vitro-in Vivo Correlations

Ideally, it should be possible to reproduce the *in vivo* situation by means of *in vitro* experiments. However, possible discrepancies between both sets of data can arise as a consequence of the following factors: a) absorption of radiation by skin components, which determines the amount of light interacting with the drug, b) deactivation of the excited drug by quenchers present under *in vivo* conditions, c) distribution of the drug and concentration reached at the skin, d) biotransformation of the drug, and e) repair processes in response to photodamage. These factors can influence the course of the photochemical reactions and, as a consequence, the associated phototoxic effects.

Structure-Activity Relationships

The properties of a drug are implicit in its chemical structure; this is the principle underlying the development of quantitative structure-activity relationships (QSARs). If the mechanistic basis for a particular property of a group of related drugs can be elucidated and the relevant physicochemical or structural parameters measured or calculated, then in principle a QSAR can be established (Barratt et al., 1996). Thus, phototoxic activity can be formulated as a function of oil/water partition (as indicator for skin permeability of the drug and/or systemic distribution) and photochemical reactivity (quantum yield of a given process such as triplet formation, singlet oxygen production or photodegradation). In principle, evaluation of the results should permit to identify the chemical substructures present in the most relevant compounds and to disclose the active chromophores with an associated high phototoxic risk.

Conclusion

Although systematic phototoxicity testing is still not included in the guidelines for safety evaluation of new drug candidates, this situation will probably change in the near future. *In vivo* methods can in principle be suitable for such a purpose; however they are expensive, time-consuming and difficult to standardize. Besides, there is a trend to restrict the use of animals for ethical reasons. By contrast, *in vitro* phototoxicity assays can be simple, inexpensive, sensitive and easily adaptable to a routine procedure.

A better understanding of the molecular basis of phototoxicity must be achieved in order to improve the predictive ability. The introduction of new phototoxicity endpoints (for instance, release of inflammatory mediators), able to detect early cellular responses prior to the advent of irreversible damage, might result in an increased sensitivity. On the other hand, the adequate use of chemical or biochemical assays specific for each possible mechanistic route might allow to find correlations between the photochemical mechanism involved and the observed biological effects. Finally, identification of the substructures (chromophores) responsible for phototoxicity and the establishment of quantitative structure-activity relationships would enable to improvements in drug design.

Acknowledgement

Financial support by the Fondo de Investigaciones Sanitarias (Grant FIS 95 1498) is gratefully acknowledged.

References

Bakri A, Beijersbergen van Henegouwen GMJ, Chanal JL (1983) Photopharmacology of the tranquilizer chlordiazepoxide in relation to its phototoxicity. Photochem Photobiol 38: 177-183

Barratt MD, Dixit MB, Jones PA (1996) The use of *in vitro* cytotoxicity measurements for the prediction of the skin corrosivity potential of acids: a QSAR method. Toxic in vitro 10: 283-290

Beijersbergen van Henegouwen GMJ (1981) The interference of light in pharmacotherapy. Pharmaceutisch Weekblad, Sci Ed 3: 85-95

Boscá F, Miranda MA, Carganico G, Mauleón D (1994) Photochemical and photobiological properties of ketoprofen associated with the benzophenone chromophore. Photochem Photobiol 60: 96-101

Boscá F, Garganico G, Castell JV, Gómez-Lechón MJ, Hernández D, Mauleón D, Martínez LA , Miranda MA (1995) Evaluation of ketoprofen (R, S, and R/S) phototoxicity by a battery of *in vitro* assays. J Photochem Photobiol B: Biol 99: 1-6

Castell JV, Gómez-Lechón MJ, Grassa C, Martínez LA, Miranda MA, Tárrega P (1993) Involvement of drug-derived peroxides in the phototoxicity of naproxen and tiaprofenic acid. Photochem Photobiol 57: 486-490

Castell JV, Gómez-Lechón MJ, Grassa C, Martínez LA, Miranda MA, Tárrega P (1994) Photodynamic lipid peroxidation by suprofen and tiaprofenic acid. Photochem Photobiol 59: 35-39

Castell JV, Gómez-Lechón, MJ, Hernández D, Martínez LA, Miranda MA (1994 a) Molecular bases of drug phototoxicity: photosensitized cell membrane damage by the major photoproduct of tiaprofenic acid. Photochem Photobiol 60: 586-590

Costanzo LL, De Guidi G, Condorelli G, Cambria A, Famà M (1989) Molecular mechanism of drug photosensitization-II. Photohemolysis sensitized by ketoprofen. Photochem Photobiol 50: 359-365

De Guidi G, Chillemi R, Costanzo LL, Giuffrida S, Condorelli G (1993) Molecular mechanisms of drug photosensitization. 4. Photohemolysis sensitized by carprofen. J Photochem Photobiol, B: Biol 17: 239-246

De Guidi G, Chillemi R, Costanzo LL, Giuffrida S, Sortino S, Condorelli G (1994) Molecular mechanism of drug photosensitization. 5. Photohemolysis sensitized by suprofen. J Photochem Photobiol, B: Biol 23: 125-133

Duffy PA, Bennett A, Roberts M, Flint OP (1989) The prediction of phototoxic potential using human A431 cells and mouse 3T3 cells. Altern Methods Toxicol 7: 327-335

Epstein JH (1983) Phototoxicity and photoallergy in man. J Am Acad Dermatol 8: 141-147

Epstein JH, Wintroub BU (1985) Photosensitization due to drugs. Drugs 30: 42-57

Girotti AW (1990) Photodynamic lipid peroxidation in biological systems. Photochem Photobiol 51: 497-509

Greenhill JV, McLelland MA (1990) Photodecomposition of drugs. Progr Med Chem 27: 51-121

Harber LC, Bickers DR (1988). Photosensitivity Diseases: Principles of Diagnosis and Treatment, Saunders, Philadelphia PA

Johnson BE, Ferguson J (1990) Drug and chemical photosensitivity. Semin Dermatol 9: 39-46

Kahn G, Fleischaker B (1971) Red blood cell hemolysis by photosensitizing compounds. J Invest Dermatol 56: 85-90

Kochevar IE (1989) Phototoxicity of nonsteroidal anti-inflammatory drugs: coincidence or specific mechanism?. Arch Dermatol 125: 824-826.

Kochevar IE, Hoover KW, Gawienowski M (1984) Benoxaprofen photosensitization of cell membrane disruption. J Invest Dermatol 82: 214-218

Ljunggren B (1985) Propionic acid-derived non-steroidal antiinflammatory drugs are phototoxic *in vitro*. Photodermatol 2: 3-9

Ljunggren B, Lundberg K (1985) *In vivo* phototoxicity of non-steroidal anti-inflammatory drugs by the mouse tail technique. Photodermatol 2: 377-382

Maurer Th (1987) Phototoxicity testing - *in vivo* and *in vitro*. Fd Chem Toxic 25: 407-414

Miranda MA (1992) Phototoxicity of drugs In: Castell JV, Gómez-Lechón MJ (eds) *In vitro* alternatives to pharmaco-toxicology. Farmaindustria, Madrid, pp 239-270

Miranda MA, Morera I, Vargas F, Gómez-Lechón MJ, Castell JV (1991) *In vitro* assessment of the phototocicity of anti-inflammatory 2-arylpropionic acids. Toxic in vitro 5: 451-455

Miranda MA, Castell JV, Gómez-Lechón MJ, Martínez LA (1993) In vitro photoperoxidation as an indicator of the potential phototoxicity of non-steroidal anti-inflammatory 2-arylpropionic acids. Toxicol in vitro 7: 523-526

Moore DE, Chappuis PP (1988) A comparative study on the non-steroidal anti-inflammatory drugs, naproxen, benoxaprofen and indomethacin. Photochem Photobiol 47: 173-180

Moysan A, Miranda MA, Morliere P, Peno-Mazzarino L, Averbeck D, Averbeck S, Clement-Lacroix P, Michel L, Bazin M, Geze M, Santus R, Jeanmougin M, Castell JV, Dubertret L (1994) Evaluation of phototoxicity: from *in vitro* to *in vivo*. In: Rougier A, Goldberg AM, Maibach HI (eds) *In vitro* skin toxicology. Methods in irritation, phototoxicity and sensitization. MA Liebert Inc, New York, pp 185-194

Navaratnam S, Parsons BJ, Hughes JL (1993) Laser flash photolysis of benoxaprofen and its analogues 1. Yields of triplet states and singlet oxygen in acetonitrile solutions. J Photochem Photobiol, A: Chem 73: 97-103

Reszka K, Chignell CF (1983) Spectroscopic studies of cutaneous photosensitizing agents-IV. The photolysis of benoxaprofen, an anti-inflammatory drug with phototoxic properties. Photochem Photobiol 38: 281-291

Spielmann H, Lovell WW, Hšlzle E, Johnson BE, Maurer T, Miranda MA, Pape WJW, Sapora O, Sladowski D (1994) *In vitro* phototoxicity testing. The report and recommendations of the ECVAM Workshop 2. ATLA 22: 314-348

Spikes JD (1989) Photosensitization. In : Smith KC (ed) The Science of Photobiology, 2nd Ed . Plenum, New York, pp 79-110

Progress in Antidotic Therapy

(Chair: A. Ferrer, Spain)

Evolution of Antidotal Therapy in Recent Decades

J. Pronczuk de Garbino
International Programme on Chemical Safety (IPCS)(WHO/ILO/UNEP), World Health Organization, CH-1211 Genève, Switzerland

Antidotes are pharmaceuticals intended to counteract the noxious effects of chemicals, either man-made or of natural origin. They may modify the kinetics of a given toxic substance, its absorption or elimination, reduce its effects on the receptors, and, through different modes of action, ameliorate the functional and vital prognosis of poisonings. The IPCS[1] definition of "antidote": "therapeutic substance administered to counteract the adverse effects of a specific xenobiotic", allows the inclusion of anti-venoms and pharmaceuticals (such as glucose, vitamin K, diazepam, or isoprenaline), widely used in the medical setting but useful too in the management of certain type of poisonings. As in the medical practice antidotes, antagonists, antivenoms and pharmaceuticals are used in conjunction with techniques intended to reduce absorption or enhance elimination of poisons, also treatment procedures are to be considered when evaluating the more specific antidotes.

The evolution of antidotal therapy in the last several decades is characterised by: (i) increasing international concern about the efficacy of antidotes, leading to coordination of activities for their evaluation, (ii) the development of a number of new antidotes, including chelating agents and immunotoxicotherapy, and (iii) new approaches in the indications and use of antidotes.

Unfortunately, no major developments have taken place concerning: (a) the development of antidotes for extremely hazardous substances such as bipyridylic herbicides and aluminium phosphide, (b) expanding the availability of antidotes, especially to areas of real need (such as developing countries), and (c) the dissemination of up-to-date information concerning the real role of antidotes in the management of poisoned patients.

---

[1] The IPCS is a joint venture of the World Health Organization, the International Labour Office, and the United Nations Environment Programme

International activities

The IPCS has undertaken the task of evaluating about 50 antidotes and agents used as emetics, cathartics, electrolyte solutions, absorption reducers and anti-foaming agents. This initiative was recommended during a meeting of the World Federation of Associations of Poisons Centres and Clinical Toxicology and a number of poisons centres, hosted by the IPCS in Geneva, Switzerland in October 1985. As a result, during the last decade antidotes became, for the first time, the subject of international evaluations (the results of which are to be presented in another communication).

The **IPCS List of Antidotes**, prepared in 1993, contains : (i) 48 antidotes considered useful in the treatment of poisoning, (ii) 12 agents used to prevent the absorption of poisons, to enhance their elimination or provide symptomatic treatment, (iii) 19 useful therapeutic agents for the management of poisoning cases, and (iv) 23 antidotes and related agents considered obsolete, and whose use is no longer recommended.

The first group of antidotes has been classified in terms of urgency of availability:

A. Required to be immediately available (within 30 minutes)
B. Required to be available within 2 hours
C. Required to be available within 6 hours

And also in terms of proven efficacy:

1. *Efficacy well documented*
2. *Widely used, but requiring further research concerning efficacy and/or indications*
3. *Questionable usefulness*

This list has been widely distributed, and has helped to increase the availability of antidotes in a number of countries and also the awareness about the real value of antidotes in clinical practice (copies may be requested from the IPCS).

A chart of **Antidotes Dosages** is being prepared, in response to the request of a number of countries, where information concerning the recommended antidotes and their indications is poor or outdated. The chart contains: name of the antidotes (and synonyms), indication(s), initial and maintenance dose(s) (for children and adults), route(s) of administration, contraindications and observations. Such a chart will facilitate an appropriate use of antidotes, mainly in these parts of the world where information is not readily available.

In 1986, the IPCS and the Commission of the European Communities (CEC) undertook a joint programme to evaluate antidotes used clinically in the treatment of poisoning. A series of **Antidote Monographs** summarise and assess the clinical use, mode of action and efficacy of antidotes. It also provides information about routes of administration, contraindications and other observations. A number of volumes have already been published and other volumes are in preparation (this is the subject of another presentation). Other international (inter-country) initiatives have been undertaken by professionals in the area of clinical toxicology. However, multicenter research is still facing a number of difficulties, mostly in the financial and regulatory aspects.

New Antidotes

Antidotes have been in use since the early history of medicine, but a large number of these so-called antidotes proved to be ineffective or toxic (eg: some years ago, strychnine salts were still reccommended for the management of barbiturate overdoses). It is only in the recent decades that antidotes of real effectiveness were developed for specific poisonings. Toxicodynamic and toxicokinetic studies have allowed a sound, scientific approach to the development of antidotes, which are designed to counteract noxious effects in specific sites, in effectors or during different processes.

The antidotes of real value developed in the recent decades are:

Developed around:	ANTIDOTE	Indicated in poisoning/ envenomation by:
1973	Naloxone	Opiates
1974	N-acetylcysteine & Methionine	Paracetamol
1976	Fab fragments	Digoxin
1981	New oximes	Organophosphorus compounds
1984	Flumazenil	Benzodiazepines
1986	Hydroxocobalamine	Cyanide
1989	New chelating agents	Pb, Hg, As, ...

Naloxone (n-allyl-noroxymorphone) has markedly improved the management of opioid overdoses, as it rapidly reverses narcosis and respiratory depression. It is a highly valuable antidote, especially in circumstances where maintenance of ventilatory functions is difficult, as is the case in some hospitals in developing countries. A similar molecule, Naltrexone (Nalmefene), is used orally in the management of drug abuse (as it has a long-acting effect). It is interesting to note that new ways of administering opiate antagonists are being tested, such as implants with naltrexone that would be triggered by exogenous morphine administration (the morphine activates an enzyme which removes the protective coating of the antagonist nalmefene) (Nakayama et al, 1995).

N-acetyl-cysteine (NAC) and methionine were proposed around 1974 in the UK and USA as glutathione precursors and protectors against paracetamol-induced hepatic and renal toxicity. They proved to be effective, especially NAC, which is also easier to administer, and more widely used. Although NAC has to be given within 8 to 10 hours after exposure, it has been postulated that even late administration might protect the liver through a hemodynamic, vasodilating mechanism (Proudfoot, 1995). A combined formulation of paracetamol/methionine tablets is being proposed in order to prevent hepatic damage in case of overdose. Animal studies demonstrated that the simultaneous administration of methionine with paracetamol in overdose protects against death and liver injury from doses up to 750 mg/kg body weight of paracetamol (Jones, 1996).

The advent of digoxin antibody fragments was a major advance in toxicological therapy. The enzymatic cleavage of IgG immunoglobulins with papain was developed around 1960, but the first immunotherapy for digitalis overdose compounds proved successful in 1976 (Smith et al, 1976). The response to F(ab) is rapid, and symptoms improve within a few hours. No serious adverse effects have been reported, (except for erythema and face urticaria), but risks of hypokalemia and exacerbation of congestive heart failure exist (Antman. et al, 1990). The main limitation to their use is the high cost. Anti-digoxin Fab fragments have also been used to reverse toxicity of cardiac glycosides present in poisonous plants such as *Digitalis purpurea* and *Nerium oleander*. Trials have also been undertaken in the treatment of *Scilla maritima* (Sabouraud et al, 1990) and *Taxus baccata* poisonings (Cummins et al, 1990).

In 1993, colchicine specific F(ab) fragments were used in humans, with extraordinary success (Bismuth et al, 1995; Baud et al, 1995). Although in 1980 it was already demonstrated that colchicine would bind to specific antibodies, inducing a restoration of the activity of tubuline *in vitro* (Wolff et al, 1980), it took more than a decade for the practical application of this knowledge in human toxicology. Extensive work is being done for developing murine antibodies against tricyclic antidepressants (Dart et al, 1996) and experimental results with ovine desipramine-F(ab) are promising. Trials with anti-paraquat and anti-endotoxin monoclonal antibodies were unsuccessful. F(ab) fragments of equine origin against venoms of *Vipera aspis*, *Vipera berus* and *Vipera ammodytes* proved to be clinically effective and safe to use (no side effects, except for mild skin rash) (de Haro et al, 1996). A new, more purified, pasteurized preparation of equine origine called F(ab)2 appears to be more effective than the existing F(ab) fragments (Scherrman et al, 1996). A crotalid polyvalent F(ab) antivenin of ovine origin has been developed in the USA (Consroe et al, 1995).

F(ab) fragments are highly promising in the therapy of intoxications as they are very specific, do not induce anaphylaxis, and have a low molecular weight that allows fast interstitial diffusion and rapid protective effect. They are not metabolized, have a short half-life, and are eliminated rapidly in the urine (reducing the possibility of immunogenic effects). The only adverse effects reported are erythema and face urticaria.

The fact that the hydroxyl group of hydroxocobalamine (OHB12) could be replaced by a cyano group, generating cyanocobalamine (an atoxic compound excreted in the urine) is known since 1950, but the concept was only applied in the clinical area around 1986 (Brouard et al, 1987). In spite of its advantages (minor adverse reactions, immediate and efficacious anti-cyanide effects), the pharmaceutical was not widely used due to difficulties in its formulation and high costs. Nowadays, OHB12 has become more widely available (as a 5% solution or a 4 gram powder preparation) and gained an important role not only as antidote to acute occupational, accidental or intentional cyanide exposures but also in the management of more specific situations such as the treatment of fire victims (Baud et al, 1988), and also in the prophylaxis of endogenous cyanide exposure resulting from sodium nitroprusside infusion (Cottrell, 1982).

Flumazenil (1,4-imidazobenzodiazepine) is valuable both in the diagnosis and treatment of benzodiazepine overdoses, especially in reversing respiratory insufficiency. As the effects are short-lived, repeat administration of the antagonist is required. Patients may relapse into coma within minutes, and require close monitoring. This, along with the possibility of withdrawal syndrome in benzodiazepine addicted patients and arrhythmias in cyclic antidepressant overdoses, imposes some precaution in the use of flumazenil. Adverse effects observed include: increased blood pressure, tachycardia, tremors, dizziness, nausea, vomiting and unpleasant "waking" episodes (anxiety, restlessness and aggressive attitudes) (Weinbroun et al, 1996). Flumazenil has also been used in reversing overdoses with promethazine (Plant & MacLeod, 1994), Zolpidem (Lheureux et al, 1990), carbamazepine and alcohol (Martens et al, 1990) and other drugs. It remains a differential diagnostic tool in unclear cases of drug overdose (Hojer et al, 1990, Burkhart & Kulig, 1990), is useful for reversal of therapeutic sedation and in selected cases of benzodiazepine overdose (Hoffman & Goldfrank, 1995).

Chelating agents have been in existence and use for half a century for the treatment of exposure to metal and metalloid derivatives. For example, British Anti-Lewisite (BAL) has been in use since 1945. Chelating agents are also used in the treatment of cyanide poisoning and in the management of some pathophysiological disorders (e.g. hypercalcemia, Wilson's disease). Interest in chelators has been raised by the significant health and developmental problems posed by chronic lead exposure (especially in children) and also by occupational diseases resulting from exposure to metals in the workplace. Special concern about the toxic effects of chelating agents exists in countries where «#chelation therapies#» are offered by doubious clinics, and patients come to be «cured from metal overcharge» and from the effects of environmental pollution ...

The main change observed in chelation therapy during recent years are has been the substitution of Dimercaprol by new thiol-providers, such as meso-2,3-dimercaptosuccinic acid (DMSA, Succimer) for lead poisoning, and by the sodium salt of 2,3-dimercapto-1-propanesulfonic acid (DMPS, Dimaval) for treating lead, mercury and arsenic intoxication (Aposhian et al, 1995). Although these chelators have been used in the Eastern countries and in China for over three decades, it is only recently that their use has become generalized in Western countries. DMSA was approved by the US Food and Drug Administration only in 1991 for treating childhood lead poisoning. It is recommended in mild to moderate pediatric lead poisoning (Besunder et al, 1995), proposed in combined therapy with $CaNa_2EDTA$ for severe lead poisoning (Flora et al, 1995; Tandon et al, 1994) and has been used successfully in organoarsenate poisonings (Shum et al, 1995). DMSA has proven to be very effective in the treatment of mercury, methyl mercury and lead poisoning. Its oral administration in combination with absorbents, is under study (Reichl, 1995). DMPS has been used in combination with haemodialysis in the management of arsenic trioxide poisoning (Kruszewska et al, 1996).

Dimercaprol is in general no longer recommended, due to a number of disadvantages, including: (i) the liposolubility of the As-chelate may increase the cerebral concentration of As (Hoover et al, 1983), (ii) it cannot be used in case of renal insufficiency, rather common in severe arsenic poisoning, and (iii) adverse effects with potentially severe consequences may occur in patients who suffer from severe hypertension and tachycardia.

New chelators are now being developed, such as derivatives of DMSA: rac-2,3-dimercaptosuccinic acid (rac-DMSA) (Fang & Fernando, 1994) and the monoesthers mono-isoamyl (Mi-ADMS) and mono-n-hexyl (Mn-HDMS) that seem to be effective in the mobilization of inorganic mercury (Kostial et al, 1995). Oral chelation therapy with 1,2-dimethyl-3-hydroxy-pyrid-4-one (Deferipone) for iron and aluminium overloaded patients has proven successful (Kontoghiorghes, 1995). A new antidote, ethane-1-hydroxy-1,1-biphosphonate (EHPB) is being tested for uranium poisoning, and might offer promising results in the area of occupational health (Ubios et al, 1994).

Further developments and new therapeutic use of chelators are expected, especially if theories postulating the involvement of circulating essential metals in the pathogenesis of fulminant hepatic failure are proven. In fact, it has been suggested that the severity of multi organ failure resulting from hepatic failure might be due to circulating Fe and Cu ions that catalyse free radical reactions. If chelators prevent the circulation of those ions, free radical reactions might be blocked and progression of hepatic failure stopped (Evans et al, 1994).

New approaches in the use of antidotes

Approaches to the indication and administration of a number of antidotes has changed in recent years. We will take as examples some changes observed in the management of poisoning by organophosphorus pesticides and by cyanide, and also in the use of calcium salts in clinical toxicology.

The use of atropine and oximes has been recommended for many decades in poisoning by organophosphate (OP) pesticides, but dosage recommendations have been changed in the last years. Atropine administration has to be titrated according to the clinical status of the patient, and higher doses are generally recommended. Concerning the oximes, a readjustement of dosages according to the type of derivative used (e.g. pralidoximes, obidoximes) is proposed in order to reach appropriate therapeutic levels. The use of diazepam is postulated not only as anticonvulsant but also for increasing the tolerance to high atropine dosages and reducing the muscular fasciculations usually observed in OP poisonings. It is also suggested that sodium bicarbonate (intravenous) be added to treatment protocols, although no confirmation of its efficacy is available. The menace of chemical warfare with OP compounds has fostered the development of new oximes that seem to reactivate cholinesterases more effectively, and in a short time, such as the HI-6 oxime, that proved to be more effective than the conventional oximes in exposures to Soman and Fosdrin in the animal (Kassa, 1995). New methods for self-administration through the use of auto-injectors (e.g. containing HI-6 oxime 500 mg and atropine 2 mg) seem to be effective, and therapeutic concentrations of the antidotes are reached in serum within one minute (Nyberg et al, 1995).

The management of cyanide poisoning represents a good example of long-lasting controversies that were solved once scientists agreed on the importance of specifiying where, when and by whom were poisoning cases diagnosed and treated. Sodium nitrite and sodium thiosulphate have been recommended since 1942, dicobalt edetate (Kelocyanor) proposed in the seventies, hydroxocobalamine and oxygen by the end of the eighties. The main advance in the therapy of cyanide poisoning have been the early diagnosis followed by the immediate administration of the antidote available at the site of exposure, plus energic supportive care, oxygen administration and correction of acidosis. According to experts involved in the IPCS activities: (i) dicobalt edetate should be used in the occupational setting, if there is certainty about exposure; (ii) hydroxocobalamine is the antidote of choice, but has to be used in the hospital setting; (iii) induction of methaemoglobinaemia with nitrites is recommended only if other antidotes are not available; and (iv) oxygen should always be administered.

It is interesting to observe that some old pharmaceuticals are gaining recognition as effective antidotes. One example is calcium gluconate (and other soluble calcium salts), very effective in the management of hydrofluoric acid burns by subcutaneous injections (10% solution), local gel (2.5%), and also intra-arterial perfusion (Velvart 1983, Vance et al 1986, Caravati 1988.) Other applications of the salts include: fluoride salt ingestion, hypermagnesemia, hyperkalemia (unrelated to digitalis toxicity), tetanic manifestations due to oxalate or ethylene-glycol poisoning, hymenoptera stings and *Lactrodectus mactans* (Black widow) envenomation (Lheureux et al, 1990).

Conclusions

In the last decades, the evolution of antidotal therapy has been characterized by a number of developments, but also by some drawbacks.

First, growing global interest about the real efficacy of antidotes promoted a more scientific approach to the validation of antidotes, and also of other pharmaceuticals and treatment procedures used in the management of poisonings. Scientific co-operation promoted by international organizations proved to be highly useful, resulting in a number of activities and publications.

Second, a number of new, promising antidotes have been developed (especially in the area of immunotoxicotherapy), and new applications of existing, older pharmaceuticals are being proposed. These developments will certainly offer new therapeutic possibilities for some specific poisoning cases.

Third, concern about the adverse effects of antidotes used for extended periods of time, or applied prophylactically, is leading to increased research in this area. One example is the Gulf War Syndrome, where it was postulated that the mysterious collection of symptoms presented by some veterans could be due to the potential toxicity of multiple chemical exposures (insecticides and repellents) and prophylactic administration of pyridostigmine bromide as anti-nerve gas agent (Pennisi, 1996).

In spite of promising findings, no antidotes have yet been found for highly hazardous chemicals (such as bipyridylic herbicides), no antidotes for organophosphorus compounds are available in some developing countries plagued by pesticide poisonings, and indiscriminate use of obsolete antidotes is still occurring.

It is expected that in the future scientific research will allow the development of new, safer antidotes. But also it can be expected that up-to-date information will be made available to all countries, that adequate regulations will increase availability of antidotes, and that cost/benefit studies will prove the savings in public health systems and the other advantages offered by the rational use of appropriate antidotes.

References

Antman EM, Wenger TL, Butler VP, Haber E, Smith TW (1990) Treatment of 150 cases of life-threatening digitalis intoxication with digoxin-specific Fab antibody fragments. Final report of a multicenter study. Circulation 81(6):1744-52

Aposhian HV, Maiorino RM, Gonzalez-Ramirez D, Zuniga-Charles M, Xu Z, Hurlbut KM, Junco-Munoz P, Dart RC, Aposhian MM (1995) Mobilization of heavy metals by newer, therapeutically useful chelating agents. Toxicol 97:23-38.

Baud FJ, Barriot P, Toffis V, Riou B, Astier A (1988) Cyanide poisoning in fire victims. Intens. Care Med. 14:S289

Baud FJ, Sabouraud A, Vicaut E et al (1995) Treatment of severe colchicine overdose with colchicine-specific Fab fragments. N Engl J Med 332:642-645.

Besunder JB, Anderson RL, Super DM (1995) Short-term efficacy of oral dimercaptosuccinic acid in children with low to moderate lead intoxication. Pediatr 96:683-687.

Bismuth Ch, Baud FJ, Scherrmann JM (1995) Application humaine de l'immunothérapie dans l'intoxication par la colchicine. La Revue du Practicien 9, 291:14-16.

Brouard A, Blaisot B, Bismuth C (1987) Hydroxocobalamin in cyanide poisoning. J. Toxicol. Clin. Exp. 7:155-168.

Burkhart KK, Kulig KW (1990) The diagnostic utility of flumazenil (a benzodiazepine antagonist) in coma of unknown etiology. Ann Emerg Med 19:319-321

Caravati EM (1988) Acute hydrofluoric acid exposures. Am. J Emerg Med 6:143-150

Consroe P, Egen NB, Russell FE, Gerrish K, Smith DC, Sidki A, Landon JT (1995) Comparison of a new ovine antigen fragment (Fab) antivenin for United States Crotalidae with the commercial antivenin for protection against venom-induced lethality in mice. Am J Trop Med Hyg 53:507-510

Cottrell JE, Casthely P, Brodie JD, Patel K, Klein A, Turndorf H (1982) Prevention of nitroprusside-induced toxicity with hydroxocobalamin. New Engl J. Med, 298:809-811.

Cummins RO, Haulman J, Quan L, Graves JR, Peterson D, Horan S (1990) Near fatal yew berry intoxication treated with external cardiac pacing and digoxin-specific FAB antibody fragments. Ann Emerg Med 19:38-43.

Dart RC, Sidki A, Sullivan JB Jr, Egen NB, Garcia RA (1996) Ovine desipramine antibody fragments reverse desipramine cardiovascular toxicity in the rat. Ann Emerg Med 27:309-315.

de Haro L, Lang J, Bedry R, Guelon D, Harry P, Marchal MF, Jouglard J (1996) Results of the tolerance study of a European F(ab)2 antivenom (Viperflow®) given as an i.v. infusion to patients poisoned by European vipers. Proceedings of the XVII International Congr. of the EAPCCT, Marseille, June 1996.

Evans PJ, Evans RW, Bomford A, Williams R, Halliwell B (1994) Metal ions catalytic for free radical reactions in the plasma of patients with fulminant hepatic failure. Free Radic Res 20:139-144.

Fang X, Fernando Q (1994) Synthesis, structure, and properties of rac-2,3-dimercaptosuccinic acid, a potentially useful chelating agent for toxic metals. Chem Res Toxicol 7:148-156.

Flora GJ, Seth PK, Prakash AO, Mathur R (1995) Therapeutic efficacy of combined meso 2,3-dimercaptosuccinic acid and calcium disodium edetate treatment during acute lead intoxication in rats. Hum Exp Toxicol 14:410-413.

Hoffman RS, Goldfrank LR (1995) The poisoned patient with altered consciousness. Controversies in the use of a 'coma cocktail'.JAMA 274:562-569.

Hojer J, Baehrendtz S, Matel G, Gustafsson LL (1990) Diagnostic utility of flumazenil in coma with suspected poisoning: a double blind, randomised controlled study. BMJ 301:1308-1311.

Hoover TD, Aphosian HV (1983) BAL increases the arsenic-74 content of rabbit brain. Toxicol Appl Pharmacol 70:160-162.

Jones A (1996) Should methionine be added to every paracetamol tablet? Proceedings of the XVII International Congr of the EAPCCT, Marseille, June 1996.

Kassa J (1995) Comparison of the effect of selected cholinesterase reactivators combined with atropine on the toxicity of soman and fosdrin in mice. Sb. Ved. Pr. Lek. Fak.Karlovy Univ. Hradci Kralove 38:63-66.

Kontoghiorghes GJ (1995) New concepts of iron and aluminium chelation therapy with oral L1 (deferiprone) and other chelators. A review. Analyst 120:845-851.

Kruszewska S, Wiese M, Kolacinski Z, Mielczarska J (1996) The use of haemodialysis and 2,3 propanesulphonate (DMPS) to manage acute oral poisoning by lethal dose of arsenic trioxide. Int J Occup Med and Env Health 9:111-115.

Lheureux P, Debailleul G, DeWitte O, Askenasi R (1990) Zolpidem intoxication mimicking narcotic overdose: response to flumazenil. Hum Exp Toxicol 9:105-107.

Lheureux P, Even-Adin D, Askenasi R (1990) Current status of antidotal therapies in acute human intoxications. Acta Clin Belg Suppl 13:29-47.

Martens F, Koppel C, Ibe K, Wagemann A, Tenczer J (1990) Clinical experience with the bonzodiazepine antagonist flumazenil in suspected benzodiazepine or ethanol poisoning. J Toxicol Clin Toxicol 28:341-356.

Nakayama GR, Roskos KV, Fritzinger BK, Heller J (1995) A study of reversibly inactivated lipases for use in morphine-triggered naltrexone delivery system. J Biomed Mater Res Nov 29 (11):1389-96

Pennisi E (1996) Chemicals behind the Gulf War Syndrome?. Science, 272:479-480

Plant JR, MacLeod DB (1994) Response of a promethazine-induced coma to flumazenil. Ann Emerg Med 24:979-982.

Proudfoot AT (1995) Antidotes: benefits and risks. Toxicol. Letters 82/83:779-783.

Reichl FX, Hunder G, Liebl B, Fichtl B, Forth W (1995) Effect of DMPS and various adsorbents on the arsenic excretion in guinea-pigs after injection with As2o3 Arch Toxicol 69(10):712-7

Sabouraud A, Urtizberea M, Cano N, Garnier R, Scherrmann JM (1990) Specific anti-digoxin Fab fragments: an available antidote for proscillaridin and scilliroside poisoning? Hum Exp Toxicol 9:191-193.

Scherrmann JM, Lang J, Pepin-Covatta S, Lutsch C (1996) Recent advances in the immunoreactivity of a new purified canine F(ab)2 preparation (Viperflow®) against European viper venom. Proceedings of the XVII International Congress of the EAPCCT, Marseille, June 1996.

Shum S, Whitehead J, Vaughn L, Shum S, Hale T (1995) Chelation of organoarsenate with dimercaptosuccinic acid. Vet Hum Toxicol 37:239-242.

Smith TW, Haber E, Yeatman L, Butler VP (1976) Reversal of advanced digoxin intoxication with Fab fragments of digoxin specific antibodies. N Engl J Med 294:797-800.

Tandon SK, Singh S, Jain VK (1994) Efficacy of combined chelation in lead intoxication. Chem Res Toxicol 7.

Ubios AM, Braun EM, Cabrini RL (1994) Lethality due to uranium poisoning is prevented by ethane-1-hydroxy-1,1-biphosphonate (EHBP). Health Phys 66:540-544.

Vance MV, Curry SC et al (1986) Digital hydrofluoric acid burns of the skin with intra-arterial calcium infusion. Ann Emerg Med 15:890-896.

Velvart J (1983) Arterial perfusion for hydrofluoric burns. Hum. Toxicol. 2:233-238.

Weinbroum A, Rudick V, Sorkine P, Nevo Y, Halpern P, Geller E, Niv D (1996) Use of flumazenil in the treatment of drug overdose: a double-blind and open clinical study in 110 patients. Crit Care Med 24:199-206.

Wolff J, Capraro HG, Brossi A, Cook H (1980) Colchicine binding to antibodies. J Biol Chem 255:7144-7148.

A Critical Review of Antidotal Immunotherapy for low Molecular Weight Toxins. Current Antidotes and Perspectives

Frédéric J. Baud[1], Stephen W. Borron[1], Jean-Michel Scherrmann[2] and Chantal Bismuth[1].

[1] Réanimation Toxicologique. Hôpital Fernand Widal - Université Paris 7. 200 rue du Faubourg Saint-Denis. F-75475 Paris Cedex 10 - France

[2] INSERM U26. Hôpital Fernand Widal. 200 rue du Faubourg Saint-Denis. F-75475 Paris Cedex 10 - France

Introduction

Immunotoxicotherapy (ITT) has been defined by Scherrmann et al. (Scherrmann et al. 1989) as a procedure able to simultaneously sequester, extract or redistribute, and eliminate the toxin by using specific active binding sites derived from different antibody molecular entities. Since the initial report in 1976 by Smith et al. of the reversal of a digoxin intoxication by an equimolar dose of digoxin-specific Fab fragments, immunotoxicotherapy (ITT) has become first-line treatment of all life-threatening intoxications by cardiac glycosides. The efficacy of ITT is such that one might pose the question as to whether all other approaches (anti-arrhythmics, hemoperfusion, etc.) are obsolete. While it appears evident that ITT rapidly reverses digitalis-induced cardiac toxicity, improved outcome, in terms of decrease in mortality rate, has not been proven. This may be in part due to the underlying health of the patients treated or to delays in treatment with Fab. However, this points to the importance of refining the place of ITT in the treatment of digitalis poisoning.

The volume of distribution of immunoglobulins is limited to the extra-cellular space, thus the accessibility to toxins by immunoglobulins and their fragments was considered a possible limiting factor for ITT. Indeed, cardiac glycosides act at the membrane level, facilitating access by antibodies to the toxin. However, the recent example of reversal of both experimental and human colchicine poisonings by colchicine-specific Fab demonstrates that the cellular localization of the toxin is not, per se, a limiting factor. In the case of colchicine, an intracellular tubular-bound toxin is successfully eliminated by extracellular Fab.

Finally, the notion of the requirement for an equimolar ratio of Fab to toxin limited the field of ITT to compounds exhibiting toxic properties in minute amounts. In the case of many toxins whose dose-effect ratios are high, such as tricyclic antidepressants, theophylline, calcium channel blockers etc., there has been a great concern that a less than equimolar dose would be inefficient and that an equimolar dose would introduce the danger of administering a large foreign protein load. However, there is now a large body of knowledge demonstrating that an inframolar dose of Fab may be able to reverse serious toxicity. This has been shown both in the case of experimental desipramine poisoning and experimental and human colchicine poisoning.

It is noteworthy that, at present, there are a very large number of toxic molecules for which antibodies of high affinity and specificity have been obtained which exhibit neutralizing properties (Scherrmann 1994). We should question why, while ITT is potentially applicable to a wide variety of drug overdoses, the therapeutic perspective since 1976 appears very limited. A major hindrance to ITT has been the elevated cost of therapy. Thus, one of the major challenges of ITT will be to establish the methodology and the rules for determining the minimally effective dose in ITT.

Current Antidotes

Digitalis-specific Fab fragments

The fundamental mechanisms of ITT were established through the experimental and clinical studies performed in the field of treatment of cardiac glycoside toxicity with digoxin-specific Fab fragments. Clinical studies dealing with a large number of poisoned patients allowed the assessment of the safety of low doses of Fab fragments. While ITT has become the first-line treatment of digitalis poisonings, several concerns have to be resolved, including the indications for ITT, the minimal efficient dose, and concurrently, the optimal mode of administration of Fab fragments.

Basic mechanisms of ITT: Fab fragments in digitalis poisoning work by inducing (1) an extra-cellular redistribution of the toxin, (2) the sequestration of the toxin in the extracellular space, and (3) renal elimination of the small molecular weight antibody-toxin complex (Smith et al. 1976, Scherrmann 1994). Indeed, administered intravenously, the fragments bind to circulating digitalis forming complexes which are unable to bind to tissue digitalis receptors. However, the ability of antibodies or their Fab fragments to form stable complexes is strongly dependent on the very high affinity of antibody-toxin binding. Provided that a dose containing equimolar binding sites has been administered, total digoxin concentrations increase 5 to 20 times, while unbound serum digitalis drops to zero within minutes. By mass action, free intracellular digitalis and receptor-bound digitalis are displaced and reactivation of membrane ATPases ensues. Finally, in patients with normal renal function, the Fab-digitalis complex elimination half-life is approximately 10 to 20 h, compared to spontaneous half-lives of 160 h for digitoxin and 39 h for digoxin. Recent data, based on the volume of distribution of immunoglobulins and Fab fragments, suggest that intracellular penetration by Fab fragments may occur. Indeed, Schaumann outlined that the apparent volume of distribution of Fab fragments is greater than the volume of extracellular water (Schaumann et al. 1986).

In an attempt to classify antidotal treatments in clinical toxicology, ITT should be considered a toxicokinetic treatment inducing an extra-cellular redistribution of the cardiac glycoside (Baud et al. 1995a). However, the redistribution is so rapid and so complete that the signs and symptoms resolve within a few hours. It should be noted that the notion of reversibility of the binding of cardiac glycosides to membrane ATP-ase was demonstrated with Fab fragments. Clinical studies showed that the onset of both pharmacokinetic and pharmacodynamic actions of Fab fragments is very rapid when the Fab are infused over 20 to 30 minutes. Fab fragments begin to reverse the toxic clinical effects of digoxin or digitoxin within the first hour after administration. Most patients who responded had clinically evident improvement by 1 hour after termination of infusion and complete response, including the resolution of hyperkalemia, by 4 hours (Antman et al. 1990).

Efficiency of Fab fragments in clinical studies: Large series of cardiac glycoside poisonings treated with Fab fragments have been published, including 717 cases by Hickey et al. (1991), 150 cases by Antman et al. (1990), 34 cases by Smolarz et al. (1985), and 28 cases by Taboulet et al. (1993). In the observational surveillance study of 717 cases, 50 % of the poisoned patients exhibited full resolution of all signs and symptoms of digitalis toxicity, 24 % exhibited a partial response, and 12 % showed no response. The response for 14 % of patients was not reported or reported as uncertain. True nonresponders are rare, and lack of response should raise the suspicion that -1- too low a dose of Fab fragments was given, -2- that the clinical signs and symptoms are not caused by digitalis toxicity, or -3- the a moribund state was already present before treatment (Hickey et al. 1991). It should be outlined that none of the patients without heart disease who ingested a single overdose were non-responders to Fab (Hickey et al. 1991). Neither concurrent cardiac disease, nor the type of digitalis, nor the age were significantly associated with a particular slow initial response to ITT (Antman et al. 1990). Concerning the initial dose, Hickey et al. (1991) were unable to establish any clear relation between initial response to treatment and percent of the estimated adequate dose actually administered. Similarly, Schaumann using a fixed initial dose of Fab fragments in 4 symptomatic poisoned patients, found no failure of treatment (Schaumann et al. 1986).

Following initial treatment with Fab fragments, the rate of recrudescent toxicity (atrioventricular block, ventricular arrhythmias, or asystole) has been reported in the range of 2.8 % in the series of 717 cases (Hickey et al. 1991) to 11 % in the series of 28 cases (Taboulet et al. 1993). Thus, the recurrence of toxicity was infrequent. The mechanism of recrudescent toxicity is not fully understood. Clinical signs compatible with reintoxication are not always accompanied by increasing free digoxin concentrations, suggesting that the clinical diagnosis of recrudescent toxicity is unreliable. Furthermore, Ujhelyi et al. (1993, 1995) have recently shown that there is no evidence to support a dissociation of the Fab-digoxin complexes over extended periods of time. Rather, recrudescent toxicity appears to be related to redistribution of free digoxin into the serum.

Patients with renal failure, on dialysis, responded to Fab treatment in a manner similar to patients with normal renal function (Antman et al. 1990). However, a theoretic possibility exists that digoxin could be released with recurrence of toxicity. Among 18 patients with a pretreatment serum creatinine concentration of more than 4.0 mg/dL, including five who where on dialysis, recurrence of toxicity was observed in one dialysed patient. However, a multivariate analysis performed on the series of 717 cases did not demonstrate an increased risk of recrudescent toxicity in patients with renal failure (Hickey et al. 1991).

Digitalis intoxication can induce gastrointestinal disturbances including ischemic colitis. Even digitoxin-induced ischemic colitis can be improved within hours after Fab administration (Bourhis et al. 1990)

Safety of a single low dose of Fab fragments: In case series involving approximately 900 patients, adverse effects of Fab fragments were rare (Smolarz et al. 1985, Antman et al. 1990, Hickey et al. 1991, Taboulet et al. 1993). Hypersensitivity reactions including pruritic rash, facial swelling and flushing, urticaria, thrombocytopenia, shaking and chills, were mild and occurred in only 0.8% of 717 adults. The likelihood of an allergic reaction was distinctly greater in patients with a history of allergy or asthma. In particular, a history of allergy to antibiotics was associated with the occurrence of allergic reaction (Hickey et al. 1991). The results of skin testing before Fab treatment have been reported by Antman et al. (1990) in 94 patients and were uniformly negative, except for one patient who had erythema without wheal or induration.

Adverse reactions caused by relatively rapid reversal of the effects of digitalis, such as accelerated ventricular rate in atrial fibrillation, worsened left ventricular function or hypokalemia were noted in 7 % among 717 treated patients (Hickey et al. 1991).

Indications for ITT in digitalis poisoning: As stated above, the efficacy of ITT, in terms of decrease in mortality rate of digitalis intoxication, has not been proven. It should be noted that the mortality rate of these intoxications prior to availability of Fab treatment is not accurately known, with estimates ranging from 4.6 to 41 % (Kelly and Smith, 1992). The global mortality rate of acute digitoxin poisoning has been reported as 13 % by Bismuth et al. (1977), and 17 % by Dally et al. (1982); the mortality rate was 23 % in acute digitalis poisonings treated by pacemaker (Bismuth et al., 1977). Since the availability of Fab fragments the mortality rates of cardiac glycoside poisonings, mainly including digoxin poisonings, have been reported in the range from 6 % by Smolarz et al. (1985) to 29 % by Antman et al. (1990). There is no doubt that digitalis Fab fragment therapy can improve the outcome. In our opinion these results raise questions regarding the accuracy of diagnosis, selection of patients to receive Fab, and the proper amount and method of administration of Fab.

Fab administration is currently restricted to patients exhibiting actual or potentially life-threatening cardiac rhythm disturbances, hyperkalemia, or both caused by digitalis intoxication when such conditions are refractory to or likely to be refractory to treatment with conventional therapeutic modalities (Hickey et al., 1991). In light of the persistent high mortality and the now-known safety of Fab, one might question the advisibility of waiting for the appearance of life-threatening symptoms before administering Fab fragments. At this point, some patients may have reached a moribund status, thus, precluding the benefit of ITT. Therefore, a prophylactic attitude should be considered, especially as the prognostic factors of acute digitalis poisoning are known (Taboulet et al. 1993). One should keep in mind that the symptomatic treatment of digitalis toxicity in the pre-Fab era (antiarrhythmics and pace-makers) had little or no effect on mortality (Gaultier et al. 1976). Taboulet et al. (1993) have since pointed out the danger in delaying the administration of Fab by attempts at standard supportive therapy.

We suggest that ITT should now be considered in two clinical settings. First, Fab administration is immediately indicated in patients with life-threatening symptoms, such as ventricular tachycardia, ventricular fibrillation, atrioventricular block of high degree, asystole, or cardiogenic shock. It is noteworthy that among 56 poisoned patients whose digitalis toxicity culminated in cardiac arrest, 30 (54%) survived hospitalization after treatment with Fab fragments (Antman et al. 1990). Furthermore, Fab administration should be strongly considered as *first-line therapy* in adult patients with poor prognostic factors, i.e. advanced age, serum potassium > 5 mmol/l, male sex, underlying cardiomyopathy, bradycardia unresponsive to atropine (Taboulet et al. 1993). In severe digitalis intoxication, supportive therapy, including pacing, is not enough. More effective use of ITT should result from earlier administration (Taboulet et al. 1993).

The dosage regimen of Fab fragments: The dosage regimen initially proposed, and which was shown to be efficient, includes the administration of the dose of Fab fragments over 15-30 minutes diluted in isotonic saline. The dose of Fab fragments is calculated to be equal on a mole-to-mole basis to the amount of digoxin or digitoxin in the patient's body (Antman et al. 1990). The amount of cardiac glycoside present in the body is calculated using either the supposed ingested dose or the serum digitalis concentration.

Experimental data have suggested the efficiency of an equimolar dose of Fab. This concept has been employed clinically as described above. It should be recalled, however, that the accuracy of such calculations in clinical practice is questionable for two reasons. First, the ingested dose is frequently not known with precision. Second, the plasma concentration of digitalis accurately reflects the body load of digoxin only after equilibration, which requires several hours following the last dose. Thus, it is unclear whether the theoretical equimolar dose is attained or exceeded. Furthermore, the necessity of administering an equimolar dose of Fab to obtain an initial beneficial response is not supported by the literature (Hickey et al. 1991, Schaumann et al. 1986). Initial response to therapy is generally favorable even when the dose of Fab administered is well below the equimolar ratio. In contrast, the risk of recrudescent toxicity is more frequent in persons receiving less than 50 % of the estimated full neutralizing dose. In 47 patients who received <50% of the estimated dose the rate of recurrence of toxicity was 8.5 %. It was 5.8 %, 2.6 %, and 2.2 % in patients who received 50-75, 75-100, and > 100 % of the estimated equimolar dose, respectively (Hickey et al. 1991). While these data suggest that recrudescent toxicity may be related to the dose of Fab administered; the incidence of recrudescent toxicity is nonetheless quite small, less than 9 % even among patients at highest risk.

Furthermore, recent data suggest that the duration of Fab infusion should be increased in order to optimize the binding of digoxin to Fab fragments. Schaumann et al. (1986) successfully treated 11 cases of severe digoxin poisoning with 230 to 480 mg of Fab infused over 30 min to 5 h. However, the percentage of Fab fragments bound to digoxin was only 45.8 ± 4.0 % (mean ± SEM). Taking into account the serum Fab concentration associated with undetectable free digoxin concentrations, they changed the dosage regimen as follows: a loading dose of 160 mg Fab fragments was infused over 0.25 h, followed by a continous infusion of 160 mg over 7 h. This revised dosage regimen was associated with an increase in the percentage of Fab fragments bound to digoxin up to 69.0 ± 6.5 %. As in many cases free digoxin concentrations reappeared in serum after 10 h, the administration of a supplemental 80 mg dose of Fab at an infusion rate of 0.1 mg/min should be considered (Schaumann et al. 1986). Thus, new pharmacokinetic investigations are required to define the optimal conditions of Fab fragment administration in the treatment of digoxin poisoning.

Colchicine-specific Fab Fragments

While digoxin-specific Fab fragments showed that an equimolar dose is efficient to reverse the highly toxic effects of membrane toxins, ITT in colchicine poisoning should be considered as a model expanding ITT to toxins of low molecular weight acting intra-cellularly. Furthermore, in digoxin poisoning there is essentially one organ at vital risk, i.e. the heart. In contrast, severe colchicine poisoning induces a multi-organ failure. ITT in colchicine poisoning suggested that the reversal of toxicity in a single organ, i.e. drug-induced cardio-vascular failure, could be associated with a favourable outcome. The colchicine model reinforced the importance of the slope of dose-reponse regarding ITT. That is, removing even a modest portion of the toxin may dramatically improve outcome (Scherrmann 1994). ITT in colchicine poisoning confirmed both in animals and in a case report the efficiency of inframolar neutralization. Finally, as proposed by Schaumann, this case report challenges the classical thinking of rapid administration of the total dose of Fab fragments (Baud 1995b).

Reversal of intracellular binding of colchicine by extracellular Fab: Although intracellular binding of colchicine to microtubules was expected to limit the efficacy of immunotoxicotherapy for colchicine poisoning, a reversible effect on microtubules was found in vitro. (Wolff et al. 1980). Rouan et al. (1990) prepared a specific high-affinity colchicine-binding monoclonal antibody and investigated its ability to reverse in vitro the effects of colchicine on Chinese hamster ovary cells. Colchicine-induced polyploidy and chromosomal aberrations were reversible even when the antibody was administered up to 6 hours after colchicine exposure. Active immunization for colchicine poisoning has been examined by administering 3 mg/kg of colchicine (above LD_{50}) to previously immunized rabbits. (Scherrmann et al. 1989b). The protective effects depend on the anticolchicine antibody titer. A study in mice showed that administration of specific anticolchicine immunoglobulin G (IgG), even after the colchicine distribution phase, significantly decreased the mortality rate. (Terrien et al. 1988, Terrien et al. 1990) In further studies, (Terrien et al. 1989) IgG colchicine antibodies were produced in goats and given to mice which had received an LD_{50} of colchicine intraperitoneally. Despite a relatively low dose of antibody, a beneficial effect was shown and colchicine pharmacokinetic parameters were altered favorably. The steady-state volume of distribution decreased significantly in the treated mice. Both IgG and Fab fragments induced sequestration of colchicine in the intravascular space and a decrease in colchicine concentrations in most tissues of antidote-treated mice, indicating a tissue extraction process (Sabouraud et al. 1992).

Organ-specific reversal of toxicity by ITT: Until recently, there was no successful specific therapy for colchicine intoxication, which is associated with a high mortality rate. The mortality rate is strongly dependent on the ingested dose with a mortality rate, in our experience, of 90 % when the ingested dose is above 0.8 mg/kg body weight (Bismuth et al. 1977b). Patients with early hemodynamic collapse due to colchicine overdose have a particularly poor prognosis. An experimental study has clearly established the cardiac toxicity of high doses of colchicine in rats. The first clinical use of ITT in human colchicine poisoning was reported in 1995 (Baud et al. 1995b). The administration of colchicine-specific Fab fragments was associated essentially with the reversal of the cardiovascular failure only. Fab fragments at the administered dose and at the time of administration, 36 hours after ingestion of 1 mg/kg of colchicine, did not prevent the delayed occurrence of bone marrow aplasia, complete hair loss, and transient peripheral neuropathy. Further studies are needed to determine the effect of Fab on bone marrow aplasia.

The importance of the slope of the dose-effect in ITT: The mortality in acute colchicine poisoning is strongly related to the onset of cardiogenic shock. In our experience, the type of organ failure observed after acute single dose colchicine poisoning is dependent on the ingested dose (Bismuth and Gaultier 1977) with bone marrow aplasia occurring at ingested doses greater than or equal to 0.5 mg/kg, and cardiogenic shock at doses greater than or equal to 0.8 mg/kg. However, the mortality rate in the former group is 10 % compared to 90 % in the latter group. Thus, removal of even a modest proportion of the toxin would be expected to be associated with a more favourable outcome.

Partial neutralization: In the case of digitalis Fab therapy, the efficacy of equimolar neutralization has been demonstrated. However, as stated above, the question as to whether an inframolar dose might be effective remains to be resolved. In the case of acute colchicine poisoning, experimental data support the hypothesis of the efficiency of inframolar neutralization. In mice intoxicated with a 100 % lethal dose (3.8 mg/kg), half-molar (M/2), quarter molar (M/4), and eighth-molar (M/8) doses were associated with a reduction of mortality of 80 %, 50 %, and 20 %, respectively. The dose-effect relationship was linear ($r = 0.99$) (Sabouraud et al. 1991).

The concept of partial neutralization has been reinforced by the demonstration of its efficiency in a clinical case. In this case, goat colchicine-specific Fab fragments were administered at a molar ratio of 0.125 of the ingested dose corresponding to about 0.4 molar ratio of persistent body burden. Based on the analysis of the prognostic factors, this patient's predicted mortality was near 100 %. In spite of the pejorative prognostic factors, the patient survived (Baud et al. 1995b). In addition to the toxicodynamic effects of colchicine Fab fragments, toxicokinetics of colchicine were significantly altered, resulting in a 6-fold increase of total colchicine concentrations 10 min after starting Fab infusion. Free colchicine concentrations fell to undetectable levels over a 7 h period. A 4.4-fold decrease of colchicine distribution volume (2.9 l/kg) under Fab treatment indicated that substantial amounts of colchicine were being removed from peripheral sites and redistributed into the extracellular space, corresponding to the distribution volume of Fab fragments. Fab infusion increased urinary colchicine excretion rate 6-fold (24 and 134 µg/h, respectively, before and after Fab administration). Colchicine was renally excreted, bound at 98 % to Fab fragments over a 16 h period. The amount of Fab-bound colchicine renally excreted was equal to the binding capacity of the Fab fragment amount recovered in urine. Thus, the neutralization yield of Fab fragments was near 100 %. The amount of neutralized colchicine was estimated at 3.7 mg out of the 9 mg of colchicine present in the body prior to Fab dosing.

Mode of administration of Fab fragments: Schaumann et al. (1986) demonstrated the advantages of split-dose Fab fragment therapy in a group 17 digoxin intoxications. Having noted that the percentage of unbound Fab fragments in patients treated with a single rapidly administered dose was 52.4 %, Schaumann decided to lengthen the duration of administration by a factor of 10, from 0.5 h to 5 h. The fraction of bound Fab fragments remained below 50 % of the total. He therefore decided to employ split doses, giving half the total dose over 0.25 h and the other half over 7 h. He succeded in improving the percentage of bound Fab fragments to a median of 70.2 % in six patients. It must be emphasized that this mode of administration allowed him to reduce the administered dose of Fab by 20 % to 33 %.

In the case of acute human colchicine poisoning we used this mode of administration, slightly modified, as the loading dose was administered over 1 h and the second dose over 7 h (Scherrmann 1994, Baud et al. 1995b). The amelioration seen within the first hour was nothing short of dramatic. At this time, the patient had received only a dose of Fab corresponding to a molar ratio of 0.06 of the ingested dose corresponding to about 0.2 molar ratio of persistent body burden (Baud et al. 1995b).

Anti-tricyclic Antidepressant Immunotherapy

Tricyclic antidepressants (TCA) compose the drug class most often responsible for death from intentional overdose in the United States (Litovitz et al. 1994). Due to the impressive potential of ITT, its efficiency had to be assessed in such a life-threatening poisoning. However, there are many particularities of TCA poisoning that preclude any extrapolation from data collected on digitalis or colchicine ITT. In particular, due to the large amount of toxin which is inducing life-threatening intoxications, only partial neutralization should be considered. As only partial neutralization is feasible, the role of ITT in TCA poisoning must be placed in perspective with conventional supportive therapy. Even taking into account inframolar neutralization, the required dose of Fab is very large in comparison with those in digitalis poisonings. Thus, the pharmacokinetics and safety of a large dose of Fab have to be determined.

Originality of the TCA model regarding ITT: The clinical usefulness of ITT has been largely established for cardiac glycosides, and some recent data suggest the interest of ITT in colchicine poisoning. However, these drugs are very atypical in that they are very potent; their toxic dose is quite low and the dose of Fab fragments required to reverse toxicity is correspondingly low. Tricyclic antidepressants have toxic doses 10-100 times higher and thus will require proportionally higher dose of Fab to treat an overdose. The major challenge in the development of specific Fab as an antidote is to know whether this therapy can be safely scaled up by several orders of magnitude in order to address these frequently encountered overdoses (Pentel and Keyler 1995).

Inframolar dose of Fab induces redistribution of TCA: Desipramine-specific IgG1 with a moderate affinity for desipramine ($K_a = 3.0 \times 10^8$ M^{-1}) can effect substantial redistribution of a subtoxic dose of either desipramine or imipramine in rats, even when the antibody dose is significantly less than equimolar to the desipramine dose (Pentel et al. 1991). In rats with toxicity from desipramine, monoclonal desipramine-specific IgG or the corresponding F(ab)'2 (Brunn et al. 1991) or the 55kD Fab' fragments (Brunn et al. 1992) bind and redistribute desipramine from tissue into serum. Indeed, administration of 1 g/kg of Fab' in rats following the injection of desipramine 30 mg/kg IP

(Fab'/tricyclic molar ratio of 0.09) rapidly increased the serum desipramine concentration 15-fold, thus demonstrating the redistribution of desipramine out of tissues (Brunn et al. 1992).

Inframolar dose of TCA-specific antibodies induces improvement in survival: In rats, toxic doses of desipramine induce hypotension, slowing heart rate and electrocardiographic QRS-interval prolongation. High affinity TCA-specific monoclonal Fab' or sheep polyclonal Fab fragments rapidly reverse the cardiovascular toxicity of desipramine (DMI) in rats (Pentel et al. 1995, Dart et al. 1996). All these drug-induced disturbances are substantially reduced by specific Fab' fragments administered at a dose corresponding to a Fab'/tricyclic molar ratio of 0.09 (Brunn et al. 1992). It should be noted that despite the relatively low Fab'/desipramine ratio, the therapeutic effect was achieved rapidly, with onset during the 10-min Fab' infusion and the maximal effect by the end of infusion. Notwithstanding, without knowledge of the effect on the outcome, the improvement in blood pressure and electrocardiographic abnormalities can be considered as only surrogates of reduction of tricyclic antidepressant toxicity and cannot be considered per se as clinically relevant. However, polyclonal desipramine specific Fab fragments at a dose of 2 g/kg corresponding to a molar Fab/desipramine ratio of 0.21 prolonged the survival time of poisoned rats by 50%. Thus, the use of drug-specific antibody fragments may represent a useful strategy for the treatment of tricyclic antidepressant overdoses.

The place of ITT in the therapy of TCA poisoning: Conventional supportive treatment of life-threatening TCA overdoses in humans includes mechanical ventilation, cardio-vascular support and the infusion of molar bicarbonate. In spite of such aggressive therapy, the mortality remains significant. Combining TCA-specific Fab with sodium bicarbonate is more effective than either treatment alone (Brunn et al. 1992). In mechanically-ventilated and anesthetized rats poisoned with desipramine, the best outcome was obtained using a combination of an inframolar dose of Fab fragments and a large dose of sodium bicarbonate. This combination produced a greater reduction in DMI-induced QRS prolongation than either treatment alone. This beneficial interaction was considered by the authors as resulting from combining pharmacodynamic (sodium bicarbonate) and pharmacokinetic (Fab) antidotes (Pentel et al. 1995). However, the effect of alkalinization, per se, on either the binding of desipramine to its cardiac target or to specific Fab fragments has not been considered.

Pharmacokinetics and safety of high doses of Fab: Several questions regarding the infusion of high doses of Fab fragments have been recently addressed (Pentel and Keyler 1995). The pharmacokinetics of a huge dose of Fab fragments in rats, 7.5 g/kg, closely approximated that reported for lower doses of digoxin-specific Fab fragments in dogs, baboons, or humans (Pentel et al. 1988). The elimination half-life of a high dose of non-specific Fab in dogs was somewhat longer and the fraction excreted in urine somewhat lower, but still in the range that should allow Fab to be an effective antidote (Keyler et al. 1991). Thus, the pharmacokinetic properties of Fab appear to be similar over a wide range of Fab doses, including those that would be needed to treat overdoses such as TCA overdose (Pentel and Keyler 1995)

High doses of Fab were generally well tolerated in rats, with no acute changes in cardiovascular parameters, normal wakening from anesthesia, and normal organ histology at 21 days (Pentel et al. 1988). One rat receiving a 2 g/kg dose of TCA specific Fab, representing a molar ratio of TFab to DMI of 0.21, died. Fab seems to be not toxic in itself, as doses higher than anticipated for clinical use (4g/kg) were well tolerated in anesthetized rats in the absence of desipramine. Possibly the combined effects of desipramine toxicity, several hours of anesthesia, and the large protein and volume load associated with antibody administration caused the observed death (Pentel et al. 1993). Dogs receiving either 5.3 or 3.2 g/Kg Fab I.V. over 1 h tolerated the Fab infusion well, but all the dogs developed transient renal insufficiency assessed by daily urine output, serum creatinine concentration and the calculation of creatinine clearance. At 2 weeks creatinine clearance had returned to normal (Keyler et al. 1991). Whether this was due to the Fab or to contaminant proteins is unclear. All animals survived for 21 days and had normal organ (including kidney) histology at that time. Thus, the administration of high doses of Fab fragments appears possible, but more safety data are needed (Pentel and Keyler 1995)

Perspectives

Toxin-dependent prerequisites for ITT: Toxin-dependent prerequisites for ITT have been previously outlined (Scherrmann et al. 1989, Scherrmann 1994). For ITT to prove maximally beneficial in relation to its high cost, the toxin should have the following characteristics:

High risk of mortality and short-term effects (digoxin, tricyclic antidepressants, colchicine, paraquat) or with *long-term effects and risk of tissue accumulation* (environmental contaminants such as hexachlorobiphenyl compounds). However the impact of Fab therapy on costs of care should also be taken into account. While ITT is quite expensive, so is admission to an intensive care unit. A retrospective cost-effectiveness analysis of the use of digoxin-specific Fab fragments indicated that such therapy is generally associated with an increase in total medical care cost for seriously toxic patients because more of them survive, requiring additional hospitalization. In contrast, savings are achieved in less serious intoxications due to decreased number of days spent in a coronary care unit and decreased use of pacemakers and other aggressive treatments (Mauskopf and Wenger 1991). We suggest that in digitoxin poisoning, the cost-benefit of ITT should be assessed not only on its ability to decrease mortality but also on its ability to decrease duration of hospitalization.

Drug addiction is a major public health problem, which has prompted an interest in antibody research in this domain. Such antibodies were initially proposed for diagnostic purposes. In 1972, evidence for active immunization against morphine was demonstrated in mice (Berkowitz and Spector 1972). Thus, the possibility of active immunization as a mechanism of tolerance was investigated. However, there was no subsequent development of ITT towards opiates as naloxone had obviated the problem of reversal of narcotic effects. In contrast, cocaine abuse continues to present a major toxicological concern, resulting in frequent admissions to critical care units for cardiac chest pain and other vascular and neurological emergencies. Cocaine is toxic in small amounts and there is no true antidote for this life-threatening intoxication. Recently, cocaine has been shown to be immunogenic in rats when conjugated to protein. Active immunization with a cocaine conjugate results in decreased tissue distribution as assessed by the measurements of cerebellar and striatal concentrations of cocaine (Carrera et al. 1995). Active immunization also prevents the increase in locomotor activity and stereotyped behaviour induced by cocaine. These promising preliminary results demand consideration of the possibility of reversal of life-threatening toxicity by ITT.

Capability of producing an antibody response in the host after conjugation to a protein: The immunological response is dependent on the chemical structure and molecular weight (haptens to macromolecules) of the toxin. Thus, it seems rather unlikely that high affinity and high specificity antibodies could be obtained in case of very low molecular weight molecules such as carbon monoxide, cyanide, et al.cohols. In contrast high affinity and specific immunoglobulins are obtained for toxicants whose molecular structure includes either homogenous or heterogenous rings such as tricyclic antdepressants, phenothiazines, barbiturates, benzodiazepines, colchicine, opiates, cocaine, etc. However, a recent study showed that rats actively immunized with an arsanilic acid conjugate were protected from a lethal dose of sodium arsenite. The antibodies were shown, in vitro, to strongly bind sodium arsenite. Arsenic has a strong affinity for thiol groups which are present in antibodies. However, a non-specific binding was excluded as immunized rats were not protected from the toxicity of $HgCl_2$ which also has a high affinity for thiols (Leikin et al. 1991). It should be noted that in case of toxic ions chelating agents have proven useful. While there are numerous similarities between chelating agents and ITT, there have been no attempts, to our knowledge, to compare their respective roles. The major differences between chelating agents and ITT are the increased ability of the former to cross cellular membrane and generally higher affinity of several orders of magnitude.

Apparent volume of distribution of the toxin: If the apparent volume of distribution of the toxin is much greater than that of the antibody or fragment (whole antibodies, for example, being restricted to the vascular compartment), the efficacy of ITT is dependent on -1- the distribution of the immunoglobulin, and -2- the degree and rapidity of redistribution of the toxin into the compartment occupied by the antibody or fragment. In the case of low molecular weight toxins, fragmentation of antibodies increases the volume of distribution, and, thus enhances the likelihood of antibody-toxin binding. However, the trend toward preparing fragments smaller than Fab would lead to a more rapid and wider distribution, but also to faster clearance of antibodies from tissue, decreasing cumulative exposure of toxin to antibody binding sites (Scherrmann 1994). These data emphasize the now demonstrated advantages of prolonged infusion of Fab fragments for toxins with large volume of distribution.

Reversibility of effects: There is no evidence that therapeutic antibodies can reverse structural lesions. This, in large part, explains their limited utility against paraquat. On the other hand, antibodies can neutralize residual unbound toxin, even if a great part of the toxin has already damaged the organism, thus potentially averting a lethal outcome. Furthermore, one must take into account the microkinetics of cellular lesions. A toxin must reside at the cellular binding site for a variable period of time before irreversible structural damage occurs. Three very different examples of toxins inducing irreversible structural lesions are iron, paracetamol, and hydrofluoric acid. Each of these possesses an efficient antidote. The real benefit depends on the duration of toxin binding at the effector site. Thus, one must not discard the possibility of successful ITT solely on the basis of a toxin's capacity to induce structural lesions.

Toxicity after absorption in the milligram range: Stoichiometric neutralization with antibodies has proven to be effective against cardiac glycosides. Now there is clear evidence that partial neutralization is effective in both colchicine and tricyclic antidepressant poisonings. Globally, it is important to consider the slope of toxicity of the toxin as ITT is likely to be of greater benefit where removal of a small proportion of the toxin results in significant pharmacodynamic improvement. On the other hand, the reversal of cardiotoxicity induced by both colchicine and tricyclic antidepressants showed that organ-specific toxicity may respond to ITT in the absence of a markedly reduced body burden of toxin (Pentel et al. 1991).

Antibody-dependent prerequisites for ITT: Antibody-dependent requisites for ITT have also been outlined (Scherrmann 1994).

Polyclonal versus monoclonal antibodies: Various experimental studies have shown that polyclonal antibodies have some important advantages as pharmaceutical agents for several reasons. First, polyclonal antibodies typically have higher affinity for a given hapten than their monoclonal counterparts. Secondly, it is possible to produce polyclonal antibodies with broad specificity. Finally, polyclonal antibodies may be less expensive to produce if high titers are achievable.

Antibody specificity: Antibodies with broad specificity are clearly an important advantage in the case of digitalis toxicity in which case the same antibodies were shown to be clinically efficient in various poisonings such as digoxin, and lanatoside C poisoning (Hess et al. 1979) (both drugs have the same genin part), digitoxin (whose genin part slightly differs from the former), or even in complex clinical settings such as those resulting from digitalis-containing plants (Shumaik et al. 1988). A broad specificity is also considered as a possible advantage in the case of TCA overdose, as many related compounds cause severe intoxications. Thus desipramine-specific polyclonal antibodies exhibit high affinities for imipramine, amitriptyline, and nortriptyline (Pentel and Keyler 1995). However, a possible limitation of the assessment of ITT in clinical toxicology can result from the fact that the majority of acute poisonings results from multiple drug ingestions which could result in unpredictable cross-reactions. Fortunately, this is usually not the case in either cardiac glycoside or colchicine poisonings. This is more often the case in tricyclic antidepressant poisonings where either several tricyclic antidepressants or structurally-related drugs, such as phenothiazines, may be present at the time of ITT. Depending on the relative affinities of the immunoglobulin for the related compounds, the degree of binding and the appropriate dose of Fab fragments cannot be predicted.

Antibody fragments: The fragmentation of antibodies (Scherrmann 1994), and specifically the removal of Fc fragments from IgG, increases the safety of antibodies, limiting the risks of hypersensitivity reactions and of serum sickness. While useful against haptens (< 1000 daltons), fragmentation must be questioned when considering immunotherapy against high molecular weight toxins (most animal venoms), whose toxin-Fab complex may largely exceed 60,000 Daltons, thus rendering them incapable of renal glomerular filtration. The advantages of fragmentation may be offset by the reduction of stability of large toxin-antibody complexes, with the risk of the subsequent release of the toxin. Nonetheless, even smaller antibody fragments may offer improved therapeutic properties. A recombinant single chain Fv fragment (sFv), one half the size of Fab has been cloned, which retains a high affinity for DMI and is able to alter DMI distribution in vivo. Because sFv has a shorter elimination half-life and more extensive renal excretion than Fab, it may have therapeutic advantages. (Huston et al. 1993) Moreover, this fragment can potentially be engineered to improve its therapeutic properties.

Conclusion

Immunotoxicotherapy is a toxicokinetic treatment basically inducing extracellular redistribution and sequestration of toxins. In comparison with other toxicokinetic treatments the redistribution is so extensive and rapid that it induces in all studied models a dramatic toxicodynamic improvement. The major scientific challenge today is to refine the knowledge regarding the minimal efficient dose and regimen of administration for the various antibodies.

Such improvements depend on a better understanding of the kinetics of redistribution of toxins and their slope of toxicity. Perhaps, an even greater challenge, however, are the barriers imposed by government regulation and economic cost of development of antibodies.

References

Antman EM, Wenger TL, Butler VP, Jr., Haber E, Smith TW (1990) Treatment of 150 cases of life-threatening digitalis intoxication with digoxin-specific Fab antibody fragments: Final report of a multicenter study. Circulation 81:1744-1752.

Baud FJ, Borron SW, Bismuth C (1995a). Modifying toxicokinetics with antidotes. Toxicol Lett 82/83: 785-793.

Baud FJ, Sabouraud A, Vicaut E, Taboulet P, Lang J, Bismuth C, Rouzioux JM, Scherrmann JM (1995b) Treatment of severe colchicine overdose with colchicine-specific Fab fragments. N Engl J Med 332:642-645.

Berkowitz B, Spector S (1972) Evidence for active immunity to morphine in mice. Science 178:1290-1292.

Bismuth C, Motte G, Conso F, Chauvin M, Gaultier M (1977) Acute digitoxin intoxication treated by intracardiac pacemaker: experience in sixty-eight patients. Clin Toxicol 10: 443-456.

Bismuth C, Gaultier M (1977) Aplasie médullaire après intoxication par la colchicine Nouv Presse Med 6: 1625-1629.

Bourhis F, Riard P, Danel V, Hostein J, Fournet J (1990) Digitalis poisoning with severe ischemic colitis: a favourable course after treatment with specific antibodies. Gastroenterol Clin Biol 14: 95-106.

Brunn GJ, Keyler DE, Ross CA et al. (1991) Drug specific F(ab')2 fragment reduces desipramine cardiotoxicity in rats. Int J Immunopharmacol 13: 841-851.

Brunn GJ, Keyler DE, Pond SM, Pentel PR (1992). Reversal of desipramine toxicity in rats using drug-specific antibody Fab' Fragment: Effects on hypotension and interaction with sodium bicarbonate. J Pharmacol Exp Ther 260:1392-1399.

Carrera RA, Ashley JA, Parsons LH, Wirsching P, Koob GF, Janda KD (1995) Suppression of psychoactive effects of cocaine by active immunization. Nature 378:727-730

Dally S, Bismuth C, Alperovitch A, Lagier G, Scherrrmann JM, Elkhouly M (1982) Facteurs pronostiques de l'intoxication digitalique. Schweiz Med Wochenschrift 112: 1113-1116.

Dart RC, Sidki A, Sullivan JB, Jr., Egen NB, Garcia RA (1996) Ovine desipramine antibody fragments reverse desipramine cardiovascular toxicity in the rat. Ann Emerg Med 27:309-315.

Gaultier M, Welti JJ, Bismuth C, Motte G, Conso F, Chauvin M (1976) Intoxication digitalique grave. Facteurs pronostiques intérêt et limites de l'entrainement électrosystolique. Ann Med Intern 127: 761-766.

Hess T, Stucki P, Baraudan J, Sholttsik G, Roese W (1979) Treatment of a case of lanatoside intoxication with digoxin specific Fab antibody fragments. Am Heart J 98:767-771.

Hickey AR, Wenger TL, Carpenter VP, Tilson HH, Hlatky MA, Furberg CD, Kirkpatrick CH, Strauss HC, Smith TW (1991) Digoxin immune Fab therapy in the management of digitalis intoxication: safety and efficacy results of an observational surveillance study. J Am Coll Cardiol 17:590-598.
Kelly RA, Smith TW (1992) Recognition and management of digitalis toxicity. Am J Cardiol 69: 108G-119G
Keyler DE, Salerno DM, Murakami MM, Ruth G, Pentel DR (1991) Rapid administration of high dose antibody fragments to dogs: pharmacokinetics and toxicity. Fund Appl Toxicol 17: 83-91
Leikin JB, Goldman-Leikin RE, Evans MA, Wiener S, Hryhorczuk DO (1991) Immunotherapy in acute arsenic poisoning J Toxicol Clin Toxicol 29: 59-70
Litovitz T, Clark LR, Soloway RA (1993) Annual report of the American Association of Poison Control Centers toxic exposure surveillance system. Am J Emerg Med 1994;12:546-586.
Mauskopf JA, Wenger TL (1991) Cost-effectiveness analysis of the use of digoxin immune Fab(ovine) for treatment of digoxin toxicity. Am J Cardiol 68: 1709-1714.
Pentel PR, Scarlett W, Ross CA, Landon J, Sidki A, Keyler DE (1995) Reduction of desipramine cardiotoxicity and prolongation of survival in rats with the use of polyclonal drug-specific antibody Fab fragments. Ann Emerg Med 26:334-341.
Pentel PR, Ross CA, Landon J, Sidki A, Shelver WL, Keyler DE (1993) Reversal of desipramine toxicity in rats with polyclonal drug-specific antibody Fab fragments. J Lab Clin Med 123:387-393.
Pentel PR, Keyler DE, Brunn GJ, Milavetz JM, Gilbertson DG, Matta SG, Pond SM (1991) Redistribution of tricyclic antidepressants in rats using a drug-specific monoclonal antibody: dose-response relationship. Drug Metabol Disp 19: 24-28
Pentel PR, Keyler DE (1995) Drug-specific antibodies as antidotes for tricyclic antidepressant overdose. Toxicol Lett 82/83: 801-806.
Pentel PR, Keyler DE, Gibertson DG, Ruth G, Pond SM (1988) Pharmacokinetics and toxicity of high doses of antibody Fab fragments in rats. Drug Metab Dispos 16:141-145
Pentel PR, Brunn GJ, Pond SM, Keyler DE (1991) Pretreatment with drug-specific antibody reduces desipramine cardiotoxicity in rats. Life Sci 91: 675-683
Rouan SE, Otterness I, Cunningham A, Holden H, Rhodes C (1990) Reversal of colchicine-induced mitotic arrest in Chinese hamster cells with a colchicine-specific monoclonal antibody. Am J Pathol 137:779-787.
Sabouraud A, Urtizberea M, Grandgeorge M, Gattel P, Makula ME, Scherrmann JM (1991) Dose-dependent reversal of acute murine colchicine poisoning by goat colchicine-specific Fab fragments. Toxicology 68: 121-132.
Sabouraud A, Urtizberea M, Cano NJ, Grandgeorge M, Rouzioux J-M, Scherrmann J-M (1992) Colchicine-specific Fab fragments alter colchicine disposition in rabbits. J Pharmacol Exp Ther 68: 121-132.
Schaumann W, Kaufmann B, Neubert P, Smolarz A (1986) Kinetics of Fab fragments of digoxin antibodies and of bound digoxin in patients with severe digoxin intoxication. Eur J Clin Phamacol 30: 527-533.
Schermann JM (1994) Antibody treatment of toxin poisoning - recent advances. J Toxicol Clin Toxicol 32: 363-375
Scherrmann JM, Terrien N, Urtizberea M, Pierson P, Denis H, Bourre JM. Immunotoxicotherapy: present status and future trends. J Toxicol Clin Toxicol 1989a; 27: 1-35.
Scherrmann JM, Urtizberea M, Pierson P, Terrien N (1989b) The effect of colchicine-specific active immunization on colchicine toxicity and disposition in the rabbit. Toxicology ;56:213-222.

Shumaik GM, Wu AW, Ping AC (1988) Oleander poisoning: treatment with digoxin-specific Fab antibody fragments. Ann Emerg Med 17: 732-735.

Smith TW, Haber E, Yeatman L, Butler VP, Jr (1976) Reversal of advanced digoxin intoxication with Fab fragments of digoxin-specific antibodies. N Engl J Med 294:797-800.

Smolarz A, Roesch E, Lenz E, Neubert H, Abshagen P (1985) Digoxin specific antibody (Fab) fragments in 34 cases of severe digitalis intoxication. Clin Tox 23:327-340.

Taboulet P, Baud FJ, Bismuth C (1993) Acute digitalis poisoning - Is pacing still appropriate ? J Toxicol - Clin Tox 31:261-273.

Terrien N, Urtizberea M, Scherrmann JM (1988) Immunotoxicotherapy for colchicine poisoning: Toxicological and pharmacological features of an experimental model simulating human poisoning. Arch Toxicol Indust Hygiene 244:322-325.

Terrien N, Urtizberea M, Scherrmann JM (1990) Reversal of advanced colchicine toxicity in mice with goat colchicine-specific antibodies. Toxicol Appl Pharmacol 104:504-510.

Terrien N, Urtizberea M, Scherrmann JM (1989) Influence of goat colchicine-specific antibodies on murine colchicine disposition. Toxicology 59:11-22.

Ujhelyi MR, Robert S, Cummings DM, Colucci RD, Green PJ, Sailstad J, Vlasses PH, Zarowitz BJ (1993) Influence of digoxin immune Fab therapy and renal dysfunction on the disposition of total and free digoxin. Ann Intern Med 119: 273-277.

Ujhelyi MR, Robert S (1995) Pharmacokinetic aspects of digoxin-specific Fab therapy in the management of digitalis toxicity. Clin Pharmacokinet 28: 483-493.

Wolff J, Capraro HG, Brossi A, Cook GH (1980) Colchicine binding to antibodies. J Biol Chem 255:7144-7148.

The Use of Antidotes in the Management of Central Nervous System Depression

Ana Ferrer, Emilia Civeira, Peña Lopez and Belen Loren
Servicio de Toxicología. Hospital Clínico. San Juan Bosco 15. E-50009 Zaragoza. Spain

Evolution of Antidotic Therapy for CNS Depression

CNS depression is the most common effect in poisoned patients on arrival in the emergency department. And it has been so since the mid 50ies when the widespread prescription of barbiturates originated the beginning of what was later to be called the "acute poisoning epidemic".

The evolution of the treatment of this clinical situation has been very interesting, showing a very strong relationship with the technical developments in the medical field. Starting with the search for antidotes when the mechanisms of action of the implicated substances were not well understood, this treatment moved towards a conservative attitude popularised as the "Scandinavian method" in the early 60's and in close relationship with the implementation of the ICU's. This method proved to be very useful for life support of patients poisoned by CNS depressants, not only barbiturates, but also other agents of increasing frequency such as benzodiazepines and opiates, all of which are functional toxics.

The epidemiological profile of acute poisoning has changed dramatically in the last 30 years, as benzodiazepines, alcohol and opiates have substituted barbiturates. Nevertheless the first places are occupied by CNS depressants which continue to produce a similar clinical picture, focusing clinical and therapeutic research on the toxic coma management.

Thus, for the last 15 years for some agents and some clinical situations there have been signs of a reversal of the treatment way previously described, from the conservative life support to the coma cocktail, including hypertonic dextrose, thiamine, naloxone and flumazenil. A systematic approach for the management of the patient with altered mental status includes assessment of the patient's vital functions. Naloxone and flumazenil can be considered antidotes in the stronger sense of the meaning because they are capable of deplacing opiates and benzodiazepines from their specific nervous receptors, reversing coma and obviating the need for intensive care measures such as intubation or mechanical ventilation. Some of the open questions about them are the specificity of their mechanism of action with the implication of the possibility of use for other toxic and non-toxic clinical situations and their safety, related to their side effects.

Flumazenil

Flumazenil belongs to a family of imidazodiazepines which selectively and potently inhibit the central effects of benzodiazepines by acting as their antagonists at CNS's benzodiazepine gamma-aminobutyric acid ($GABA_A$) receptors (Hunkeler et al., 1981). Under the name Ro 15-1788 it was selected for clinical trials in 1981. Its safety range is wide with an LD_{50} between 4,300 and 6,000 mg/kg. In experimental animals it has failed to produce any of the characteristic benzodiazepine effects: It is without sedative, anticolvulsant, muscle relaxant or anxiolytic effects and so it can be considered as being a pure antagonist. Its ability to revert sedation induced by benzodiazepines in man was demonstrated in healthy volunteers in 1981 (Darragh et al., 1981).

It has been used largely since the mid 80's, first in controlled clinical trials and then in open therapeutic applications (Hofer and Scollo Lavizzari, 1985). As an example of a series of them, a double-blind, placebo-controlled trial, including 326 patients from thirteen emergency departments in the US in 1992 confirmed the efficiency of flumazenil in reversing benzodiazepine-induced sedation in patients with benzodiazepine overdose (Flumazenil study group, 1992).

It is used by the parenteral route, with doses of 0.25 mg up to 3 mg, starting with a bolus iv administration followed when necessary by iv infusion.

It has been also used in terminating the sedative actions of benzodiazepines in short-term anesthesia, for the diagnosis of coma of unknown etiology, other toxic coma as produced by alcohol, cannabis, carbamazepine and prometazine and other clinic situations relating to $GABA_A$ receptors such as idiopathic recurring stupor and hepatic encephalopathy (Scollo-Lavizzari and Matthis, 1985; Rothstein et al., 1992; Winkler et al., 1993; Rubio et al., 1993; Pomier-Layargues et al., 1994). In the report of the first case of a patient with prometazine-induced coma responding to treatment with flumazenil, the authors brought this reversal into relationship with the effect of prometazine, similar to other antihistaminics, on GABA receptors (Plant and MacLeod, 1994).

Most frequent adverse experiences related to its use are agitation, vomiting, abnormal crying, dizziness, headache and tachycardia. The severest are seizures, cardiac arrhythmias and psychotic disorders.

In 1992 a total of 42 cases of seizures related to the administration of flumazenil had been reported (Spivey, 1992). They are considered to be related to reversal of the anticonvulsive activity of the benzodiazepine in the presence of convulsive toxic agents. Patients at high risk of complications include those with mixed benzodiazepine/other convulsive drugs (mainly cyclic antidepressives), patients with a history of epilepsy, patients treated with benzodiazepines for seizures and patients receiving benzodiazepines on a chronic basis. A pediatric patient, without any of the forementioned risk factors, developing a generalised seizure following flumazenil administration has been lately reported (McDuffee and Tobias, 1995).

The authors go so far as to deny the indication of the use of flumazenil as a diagnostic tool in the coma of unknown cause, which in the light of the existing data seems to be somewhat exaggerated.

What seems better substantiated is to contraindicate the use of flumazenil if there is a history of seizures or a suspicion of the presence of cyclic antidepressants or other convulsant toxics.

Naloxone

Naloxone hydrochloride is a synthetic N-allyl derivative of oxymorphone which acts essentially as a specific pure competitive antagonist at several receptors, with effects on pain perception, mood and feeling state, and respiration, due to its ability to displace or prevent the binding of agonistic narcotics. It is devoid of agonistic activity.

The pharmacodynamics of naloxone, relating to endorphin receptor sites, focus its potential use in a variety of clinical situations as continuing research illustrates the association of endogenous opioid compounds with various disease states (Handal et al, 1983).

First of all it has been used for reversal of the cardiovascular and respiratory depression associated with narcotic and non-narcotic overdoses (morphine, propoxyphene, methadone, dextromethorphan, lomotil, codeine, fentanyl, chlorpromazine), reversing surgical analgesia and in the treatment of opiate dependence.

It rapidly reverses hypotension caused by endotoxin, hypovolemia, and spinal transection (Faden and Holaday, 1979). From these findings, it may be inferred that endorphins may play a role in the pathophysiology of several types of shock, conditions in which endogenous opioids appear to be mobilised, resulting in increased opioid receptor occupation (McNicholas and Martin,1984). It has also shown an antipruritic effect. As it has been shown to block or reverse the hypotensive actions of captopril (Varon and Duncan, 1991) it can be hypothesized that the hemodynamic effects of angiotensin-converting enzyme inhibitors may be mediated by the endogenous opioid system. It has also shown a protective effect on the ischemically damaged small bowel and reverses chronic idiopathic constipation. As demonstrated in animal experiments the beneficial effects of naloxone on circulatory dynamics may have clinical implications in the prevention and treatment of ischemic heart disease (Lee et al., 1993).

On the basis of the hypothesis that the opiate-like neuropeptides may be involved in the etiology of schizophrenic symptoms and that opiate antagonists would alter mood level, naloxone has been tested in a number of neuropsychiatric disorders such as schizophrenic and manic syndromes, depression, tardive dyskinesia, nervous anorexia and senile dementia with controversial results. Some studies suggest that naloxone and other opiate antagonists may have antipsychotic properties. It seems to be a potential treatment for psychiatric patients displaying self-induced water intoxication indicating that endogenous opioid systems are involved in the compulsive drinking behaviour of this syndrome (Nishikawa et al., 1994). However the literature describing the use of naloxone to treat schizophrenic patients has shown contradictory results (Welch and Thompson, 1994). It has also been used in the treatment of pure neurological disorders such as acute ischemic stroke and acute hemiplegia after intracranial neurological procedures.

It has been demonstrated that some encephalopathies respond to opiate receptor blockade. This effect may be secondary to increased opiate-binding sites or to elevated central nervous system levels of endogenous opiates (Mattana et al., 1993).

In the field of Clinical Toxicology the use of naloxone has become the standard way of treating narcotic-induced coma and respiratory depression. As early as 1973 it was stated that naloxone when available should be regarded as the treatment of choice in all opiate poisonings (Evans, 1973). According to the data of the 1992 Annual Report of the American Association of Poison Control Centers naloxone is the most frequently prescribed specific antidote for human poisonings having been used in 1,5 % of all poisonings seen at health care facilities (Chamberlain and Klein,1994).

Due to naloxone's pharmacokinetic characteristics it is generally administered parenterally (iv, im and sc) and occasionally it has also been used endotracheally. The usual dose of naloxone is 0.4 to 1.2 mg iv although in some cases doses of up to 3.2 mg are necessary (Jefferys and Volans, 1983). A continuous infusion protocol is recommended when prolonged narcotic antagonist effects are required. In paediatric narcotic poisoning an initial dose of 0.01 mg/kg given as a bolus has been recommended. Constant naloxone hydrochloride infusion at an initial rate of 0.4 mg/hr should be considered if the patient responds inadequately to this dose, requires repeated administration to reverse narcotic-induced effects, or has ingested long-acting agents (Lewis et al., 1984). The brief duration of action of naloxone has been considered to recommend the careful observation of overdosed patients in the emergency room. Nevertheless this kind of patients usually discharged themselves as soon as they are aroused by the antidote therapy. In our experience this fails to produce a relapse of the overdose symptoms.

As shown by the analysis of the records of 188 consecutive patients admitted for acute opiate intoxication this life compromising situation has by now an excellent prognosis when treated in hospital owing to the availability of the specific antagonist. Only one died because of cardiac arrest before admission and three quarters of the patients, even when in deep coma on admission, spend less than 1 day in hospital (Larpin et al., 1990).

Other opiate related symptoms such as acute colonic pseudo-obstruction caused by opiates have also been successfully treated with naloxone (Bianchi and Ubach, 1994).

Cardiac toxicity following dextropropoxyphene overdose seems related to its potent local anaesthetic properties rather than its narcotic properties. The usefulness of naloxone to treat these cardiac problems is still debated. Hantson et al. (1995) presents two cases in which cardiovascular depression caused by dextropropoxyphene overdoses seems to be improved after naloxone infusion and suggest that naloxone can reverse hypotension as in the case of captopril overdose through a blockage of the beta-endorphin inhibition of angiotensin II.

Some reported side effects can be related to an effect of the drug on cardiac excitability, so ventricular irritability, cocaine-induced ventricular arrhythmias and rapid atrial fibrillation temporally related to naloxone administration (Merigian, 1993) and even cardiac arrest after reversal of effects of opiates with naloxone (Cuss et al., 1984) have been reported in isolated cases. Other reports include pulmonary edema, hypertension and cardiac arrest following postsurgical narcotic antagonism with naloxone at standard doses (Neal and Owens, 1993).

It has been found to be capable of inducing withdrawal symptoms, the intensity of which being a function of the amount of naloxone administered and of the level of physical dependence (Kurland et al., 1976). Naloxone-induced withdrawal syndrome has been studied in patients dependent on buprenorphine, within 3 to 6 hours of their last dose. Withdrawal severity began 5 minutes after 1.2 mg naloxone given intravenously and reached a peak at 60 minutes. Withdrawal signs and symptoms included mydriasis, systolic hypertension, tachypnoea, muscle pains, yawning, anxiety, restlessness and craving (Nigam et al., 1994).

Naloxone has been said to be an antidote of acute ethanol intoxication in man. However, experimental and clinical studies are not convincing and contradictory and the latest clinical trials tend to confirm that naloxone lacks effectiveness in human ethylic coma (Garces et al., 1993).

Despite previous descriptions of success using naloxone in acute clonidine poisoning, five cases of acute pediatric ingestions of clonidine ranging from mild to severe occurred in which naloxone hydrochloride in doses up to 0.1 mg/kg was unsuccessful in reversing the signs and symptoms of this toxicant. Although naloxone can be safely administered to children who ingested clonidine to distinguish them from those who are intoxicated by opiates, it is not therapeutically useful to use naloxone to reverse poisoning with this class of agents (Banner et al., 1983). In a review of 47 consecutive inpatient records to determine the clinical course, role of supportive measures, and response to naloxone in children with clonidine poisoning the authors conclude that naloxone is an inconsistent antidote for clonidine poisoning (Wiley et al., 1990). Other authors (Bamshad and Wasserman, 1990) state that naloxone may be an important adjunct to therapy in pediatric clonidine intoxications as one of its postulated mechanisms of action is the release of endogenous opioids and/or the stimulation of opioid receptors in the central nervous system (CNS)

The actual use of Naloxone and Flumazenil in the Emergency Room

In order to establish how these antidotes are being actually used we have studied in a prospective way its use in the 467 poisoned patients admitted to the Emergency Unit of the Clinical Hospital of Zaragoza between January and June 1996.

The number of cases in which one or both of them have been used is 84 (18%) which means 94,4 % of the total 89 cases in which some antidote has been used. The cases distributed by age (mean 33,8±13,8) and sex (m 61,9 %; f 38,1 %) is practically the same as in the all cases. Nearly fifty percent of the cases are overdoses (40) and the other fifty percent are self-poisoning (41) which shows a neat difference with the total cases in which overdoses accounts for 57 % and suicide attempts for 27 %.

In table 1 the main implicated agents in the total and the antidote cases with an effect of depression of the CNS suspected and confirmed analytically are shown. They have been used this kind of antidote in 52% of the 100 cases in which benzodiazepine poisoning was suspected and in the 77,5% of the 40 opiate overdoses suspected. Its use in alcohol overdoses (11,6 %) can be explained by the suspicion of an association with benzodiazepines (20 cases) or opiates (7 cases). 94% of the antidote cases had symptoms at admission which leaves 5 cases in which perhaps there have been used to try to prevent the development of symptoms in banal cases.

Table 1: Toxic agents in total and antidote cases

	Total cases		Antidote cases	
	Suspected	Confirmed	Suspected	Confirmed
Benzodiazepines	100 (21,4%)	78 (16,7)	52 (61,9%)	40 (47,6%)
Antidepresives	26 (5,6%)	17 (3,6%)	13 (15,5%)	10 (11,9%)
Neuroleptics	14 (3%)	7 (1,5%)	4 (4,8%)	4 (4,8%)
Alcohol	259 (54,8%)	147 (31,5%)	30 (35,7%)	20 (23,8%)
Opiates	40 (8,6%)	19 (4,1%)	31 (36,9%)	14 (16,7%)

In table 2 a relation of the relevant symptoms to evaluate the toxic action over CNS can be seen in total toxic and antidote cases. Rousing antidotes have been used in 21,21% of toxic coma cases with a Glasgow > 10 and in the 53,8 % of toxic coma cases with a Glasgow <10; in 10 out of 13 cases of respiratory depression (76,9%) and in 7 out of 9 cases of respiratory arrest (77,8%); and in 29 out of 44 cases of miosis (65,9%).

Table 2: CNS symptoms in total and antidote cases

	Total cases	Antidote cases
Coma Glasgow >10	198 (42,4%)	42 (50%)
Coma Glasgow < 10	26 (5,6 %)	14 (16,7%)
Respiratory depression	13 (2,8%)	10 (11,9%)
Respiratory arrest	9 (1,9%)	7 (8,3%)
Miosis	44 (9,4%)	29 (34,5%)

Some data indicate a greater severity in the cases treated with rousing antidotes: 41% of these cases had been admitted to Hospital (11,9% in the ICU) vs 25,1% of the total cases (4,1% in the ICU).

The time spent in treatment doesn't show any difference between either group. Another difference is the number of cases released against medical criteria which represents 20,23% in the antidote cases and 8,6% of the total cases.

There have been 2 fatalities (2,4%) in the antidote cases and 7 (1,5%) in total cases.

The antidotes used are shown in Table 3. Mean doses of flumazenil were 1,6 mg ± 1,9 [0,25-12,5] and of naloxone 0,91 mg ± 0,71 [0,2-3,2]. Only one intravenous dose has been administered in 25 cases (29,8%), whereas more than one iv dose or an infusion has been necessary in 59 cases (70,2%).

Table 3: Antidotes used

	n	%
Naloxone	18	21,4%
Flumazenil	50	59,5%
Nal/Flu	16	19,1%

The indicated antidote for the suspected toxic has been used in 58 cases (69%): flumazenil alone in 37 cases (74% in the total cases of use of flumazenil alone); naloxone alone in 15 cases (83,3% in the total cases of use of naloxone alone); and both in 6 cases (37,5% in the total cases of use of both antidotes together). The analytical results show a somewhat lower degree of concordance: 65,6% of the total cases with 70,2% for flumazenil, 71,4% for naloxone and 40% for both of them.

A clearcut response to antidotes has been reported in 51 cases (60,7%), with 9 (10,7%) non-response and 24 (28,6%) dubious cases.

The response has been graduated in four groups (Table 4): A stable return to Glasgow 15; a return to Glasgow 15 with a relapse; a stable rise of at least two points on the Glasgow scale; and a rise of at least two points on the Glasgow scale with a relapse. In all cases but two there has been a stable recovery of the respiratory function.

Table 4: Response to antidotes

	n	type of response
yes	5	stable return to Glasgow 15: 27 cases (32,1%) return to Glasgow 15 with a relapse: 5 cases (5,9%) stable rise on the Glasgow scale: 16 cases (19%) rise on the Glasgow scale with a relapse : 3 cases (3,6%)
no	9	
dubious	24	

In 4 cases some side effects have been reported. In 3 of them flumazenil alone and in the fourth flumazenil and naloxone had been used. Three cases presented severe agitation. The other case was a self-poisoned patient having ingested high doses of benzodiazepines and amitriptyline (plasmatic concentration of amitriptyline at admission of over 1000 ng/ml), who developed convulsions after administration of 1,5 mg of flumazenil.

In 36 cases the antidotes have been used in combination with an infusion of glucose and in 7 with tiamine.

Conclusions

Flumazenil and Naloxone are by far the most commonly used antidotes in acute poisoning.

The most frequent use of flumazenil is related to the most frequent implication of benzodiazepines in acute poisoning and also with its extensive use in an unspecific way, without a clearcut indication.

Naloxone is used in a more specific way but nevertheless there is an appreciable degree of use without indication.

The highest unspecificity corresponds to the use of both antidotes together when they are used with glucose infusion in a "coma cocktail" fashion.

In most cases more than one dose of the antidote has been needed.

Only one severe adverse effect has been found confirming the risk of the use of flumazenil when a cyclic antidepressive plus benzodiazepine poisoning exists.

References

Bamshad MJ, Wasserman GS (1990) Pediatric clonidine intoxications. Vet Hum Toxicol 32:220-223

Banner W, Lund ME, Clawson L (1983) Failure of naloxone to reverse clonidine toxic effect. Am J Dis Child 137:1170-1171

Bianchi A, Ubach M (1994) Acute colonic pseudo-obstruction caused by opiates treated with naloxone. Med Clin Barc 103:78

Cuss FM, Colaco CB, Baron JH (1984) Cardiac arrest after reversal of effects of opiates with naloxone. Br Med J Clin Res Ed 288:363-364

Chamberlain JM, Klein BL (1994) A comprehensive review of naloxone for the emergency physician. Am J Emerg Med 12:650-660

Darragh A, Lambe R, Scully M, Brick I, O'Boyle C, Downe WW (1981) Investigation in man of the efficacy of a benzodiazepine antagonist, Ro 15-1788. Lancet 2:8-10

Evans LE (1973) Use of naloxone in opiate poisoning. Bri Med J 2:717

Faden AI, Holaday JW (1979) Opiate antagonists: a role in the treatment of hypovolemic shock. Science 205:317-318

Flumazenil Study Group(1992) Treatment of benzodiazepine overdose with flumazenil. The Flumazenil in Benzodiazepine Intoxication Multicenter Study Group. Clin Ther 14:978-995

Garces J, de la Torre R, Gutierrez J, Miralles R, Torne J, Cami J (1993) Clinical effectiveness of naloxone in acute ethanol intoxication. Rev Clin Esp 193:431-434

Handal KA, Schauben JL, Salamone FR (1983) Naloxone. Ann Emerg Med 12:438-445

Hofer P, Scollo Lavizzari G (1985) Benzodiazepine antagonist Ro 15-1788 in self-poisoning. Diagnostic and therapeutic use. Arch Intern Med 145:663-664

Hunkeler W, Mohler H, Pieri L, Polc P, Bonetti EP, Cumin R, Schaffner R, Haefely W (1981) Selective antagonists of benzodiazepines. Nature 290:514-516

Jefferys DB, Volans GN (1983) An investigation of the role of the specific opioid antagonist naloxone in clinical toxicology. Hum Toxicol 2:227-231

Kurland AA, McCabe L, Hanlon T (1976) Contingent naloxone treatment of the narcotic addict: a pilot study. Int J Addict 11:131-142

Larpin R, Vincent A, Perret C (1990) Hospital morbidity and mortality of acute opiate intoxication. Press Med 19:1403-1406

Lee AY, Chen YT, Lin CJ, Wu JP, Chen KS (1993) Beneficial effects of the opiate antagonist naloxone on hemodynamics and ventricular function following coronary artery occlusion and reperfusion in the dog. Proc Natl Sci Counc Repub China B 17:29-34

Lewis JM, Klein Schwartz W, Benson BE, Oderda GM, Takai S (1984) Continuous naloxone infusion in pediatric narcotic overdose. Am J Dis Child 138:944-946

Mattana J, Ahn J, Desroches L, Fitzmaurice S, Singhal PC (1993) Naloxone-responsive encephalopathy in end-stage renal disease. Am J Kidney Dis 21:669-672

McDuffee AT, Tobias JD (1995) Seizure after flumazenil administration in a pediatric patient. Pediatr Emerg Care 11:186-187

McNicholas LF, Martin WR (1984) New and experimental therapeutic roles for naloxone and related opioid antagonists. Drugs 27:81-93

Merigian KS (1993) Cocaine-induced ventricular arrhythmias and rapid atrial fibrillation temporally related to naloxone administration. Am J Emerg Med 11:96-97

Neal JM, Owens BD (1993) Hazards of antagonizing narcotic sedation with naloxone. Ann Emerg Med 22:145-146

Nigam AK, Srivastava RP, Saxena S, Chavan BS, Sundaram KR (1994) Naloxone-induced withdrawal in patients with buprenorphine dependence. Addiction 89:317-320

Nishikawa T, Tsuda A, Tanaka M, Nishikawa M, Koga I, Uchida Y (1994) Naloxone attenuates drinking behavior in psychiatric patients displaying self-induced water intoxication. Prog Neuropsychopharmacol Biol Psychiatry 18:149-153

Plant JR, MacLeod DB (1994) Response of a promethazine-induced coma to flumazenil. Ann Emerg Med 24:979-82

Pomier Layrargues G, Giguere JF, Lavoie J, Perney P, Gagnon S, D'Amour M, Wells J, Butterworth RF (1994) Flumazenil in cirrhotic patients in hepatic coma: a randomized double-blind placebo-controlled crossover trial. Hepatology 19:32-37

Rothstein JD, Guidotti A, Tinuper P, Cortelli P, Avoni P, Plazzi G, Lugaresi E, Schoch P, Montagna P (1992) Endogenous benzodiazepine receptor ligands in idiopathic recurring stupor. Lancet 340:1002-1004

Rubio F, Quintero S, Hernandez A, Fernandez S, Cozar L, Lobato IM, Pantoja S (1993) Flumazenil for coma reversal in children after cannabis. Lancet 341:1028-1029

Scollo Lavizzari G, Matthis H (1985) Benzodiazepine antagonist (RO 15-1788) in ethanol intoxication: a pilot study. Eur Neurol 24:352-354

Spivey WH (1992) Flumazenil and seizures: Analysis of 43 cases. Clin Ther 14: 292-305

Varon J, Duncan SR (1991) Naloxone reversal of hypotension due to captopril overdose. Ann Emerg Med 20:1125-1127

Welch EB, Thompson DF (1994) Opiate antagonists for the treatment of schizophrenia. J Clin Pharm Ther 19:279-283

Wiley JF, Wiley CC, Torrey SB, Henretig FM (1990) Clonidine poisoning in young children. J Pediat 116:654-658

Winkler E, Almog S, Kriger D, Tirosh MS, Halkin H, Ezra D (1993) Use of flumazenil in the diagnosis and treatment of patients with coma of unknown etiology. Crit Care Med 21:538-542

Antidotes: Availability, Use and Cost in Hospital and Extra-Hospital Emergency Services of Catalonia (Spain).

S Nogué, D Soy, P Munné and J Millá
Clinical Toxicology Unit, Emergency Department and Pharmacy Service. Hospital Clínic. Villarroel 170. E-08036 Barcelona. Spain.

Introduction

Antidotes are, together with general supportive treatment, the measures to prevent absorption and the methods to increase elimination of poisons, one of the bases of the treatment of acute intoxications.

We have carried out a study in Catalonia to determine the availability of these antidotes in Hospitals and medically equipped ambulances, here named Extra-Hospital Emergency Services, to investigate the real use of these antidotes at hospital level and to determine the economic cost for this comsumption of antidotes.

Patients and Methods

We have considered as antidotes those compounds defined as such by the latest list of the International Programme for Chemical Safety (IPCS/CEC, Working Group on Prevention of Absorption in the Management of Poisoning Cases, Milan, Italy, 18-22 May 1993) which is a satellite Institution of the World Health Organization.

To study the availability of these antidotes in Hospital Emergency Services, collaboration was sought from 24 Hospitals in Catalonia by sending them the list of antidotes drawn up by the IPCS, with the object of determining whether or not the antidotes were available on the day of the survey, the 25th May 1996.

These same Hospitals had carried out a previous study (from June 1994 to May 1995) in which intoxications treated during the first 7 days of each month were noted for 12 months, and in which one of the variables was the real use of these antidotes.

To study the availability of these same antidotes in Extra-Hospital Emergency Services, collaboration was sought from 3 medically equipped teams performing these functions in Barcelona city (the "061" and Firemen) and in the rest of Cataluña (the "S.E.M."), using the same list of the IPCS and the same method, on the 31 of July 1996.

Finally, the economic evaluation of the use of these antidotes was carried out in only one of the participating hospitals (the "Clinic Hospital" of Barcelona) and consisted in accounting for the cost of their use during the first 6 months of 1996, and comparing this with the rest of the medications used in other medical emergencies and with the total cost of pharmaceuticals in the Hospital.

Results

The first part of the study refers to the availability of antidotes. Listed in Table 1 there are the 26 antidotes considered to be Type "A" by the IPCS, which means that they must be available in less than 30 minutes, and the percentage of Hospital and Extra-Hospital Emergency Services where they were indeed available.

In none of the Hospitals were all of the 26 antidotes available for immediate use. Only 12 antidotes were available in all the Hospitals, but 3 were not found in any of the 24 Hospitals. For example, hydroxicobalamin was only available in 29% of the Hospitals, and in none of the Extra-Hospital Emergency Services. Another example: The Extra-Hospital Services did not always carry beta-blockers, calcium salts or glucagon. No-one has methylpyrazole or prenalterol, only some, not all, of the Hospital Emergency Services have phentolamine or physostigmine. Furthermore, the Extra-Hospital Emergency Services do not carry either methylene blue or protamine or pyridoxine.

In Table 2 there are those 13 antidotes listed which are considered Type "B" by the IPCS, which means that they must be available in less than 2 hours. Again, no Hospital had all 13 antidotes available: Only 3 antidotes were found in all the Hospitals, and 2 in none. For example, not all Hospitals had antagonists of the organophosphate insecticides (pralidoxime or obidoxime), only 33% had silibinin and only 88% had deferoxamine.

Table 3 shows the 8 Type "C" antidotes which must be available in less than 6 hours. In no hospital were all of these 8 antidotes available. Only one antidote (phytomenadione) was found in all hospitals and 3 were not found in any Hospital. Thus, for example, not all hospitals had edetate calcium or penicillamine.

Table 1 Antidotes considered to be Type "A" (available in less than 30 minutes)

Antidote	Hospital[a]	Extra-Hospital[b]
Amyl nitrite	38	0
Atropine	100	100
Beta-blockers	100	67
Calcium salts	100	33
Dantrolene	79	0
Diazepam	100	100
Dicobalt edetate	54	0
Digoxin-specific antibodies	4	0
Dimethylaminophenol (4-DMAP)	0	0
Ethanol	83	0
Glucagon	100	33
Glucose	100	67
Hydroxicobalamin	29	0
Isoprenaline	83	33
4-Methylpyrazole	0	0
Methylene blue	100	0
Naloxone	100	100
Oxygen	100	100
Phentolamine	38	0
Physostigmine	83	0
Prenalterol	0	0
Protamine	100	0
Pyridoxine	100	0
Sodium nitrite	21	0
Sodium nitroprusside	100	33
Sodium thiosulfate	38	0

[a] Percentage of Hospital Emergency Departments where antidote was available
[b] Percentage of Extra-Hospital Emergency Services where antidote was available

Summing up the first part, we can consider that with respect to the availability of antidotes, on average, 64 % of Type "A" are available in Hospital Emergency Services and 26% in Extra-Hospital Emergency Services, and that 53% of Type "B" and 46% of Type "C" are available in Hospital Emergency Services.

Table 2 Antidotes considered to be Type "B" (available in less than 2 hours)

Antidote	Hospital[a]
Acetylcysteine	100
Benzylpenicillin	100
Deferoxamine	88
Dimercaprol	67
Flumazenil	100
Folinic acid	96
Methionine	0
Neostigmine	0
Obidoxime	8
Prussian blue	17
Pralidoxime	67
Silibinin	33
Succimer (DMSA)	13

[a] Percentage of Hospital Emergency Departments where antidote was available

Table 3 Antidotes considered to be Type "C" (available in less than 6 hours)

Antidote	Hospital[a]
Edetate calcium	88
N-Acetyl penicillamine	88
Oxygen, hyperbaric	4
Penicillamine	88
Pentetic acid (DTPA)	0
Phytomenadione	100
Trientine	0
Unithiol (DMPS)	0

[a] Percentage of Hospital Emergency Departments where antidote was available

The second part of the study concerned the real use of these antidotes. The base is a multicentre study carried out in Hospital Emergency Services in Cataluña in which 24 Hospitals participated. All the acute intoxications treated during the first week of each month were collected from June 1994 to May 1995.

Table 4 shows the distribution of the 1,293 intoxications thus collected by the type of toxic agent responsible, which highlights the fact that two thirds of them were caused by medicaments or alcohol (ethanol). Less important, following this order, were household products, abusive drug use, animal bites or stings, industrial products, plants and fungi and agricultural products.

Table 4 Epidemiology of acute intoxication in Catalonia (n=1.293). Distribution by type of toxin

Toxin	Cases (%)
Medicaments	40.5
Alcohol	27.0
Household products	10.9
Drugs of abuse	10.6
Animal bite/sting	4.3
Industrial products	3.7
Plants and fungi	0.5
Agricultural products	0.3

In 185 of these 1,293 acute intoxications, antidotes were used, that is, in 14% of the cases, although only 9 different antidotes were used. Note that 2 antidotes, flumazenil and naloxone, account for 93% of the cases in which antidotes were used. Also used were Acetylcysteine, phytomenadione and others represented in Table 5.

The third part of the study had an economic goal. The data were collected during the first six months of 1996 in the Pharmacy Service of the Hospital Clinic of Barcelona, a 900-bed Hospital, which annualy receives around 136.000 emergencies of all types, among which are 41.000 medical emergencies, of which 700 are toxicological emergencies.

During the cited period of 6 months, the total expenditure of the Hospital on pharmaceutical products was more than 7 million dollars, of which 73.000 $ were used in the treatment of 20.000 medical emergencies and 6.000 $ was the cost of antidotes for 348 toxicological emergencies. This consumption of antidotes represents less than 0.1% of the total Hospital pharmaceutical expenditure. In other words, of each 1.185 $ of general pharmaceutical expenditure by the Hospital, 12 $ are for medical emergencies, and of this only 1 $ is for antidotes for toxicological emergencies.

Table 5 Drugs used in 185 cases of acute intoxications treated by antidotes

Antidote	Number of cases
Flumazenil	103 / 185
Naloxone	66 / 185
N-Acetyl-Cysteine	8 / 185
Phytomenadione	2 / 185
Atropine	1 / 185
Ethanol	1 / 185
Glucose	1 / 185
Penicillin	1 / 185
Çalcium salts	1 / 185

Conclusion

None of the 24 hospitals surveyed had 100% of the antidotes recomended by the IPCS available, and the Hospital and Extra-Hospital Emergency Services of Cataluña lack some antidotes that they might need extremely urgently.

Only in 14% of acute intoxications were antidotes used, and the economic cost of these represents less than 0.1% of the total pharmaceutical expenditure.

Finally, in Catalonia, the lack of Clinical Toxicology Services which can assess antidotes, the absence of specific recommendations by the scientific societies most closely related to the subject (the Spanish Society of Toxicology and the Spanish Society of Emergency Medicine) and the lack of a legal directives are all contributing to the fact that antidotes are not at present subject to a rationalization of their availability and use.

Reference

IPCS/CEC (1993) Working group on prevention of absorption in the management of poisoning cases. World Health Organisation 93.36

The Relative Efficacy of Antidotes: The IPCS Evaluation Series

Dag Jacobsen[1] and John A. Haines[2]
[1] Clinic for Acute Medicine, Ullevål University Hospital, N-0407 Oslo
[2] International Programme of Chemical Safety (IPCS), CH-1211 Geneva

Abstract

Antidotes may play an important role in the treatment of poisoning. For many physicians and toxicologists an antidote is an antidote. According to the IPCS definition, an antidote is a therapeutic substance used to counteract the toxic action(s) of a specified xenobiotic. Given this wide definition, the efficacy of an antidote may vary considerably depending on which toxic action(s) is/are being counteracted and on the level of counteracting power: An almost 100% efficacy is seen using specific antagonists, such as naloxone in opiate poisoning or flumazenil in benzodiazepine poisoning, e.g. resulting in complete reversal of opiate toxicity unless complications, such as anoxic brain damage, have developed. At the other end of the efficacy scale, we may place chelating agents for heavy metal poisoning and diazepam for organophoshorus insecticide poisoning. Therefore, in the IPCS/EC evaluation series of antidotes, some chelating agents are considered only to be an adjunct to supportive care which is the cornerstone of treatment. When teaching clinical toxicology or recommending the use of antidotes in poisoned patients, the expected efficacy level of the antidote in question should be stressed. This may be particularly important in severe poisonings when the antidote may only be considered as an adjunct to supportive care, e.g. deferoxamine in acute iron poisoning. Unless this is stressed, the unexperienced physician may rely too much on the antidote and may not pay sufficient attention to the supportive care. In this presentation, the varying efficacy levels of antidotes are discussed as based on the presently ongoing IPCS/EC evaluation programme on antidotes.

Introduction

While good supportive care and elimination techniques may, in many cases, restore a poisoned patient to good health and stabilize body functions, the appropriate use of antidotes and other agents may greatly enhance elimination and counteract the toxic actions of the poison. In certain circumstances they may significantly reduce the medical resources otherwise needed to cure a poisoned patient, shorten the period of therapy and, in some cases, save a patient from death; thus reducing the overall burden on the health service of poisoned patient management. In areas remote from good hospital services, and particularly in developing countries with a lack of adequate facilities for supportive care, antidotes may be even more essential in the treatment of poisoning, as well as in primary health care use. In rural areas of these countries the workers and their families, especially the children, may be at greater risk of poisoning, for example, from the misuse of pesticides, or from poisonous plants and venomous animals, than in developed countries.

Individual physicians have frequently expressed concern about the difficulties in obtaining certain antidotes in an emergency for treating poisoned patients. The need for an international evaluation of the efficacy in clinical use of antidotes for specific types of poisoning cases in order to encourage their availability was first recognized at the joint meeting of the World Federation of Associations of Poisons Control Centres and Clinical Toxicologists, the International Programme on Chemical Safety (IPCS) and the Commission of the European Union (EC), held at WHO headquarters, Geneva, 6-9 October 1985. Subsequently, the IPCS and the EC jointly instituted a project, which is being undertaken in consultation with the World Federation, designed to evaluate the use of antidotes. The first three volumes of this series have recently been published (Meredith et al. 1993a; 1993b; 1995) and further volumes will follow shortly.

In a preparatory phase of this project, an antidote was defined as *a therapeutic substance used to counteract the toxic action(s) of a specified xenobiotic.* A preliminary list of antidotes, as well as other agents used to prevent the absorption of poisons, to enhance their elimination and to treat their effects on body functions, was established and they were preliminarily classified according to urgency of treatment and efficacy in practice.

The efficacy of a substance used as an antidote must be validated, initially with animal experiments and preferably using species which show a pattern of toxicity similar to humans. It may be more difficult to document the clinical efficacy of an antidote in humans compared to other pharmaceutical agents. Since there is less opportunity for clinical trials, the potential toxicity of an antidote is also important in deciding its use; and the possibility of an adverse reaction to the antidotes should be considered. A non-toxic antidote may be employed in cases of poisoning where efficacy is uncertain, while an antidote with more pronounced side-effects should only be used if the therapeutic effect is known and the diagnosis is certain. Long-term adverse effects and chronic toxicity may be less important than for an ordinary pharmaceutical agent, since an antidote is only likely to be used once. It is important that increased toxicity does not result from mobilization of the toxin from tissue stores or changes in tissue distribution, which have been discussed using older chelating agents in heavy metal toxicity (Aaseth 1983).

Classification of antidotes according to efficacy

Specific antagonists, such as naloxone and flumazenil, may counteract the toxic effects of opiates and benzodiazepines, respectively, within minutes (Evans et al. 1973; Aarseth et al. 1988). Therefore these antidotes may be considered to be the most effective antidotes available as indicated in Table I. However, efficacy of an antidote is still a relative parameter as the time lapsing between ingestion of the toxic agent and the antidote may be critical on the level of minutes. For example, in opiate poisoning naloxone may reverse respiratory arrest and may also help in restoring pulse in patients also suffering from cardiac arrest. Of the 178 opiate poisoned patients in our study in Oslo, 31 suffered from respiratory arrest of whom 2 also had cardiac arrest (Jacobsen et al. 1984). In all these patients some sort of primary resuscitation had been started by bystanders. Paramedics continued ABC resuscitation and gave naloxone to some of the patients. Of these 31 patients, 29 were discharged without any sequelae, one suffered from permanent hypoxic brain damage and one died. As illustrated in this study, a delay of antidote administration of more than 5-10 minutes may result in hypoxic brain damage or even brain death if early primary resuscitation had not been initiated by lay people.

Although effective from a pharmacologic point of view, i.e. in reversing the toxic features of benzodiazepine poisoning (Meredith et al. 1993), there may not be much need for flumazenil due to the very low morbidity and lethality in benzodiazepine poisoning. However, flumazenil has also proven to be useful in triage and diagnosis in comatose patients as many of these are poisoned with benzodiazepines.

The chelating agents may serve as a good example of a group of antidotes with only moderate efficacy in counteracting the toxic effects of various heavy metals and metaloids. Most of the documentation on the effect of chelating agents is based on experimental work in animals (Inns & Rice 1993; Xu & Jones 1988). In these experiments, various end points such as lethality, reduced amount of toxic agent in target organ(s) and increased excretion of toxic agent, have been employed. Of these, lethality is the only hard endpoint and, therefore, the studies using the other softer endpoints may seem inappropriate. On the other hand, patients with heavy metal poisoning are most often presenting with diffuse neurological signs or symptoms and are usually not in a primordial or critically ill condition. One could therefore argue that a significant reduction of the toxic compound in the target organ(s), e.g. the central nervous system, should benefit the patient.

In contrast to the extensive amount of controlled experimental data on heavy metal poisoning treated with various chelators, the clinical literature is almost devoid of such data. Our clinical knowledge is almost solely based on more or less well-performed case studies, often with inapproprate historical controls (Ennis & Harrison 1950), although some good clinical studies addressing the use of chelators in childhood plumbism have recently been performed (O Connor 1992; Markowitz et al. 1993). So far, the chelating agents have not demonstrated any clear life-saving effect in the clinical situation and it may therefore be correct to place them somewhere in the middle of the efficacy scale shown in Table I. The group of chelating agents, however, are not homogenous and some may therefore be more efficacious than others as indicated in Table I.

Table I Proposed efficacy grading of some antidotes as scientifically based on their ability to counteract toxic effects.

Atropine (nicotinic features)
Diazepam (OP-poisoning)
Silibinine
Dantrolene
(BAL) Chelating agents (DMSA/DMPS)
Obidoxime
Pralidoxime
Cyanide antidotes
Atropine (muscarinic)
Ethanol (4-MP)
Fab-fragments
N-acetylcysteine
Flumazenil
Naloxone

0% ← Relative Efficacy Scale of Antidotes → 100%

OP = organophosphate
4-MP = 4-methylpyrazole

In many countries silibinin is used as an antidote for *Amanita phalloides* and *A. virosa* poisoning despite a considerable lack of both experimental and clinical documentation. There are a few experimental studies indicating that silibinin reduces liver damage and increases survival in dogs and mice poisoned with these mushrooms (Floersheim et al. 1978; Vogel et al. 1984). Silibinin also prevents amanitin-induced nephrotoxicity in rats (Vogel 1980). Clinical data are also indicative of some effect on lethality (Hruby 1987) but conclusive data are still lacking which is why silibinin is positioned between 0 and 50% on the efficacy scale in Table I. However, given the very low toxicity of this compound, it may be more useful than this from a more pragmatic point of view. Nevertheless, antidote therapy in these mushroom poisonings should be considered only as an adjunct to supportive therapy aimed particularly at the renal and liver failure of these toxic emergencies.

Due to the species differences in methanol poisoning, limited information is derived from experimental studies. Based on one extensively studied case of human methanol poisoning (Jacobsen et al. 1983), however, and all the other indicative clinical reports (Jacobsen & McMartin 1986), the positioning of ethanol as a very efficient antidote in methanol (and ethylene glycol) poisoning in Table I will not cause much dispute. Certain disadvantages with this antidote may, however, make 4-methylpyrazole a more desirable antidote, at least in ethylene glycol poisoning (Baud et al. 1988; Jacobsen et al. 1996). So far, however, its use outside France has to await a commercially available pharmaceutical formulation.

Conclusions

Few antidotes are "absolute" antidotes such as naloxone and flumazenil. Most antidotes are therefore only adjuncts to, and not a substitute for, supportive care. Whenever toxicology professionals are recommending the use of an antidote to the physician treating a poisoned patient, the expected efficacy level of the antidote should be clearly explained in order to avoid less emphasis being placed on supportive therapy which remains the cornerstone in the treatment of the vast majority of the poisoned patients.

Acknowledgements

We would like to acknowledge the IPCS editoral staff in Geneva and the many colleagues worldwide who have contributed to this evaluation programme which is part of the IPCS INTOX Package: An interactive computerized database and information management software designed to enhance the services of poison centres worldwide. The IPCS is a joint programme of WHO, UNEP and ILO established in 1980 and is under the administration of WHO.

References

Aarseth HP, Bredesen JE, Grynne B, Lyngdal PT, Storstein L, Wiik-Larsen E (1988) Benzodiazepine-receptor antagonist, a clinical double-blind study. J Toxicol Clin Toxicol 6: 283-292

Aaseth J (1983) Recent advances in the therapy of metal poisonings with chelating agents. Hum Toxicol 29: 257-272

Baud FJ, Galliot M, Astier A, Bien DV, Garnier R, Likforman R, Bismuth C (1988) Treatment of ethylene glycol poisoning with intravenous 4-methylpyrazole. N Engl J Med 319: 97-100

Ennis JM, Harrison HE (1950) Treatment of lead encephalopathy with BAL. Pediatrics 5: 853-868

Evans LEJ, Roscoe P, Swainson CP, Prescott LF (1973) Treatment of drug overdose with naloxone, a specific narcotic antagonist. Lancet 1: 452

Floersheim GL, Eberhard M, Tschumi P, Duckert F (1978) Effects of penicillin and silymarin on liver enzymes and blood clotting factors in dogs given a boiled preparation of *Amanita phalloides*. Toxicol Appl Pharmacol 6: 455-462

Hruby K (1987) Knollenblätterpilzvergiftung. Intensivmed 24: 269-274

Inns RH, Rice P (1993) Efficacy of dimercapto chelating agents for the treatment of poisoning by percutaneously applied dichloro(2-chlorovinyl)arsine in rabbits. Hum Exp Toxicol 12: 241-246

Jacobsen D, Øvrebø S, Arnesen E, Paus PN (1983) Pulmonary clearance of methanol in man. Scand J Clin Lab Invest 43: 377-379

Jacobsen D, Frederichsen PS, Knutsen KM, Sørum Y, Talseth T, Ødegaard OR (1984) Clinical course in acute self-poisonings: A prospective study of 1125 consecutively hospitalised adults. Hum Toxicol 3: 107-116

Jacobsen D, McMartin KE (1986) Methanol and ethylene glycol poisonings. Med Toxicol 1986; 1: 309-334

Jacobsen D, Ostensen J, Bredesen L, Ullstein E, McMartin KE (1996) 4-methylpyrazole (4-MP) is effectively removed by hemodialysis in the pig model. Hum Exp Toxicol 13: 131-4

Markowitz ME, Bijur PE, Ruff H, Rosen JF (1993) Effects of calcium disodium versenate ($CaNa_2EDTA$) chelation in moderate childhood lead poisoning. Pediatrics 92: 265-271

Meredith TJ, Jacobsen D, Haines T, Berger JC, eds. (1993a) Naloxone, flumazenil and dantrolene as antidotes. Cambridge University Press, Cambridge

Meredith TJ, Jacobsen D, Haines J, Berger JC, Van Hejist ANP, eds. (1993b) Antidotes for poisoning by cyanide. Cambridge University Press, Cambridge

Meredith TJ, Jacobsen D, Haines J, Berger JC, eds. (1995) Antidotes for poisoning with paracetamol. Cambridge University Press, Cambridge

O'Connor ME (1992) CaEDTA vs CaEDTA plus BAL to treat children with elevated blood lead levels. Clin Pediatr Phila 31: 386-390

Vogel G (1980) The anti-amanita effect of silymarin. In: *Amanita* toxins and poisoning. Faulstich H, Kommerell B, Wieland T, eds. New York: Wiltzstrock, Baden-Baden, Köln, New York, pp 180-187

Vogel G, Tuchweber B, Trost W, Mengs U (1984) Protection by silibinin against *Amanita phalloides* intoxication in beagles. Toxicol Appl Pharmacol 73: 355-362

Xu ZF, Jones MM (1988) Comparative mobilization of lead by chelating agents. Toxicology 53: 277-288.

In Vitro Cell Models for Investigating Molecular Mechanisms of Toxicity

(Chairs: J.V. Castell, Spain,
and G. Repetto, Spain)

In vitro Investigation of the Molecular Mechanisms of Hepatotoxicity

José V. Castell, María José Gómez-Lechón, Xavier Ponsoda and Roque Bort
Unidad de Hepatología Experimental Centro de Investigación. Hospital Universitario La Fe. Avda. Campanar 21. E-46009 Valencia (Spain)

Drug-induced Hepatic Injury

Iatrogenic Hepatitis: Intrinsic and Idiosyncratic Toxicity. Substances capable of producing liver damage and, more specifically, hepatocyte damage are known as *hepatotoxins*. They are classified (Zimmerman and Ishak, 1995, Castell et al., 1992) according to whether they exert their effects in all individuals, in a dose-dependent and hence predictable manner (*intrinsic hepatotoxins*), or in certain individuals, occasionally after several contacts, in a non-dose dependent and therefore unpredictable way (*idiosyncratic hepatotoxins*). These substances can act directly on cells (*active* hepatotoxins), or become toxic after biotransformation (*latent* hepatotoxins). Idiosyncratic hepatotoxicity is the consequence, either of an unusual metabolism of the drug by susceptible individuals which produce too large amounts of toxic metabolites (*metabolic idiosincrasy*), or is due to an immune-mediated attack to sensitised hepatocytes (*drug hypersensitivity*).

The damage caused to hepatocytes can be *cytotoxic*, *genotoxic* or *metabolic* (Zimmermann and Ishak, 1995). The first type of injury is a common feature of many intrinsic hepatotoxins. It is evidenced by important morphological changes in the structure of hepatocytes (vacuolization, steatosis, acidophilia, necrosis, etc.), and is accompanied by increases in the serum activity of hepatic enzymes. Genotoxins are substances that produce, in first term, DNA damage and show tendency to induce primary hepatocarcinomas. Finally, drugs can also alter the cellular metabolism of hepatocytes without causing cell death. This usually takes the form of alterations of the functional capacity of hepatocytes and, in particular, of the uptake, conjugation and secretion of bile acids, which, in turn, results in an impaired bile flux (*cholestasis*).

The liver is very active in metabolizing foreign compounds. Although the biotransformation reactions are generally involved in detoxification processes, there are many cases in which the metabolism of a drug is the cause of deleterious effects to cells (Zimmerman and Ishak, 1995). This is the case of reactions catalyzed by cytochrome P450 enzymes (CYP) that can generate more toxic or more reactive metabolites capable of reacting with nucleophiles, of covalently binding to macromolecules or of initiating radical-chain reactions within cells. Against these potential hazards, hepatocytes have effective defence mechanisms (specialized enzymes, reducing molecules like GSH, and DNA and protein repair mechanisms). Ultimately, it is the balance between *bioactivation*, *detoxification* and *defense mechanisms* that determines whether a reactive metabolite will or will not elicit a toxic effect.

Studies on hepatotoxicity most frequently deal with a type of cell death which is commonly defined as *lytic necrosis*, characterized by a rapid collapse of the cell's internal homeostasis and evidenced by alterations of cytoplasmic organelles, cell swelling, membrane lysis, release of cellular content and inflammation of the surrounding areas (Columbano, 1995). The other type of cell death, *apoptosis*, is characterized by a progressive condensation of the chromatin on the inner face of the nuclear membrane, cell shrinkage with subsequent loss of membrane contact with neighbouring cells, and fragmentation of the cell with formation of membrane-bound acidophilic globules (apoptotic bodies). There is now accumulating evidence showing that most of the agents capable of inducing death of liver cells can induce both apoptosis and necrosis. Interestingly, there is very little evidence of the occurrence of necrosis in the liver in the total absence of a preceding or concomitant apoptosis (Columbano, 1995).

Molecular Mechanisms of Toxicity

Impairment of Cellular Metabolism. The alteration of the metabolic functions of hepatocytes by a drug or any of its stable metabolites, is frequently at the origin of hepatotoxicity. Certain compounds can directly act as enzyme or ion transport inhibitors, or compete with cellular metabolites in hepatocyte metabolic pathways (*antimetabolites*, e.g., galactosamine, ethionine). Conversely, many hepatotoxins can indirectly alter the energetic balance of cells by dramatically increasing the energy demand (i.e. increased ATP consumption), reducing ATP production, or both (Swartz, 1995). ATP depletion is, in fact, a common event in the course of drug-induced cell damage, frequently preceding the irreversible stages of cell injury. An interesting example of increased energy demand constitutes the *de novo* synthesis of GSH. Intracellular oxidized glutathione concentration can not go beyond a certain level. Glutathione disulfide, if not effectively reduced by glutathione reductase, is eliminated from the cell. This results in a decrease of the GSH pool, requiring *de novo* synthesis of this metabolite and, concomitantly an important ATP demand. A similar situation can occur if a metabolite is eliminated mainly by conjugation with GSH, and the recovery mechanisms (hydrolysis of glutamic acid and glycine from conjugates, and elimination as mercapturic acid conjugates) are overwhelmed (Reed, 1990).

Mitochondria constitute an important target of drug hepatotoxicity. Mitochondrial dysfunction has immediate effects on the energetic balance of cells becoming a crucial event in the onset of hepatocyte necrosis (Rosser and Gores, 1995). Several mechanisms can be involved in mitochondrial injury: a) direct inhibition of mitochondrial metabolism, for instance electron transport and/or oxidative phosphorylation (Nieminen et al., 1995); b) alteration of the physicochemical properties of the mitochondrial membrane as a result of oxidative damage (i.e. lipid peroxidation after mitochondrial GSH depletion; Nieminen et al., 1995); c) intercalation of compounds in the lipidic membrane that alter its physicochemical properties, the mitochondrial membrane potential and, hence, the driving force of ATP production; and d) damage to mitochondrial DNA.

Drug-induced Lipid Peroxidation. Lipid peroxidation is also frequently involved in the mechanism of hepatotoxicity (Poli et al., 1987). It is a free radical process leading to the oxidative degradation of the lipids present in cell membranes, and it is easily propagated to other unsaturated molecules when oxygen is present. The oxidation of lipids ends with the formation of a wide range of degradation products, i.e. hydroperoxides, aldehydes, including malondialdehyde, ketones etc. The first consequence of this process is the profound alteration of the physico-chemical properties of the membrane and of the functionality of membrane-allocated enzyme activities (Ross, 1989).

Cells use several strategies to protect themselves against uncontrolled lipid peroxidation: a) inactivation of active oxygen species, b) trapping of eventually formed radicals, c) inhibition of the radical chain propagation, and d) repair of damaged lipids. Superoxide dismutase, catalase and glutathione peroxidase, together with reduced glutathione (GSH) are the most efficient cellular agents against oxygen species and radicals. Lipid hydroperoxides can be efficiently reduced by glutathione peroxidase. Finally, natural antioxidants, like vitamin E present in biological membranes, act by inhibiting the propagation step of lipid peroxidation.

Drug-induced Oxidative Stress. The term *oxidative stress* is defined as a disturbance in the prooxidant-antioxidant balance of a cell (Sies, 1991). In active aerobic cells, a certain oxidative challenge occurs during the oxidation of substrates taking place in the cell to produce ATP. This may also happen as a consequence of a decrease in the cellular antioxidant elements because of nutritional or physiological changes but it does not constitute an oxidative stress. However, when as a result of the metabolism of a xenobiotic there is an increased formation of active oxygen species accompanied by a loss of GSH equivalents, the balance is altered in favour of *oxidative damage* (Sies 1991). The immediate result is an increased lipid peroxidation, alteration of membrane functionality and, indirectly, the inability of the cell to generate its own energy or to maintain the ionic homeostasis.

The most active substances in eliciting oxidative stress are compounds able to undergo repeated oxidation and reduction cycles within the cell. A representative example of this type of toxins are quinones (Figure 1), which cause cell injury by two mechanisms. First, quinones are electrophiles, able to react with nucleophiles such as the thiol group of GSH and proteins. Since GSH plays a key protective role within the cell, depletion of GSH renders the cell more sensitive to physiologically-generated oxygen active species. Second, a number of flavoproteins (e.g. cytochrome P450 reductase) catalyze the NAD(P)H-dependent reduction of quinones to semi-quinones which can be detoxified by reaction with nucleophiles (GSH), or spontaneously react with molecular oxygen in a one-electron oxidation process to regenerate the parent quinone and superoxide anion. The H_2O_2 which is subsequently produced by superoxide dismutases can lead to the formation of oxygen-derived radicals ($OH^{\cdot}$, $O_2^{\cdot-}$, $^1O_2$) by the $Fe^{2+/3+}$ catalyzed Haber-Weiss and Fenton reactions. The quinone can again be reduced, thus becoming trapped in a series of redox reactions. The stoichometric balance of the redox cycling is the continuous one-electron reduction of molecular oxygen by cellular NAD(P)H. The consequences are a large amount of reactive oxygen species produced and a depletion of reduced thiol and nicotinamide nucleotide pools, which makes the biomolecules within the cell more susceptible to damage by active oxygen species (Boobis et al., 1989). Oxidative stress can also lead to S-thiolation of proteins, a phenomenon that has been shown to occur with several xenobiotics. Ca^{2+}-dependent ATPases are among the most sensitive proteins to these modifications (Bellomo et al., 1983).

Mitochondria have no catalase and they rely totally on GSH peroxidase, and hence, on mitochondrial GSH, to prevent the effects of active oxygen species generated during oxidative phosphorylation. Since no transport system for oxidized glutathione out of the mitochondria exist, GSSG must be reduced inside mitochondria making this organelle particularly susceptible to GSH-depleting xenobiotics.

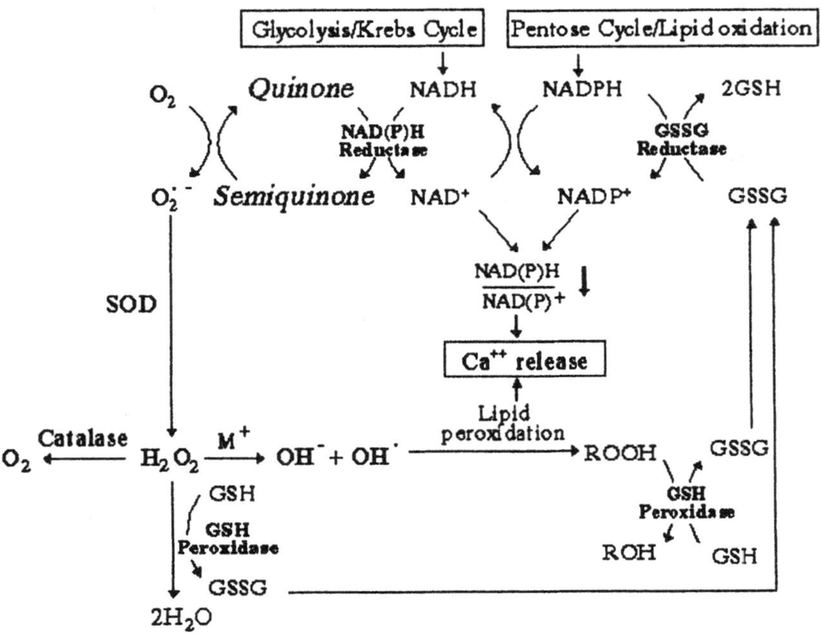

Figure 1. Drug redox-cycling and oxidative stress.

Calcium and Drug-induced Cell Injury. Intracellular calcium plays a fundamental role as a regulator of many enzymes and as an effector for hormones and growth factors controlling a wide variety of physiological processes. The cytosolic free Ca^{2+} concentration in mammalian cells is very low (~0.1 μM) compared with the concentration of extracellular fluids (1-2 mM). The major pathways for Ca^{2+} entry into cells are voltage-dependent channels gated by the electrical potential accross the plasma membrane, and receptor-operated channels (Spedding and Paoletti, 1982). The cytosolic Ca^{2+} concentration is controlled by active sequestration into the mitochondria and the endoplasmic reticulum. The driving force for Ca^{2+} uptake in the mitochondria is the electrochemical gradient across the inner membrane generated during oxidative phosphorylation, while in the endoplasmic reticulum the uptake of Ca^{2+} is done by a Ca^{2+} -ATPase pump. Other Ca^{2+} extrusion mechanisms occur in cells associated with the influx of Na^{+} by membrane pumps acting as Na^{+}-driven antiports.

Toxins can alter Ca^{2+} homeostasis either by: a) increasing the external influx or release from intracellular stores; b) impairing the uptake by organelles, or ion extrusion to the external medium. Although xenobiotics able to inactivate Ca^{2+}-ATPases are known, most toxic xenobiotics act indirectly by modifying membrane properties (e.g. lipid peroxidation), reducing the energy status of cells or depleting GSH.

As a consequence of the toxic action of xenobiotics, a sustained rise in cytosolic Ca^{2+} can occur. An immediate consequence of elevated cytosolic Ca^{2+} is plasma membrane blebbing, activation of phospholipases and non-lysosomal proteases which can cause irreparable damage and functional injury to cell components (Boobis et al.,1989; Shen et al., 1992). The elevated cytosolic Ca^{2+} can also trigger endonucleases activation and a concomitant DNA fragmentation, similar to that observed in apoptosis (Shen et al., 1992).

Covalent Binding. Biotransformation of xenobiotics may result in the formation of chemical intermediates (electrophiles, radicals, conjugates) able to react with cell macromolecules forming stable drug adducts (Hinson and Roberts, 1992). Unsaturated carbonyl compounds (i.e. α, β unsaturated ketones), aldehydes, epoxides and glucuronides are among the most frequently involved reactive groups. Proteins, DNA and RNA are the most frequent targets. Covalent binding is dependent on the proportion of the chemical converted into a reactive metabolite, the half-life of the reactive intermediate, and its ability to react with the target biomolecule. Depending both on the site of generation and on the nature and reactivity of the intermediate, differences in the subcellular localization of adducts occur. The endoplasmic reticulum is where more often protein-drug adducts are found; a fact that probably reflects the proximity of the site of bioactivation (CYP) and of binding (Boelsterli, 1993). Drug-protein adducts have also been identified in the plasma membrane (Hinson and Roberts, 1992) a finding that became relevant in relation to the immune-mediated drug hepatotoxicity.

Few examples are known where the drug-protein adduct could be fully identified. CYPs giving rise to the formation of reactive metabolites are among the most frequent protein targets (Boelsterli 1993; Holtzman, 1995), although other enzymes, (i.e. Ca^{2+}-ATPase, carboxyl esterase, GSH-transferase) have also been reported as targets (Boelsterli 1993; Kenna et al., 1993). Halothane (Van Pelt and Kenna, 1994) and diclofenac (Kretz-Rommel and Boelsterli 1994) have been studied in detail, and the presence of several protein bands bearing the trifluoroacetyl or diclofenac moiety have been identified.

Covalent binding is a phenomenon that, although frequently appearing in the course of toxic phenomena, has only in a few cases been associated as the possible cause of a well defined cell injury. For certain drugs (acetaminophen, bromobenzene), there is a clear correlation between the extent of covalent binding to proteins and the severity of hepatocyte injury (Boelsterli, 1993; Holtzman, 1995). However, with other xenobiotics, there is a lack of correlation with hepatotoxicity. This is the case with bromophenol and with the paracetamol analog N-acetyl-m-aminophenol, which are able to covalently bind to hepatocytes, yet do not result in liver damage (Boelsterli, 1993). On the opposite, covalent binding is inversely correlated with the extent of hepatotoxicity in the case of diclofenac. Binding of the drug to proteins takes place mainly via transacylation of the diclofenac glucuronide moiety (Hargus et al., 1994). This process does not per se cause cytotoxicity, while CYP metabolism of the drug produces toxic metabolites that ultimately cause cell injury (Ponsoda et al., 1995). These two metabolic routes co-exist. Thus, the more diclofenac is metabolized by P450, the less it becomes available for covalent binding through the formation of the glucuronide (Kretz-Rommel and Boelsterli, 1994).

Immunological Mechanisms of Drug Hepatotoxicity

Allergic hepatitis. Among the various forms of drug-induced hepatitis, some of them show clinical features compatible with an hypersensitivity reaction against the drug. *Allergic hepatitis* is a type of idiosyncratic toxicity that occurs in certain individuals in an unpredictable and dose-independent manner and is very difficult to anticipate with animal models.

Drugs are small organic molecules unable to elicit an immune response by themselves, unless they are bound to a macromolecule (hapten). Covalent binding of the drug (or of a metabolite) to proteins becomes thus a necessary stage in the mechanism of toxicity (Figure 2). In addition, the neoantigen has to be accessible to the immune surveillance. This may occur by different ways: a) binding to the CYP that has interacted with the drug and transport to the outer cellular membrane (Robin et al., 1995); b) diffusion of the reactive metabolite out of the site of formation, and reaction with membrane proteins or proteins that will be transported to the membrane; c) drug-protein adduct-formation by membrane located CYPs (Loeper et al., 1993).

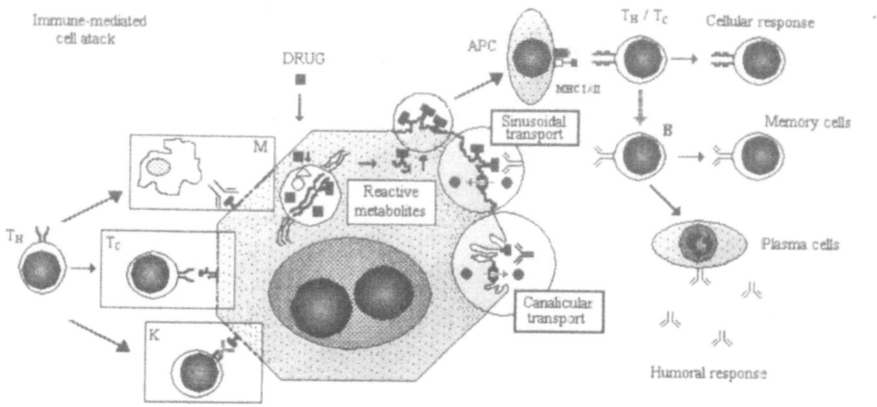

Figure 2. Key events in the course of an allergic hepatitis

Immune-mediated mechanisms of toxicity. Drug-induced allergic hepatitis becomes clinically evident either as *necrosis* (cytolytic hepatitis with increased levels of serum enzyme markers), *cholestasis* (with minor signs of necrosis) or both (mixed hepatitis).

The immune response in allergic hepatitis has an humoral and a cellular component (Homberg et al., 1985; Tsutsui et al., 1992). From the review of a significant number of clinical cases of allergic hepatitis, Homberg et al., (1985) could describe four types of response in patients: a) individuals having drug-directed antibodies in serum; b) those with antibodies directed against CYP isoforms; c) those with the presence of auto-antibodies; and d) patients with undetectable presence of antibodies against hepatic or drug antigens.

Contrary to what it could be expected, the presence of drug-directed antibodies is not the most frequent finding. Antibodies directed against either CYPs or autoantigens are more frequently reported. One possible explanation to this observation is the fact that the immune response may have arisen against a metabolite rather than to the parent molecule. Recently (Gómez-Lechón et al., 1996), antibodies against erythromycin could be demonstrated in a patient suffering from an acute drug-induced hepatic cholestasis. The antibodies were able to recognize only erythromycin that became covalently bound to human hepatocytes after prior incubation and metabolism of the antibiotic with human cells.

Little is known about the linkage between circulating antibodies and the observed liver dysfunction. Antibodies are large macromolecules and can not readily enter the cell. Instead, they could bind to haptenized hepatocyte membranes, thus altering their physicochemical properties and, indirectly, the functionality of membrane-allocated proteins (i.e enzymes, ion pumps, etc.), ion transport and ultimately bile production (Figure 2).

T-lymphocyte sensitization to drugs (or metabolites) is also considered to be a mechanism involved in allergic hepatitis (Tsutsui et al., 1992). The existence of sensitized T-cell clones can be demonstrated in (sensitized) patients by means of appropriate in vitro tests; these are based on a proliferation assay of peripheral lymphocytes incubated with the suspected drug and/or its metabolites, and a prostaglandin inhibitor. With this combined approach, a positive response for a drug could be demonstrated in ca. 80% of patients investigated (Maria and Victorino, 1994).

References

Bellomo G., Mirabelli F., Richelmi P. and Orrenius S. (1983). Critical role of sulfhydryl groups in the ATP-dependent Ca^{2+}-sequestration by the plasma membrane fraction from rat liver. FEBS letters 163, 136-139.

Boelsterli U. A. (1993). Specific targets of covalent drug-protein interactions in hepatocytes and their toxicological significance in drug-induced liver injury. Drug Metabolism Reviews 25, 395-451.

Boobis A. R., Duncan J., Fawthrop J. and Davies D. S. (1989). Mechanisms of cell death. Trends in Pharmacological Sciences 10, 275-280.

Castell J.V. and Gómez-Lechón M.J. (1992). The in vitro evaluation of the potential risk of hepatotoxicity of drugs. In: In vitro alternatives to animal pharmaco-toxicology. (J. V. Castell and M. J. Gómez-Lechón, Eds), pp. 179-204. Farmaindustria, Barcelona.

Columbano A. (1995). Cell death: Current difficulties in discriminating apoptosis from necrosis in the context of pathological processes in vivo. Journal of Cellular Biochemistry 58, 181-190.

Gómez-Lechón M. J., Carrasquer J., Berenguer J. and Castell J. V. (1996). Evidence of antibodies to erythromycin in serum of a patient following an episode of acute drug-induced hepatitis. Clinical and Experimental Allergy 26, 590-596.

Hargus, S. J., Amouzedeh H. R., Pumford N. R. et al. (1994). Metabolic activation and immunochemical localization of liver protein adducts on the nonsteroidal anti-inflamatory drug diclofenac. Chemical Research in Toxicology 7, 575-582.

Hinson J.A., and Roberts D.W. (1992). Role of covalent and noncovalent interactions in cell toxicity: effects on proteins. Annu. Reviews in Pharmacology and Toxicology 32, 471-510.

Homberg J. C. N., Abauf S., Helmy-Khalil M. et al., (1985). Drug induced hepatitis associated with anticytoplasmic organelle autoantibodies. Hepatology 5, 722-727.

Holtzman J. L. (1995). The role of covalent binding to microsomal proteins in the hepatotoxicity of acetaminophen. Drug Metabolism Reviews 27, 277-297.

Kretz-Rommel A. and Boelsterli U. A. (1994). Mechanism of covalent adduct formation of diclofenac in rat hepatic microsomal proteins. Retention of the glucuronic acid moiety in the sdduct. Drug Metabolism and Disposition 22, 956-961.

Loeper J. Descatoire V., Maurice M. et al. (1993). Cytochromes P-450 in human hepatocyte plasma membrane: recognition by several autoantibodies. Gastroenterology 104, 203-216.

Maria V. A. J. and Vitorino R. (1994). Lymphocyte proliferative response to drugs: Analysis of the value of a 24-well lymphocyte culture system. Toxicology in Vitro 5, 1041-1044.

Nieminen A. L., Saylor A. K., Tesfai S. A., Herman B. and Lemasters J. J. (1995). Contribution of the mitochondrial permeability transition to lethal injury after exposure of hepatocytes to t-butylhydroperoxide. Biochemical Journal 307, 99-106.

Poli, G., Albano, E., and Dianzani, M.U. (1987). The role of lipid peroxidation in liver damage. Chemistry and Physics of Lipids 45, 117-142.

Ponsoda X., Bort R., Jover R., Gómez-Lechón M. J. and Castell J.V. (1995). Molecular mechanisms of diclofenac hepatotoxicity: cell injury is associated to the metabolism of the drug and is preluded by a decrease in ATP levels. Toxicology in Vitro 9, 439-444.

Reed J. D. (1990). Glutathione: Toxicological implications. Ann. Rev. Pharmacological Toxicology 30, 603-631.

Robin M. A., Maratrat M., Loeper J., et al. (1995). Cytochrome P4502B follows a vesicular route to the plasma membrane in cultured rat hepatocytes. Gastroenterology 108, 1110-1123.

Ross D. (1989). Mechanistic toxicology: A radical perspective. Journal of Pharmacy and Pharmacology 41, 505-511.

Rosser B. G. and Gores G. J. (1995). Liver cell necrosis: cellular mechanisms and clinical implications. Gastroenterology 108, 252-275.

Shen W., Kamendulis L.M., Ray S.D., and Corcoran G.B. (1992). Acetaminophen-induced cytotoxicity in cultured mouse hepatocytes: effects of Ca^{2+}-endonuclease, DNA repair, and glutathione depletion inhibitors on DNA fragmentation and cell death. Toxicology and Applied Pharmacology 112, 32-40.

Sies H. (1991). Oxidative stress: Introduction. In: Oxidative Stress. Oxidants and Antioxidants. (H. Sies, Ed), Academic Press. London..

Swartz M. N. (1995). Mitochondrial toxicity: New adverse drug effects. New England Journal of Medicine. 333, 1146-1148.

Tsutsui H., Terano Y., SAkagami C. et al. (1992). Drug-specific T cells derived from patients with drug-induced allergic hepatitis. Journal of Immunology 149, 706-716.

Van-Pelt F. N. and Kenna J. G. (1994). Formation of trifluoroacetylated protein antigens in cultured rat hepatocytes exposed to halothane in vitro. Biochemical Pharmacology 48, 461-471.

Zimmerman H. J. and Ishak K. G. (1995). General aspects od drug- induced liver diseases. Gastroenterology Clinical North America 24, 739-758.

Protein Targets of Neurotoxicity

(Chair: M.K. Johnson, United Kingdom)

Neuropathy Target Esterase (NTE): Molecular Characterisation and Cellular Localisation

Paul Glynn
MRC Toxicology Unit, University of Leicester, Hodgkin Building, P.O. Box 138,Lancaster Road, Leicester, LE1 9HN, UK.

Organophosphate-induced Neuropathy and Neuropathy Target Esterase (NTE): Historical Perspective

Organophosphorus esters (OP) are covalent inhibitors of serine esterases. Those OPs which inhibit acetylcholinesterases and are thus acutely toxic have been used worldwide as insecticides. Different physicochemical properties of the tri-aryl phosphates have lead to their widespread use as plasticizers and flame-retardants. Certain OPs, for instance some saligenin cyclic phosphates which are metabolites of tri-aryl phosphates, have relatively little acute anti-cholinesterase directed toxicity, but cause a paralytic neuropathy which is manifest 1-3 weeks after exposure. Two major epidemics of human OP-induced neuropathy have occurred in the past 60 years.

The syndrome is characterised by degeneration of the distal portions of long axons in the spinal cord and peripheral nerves. An early morphological change is the formation of varicosities in the distal (but not the terminal) portions of axons; the degeneration subsequently spreads towards the axon terminal in a Wallerian-like fashion, leading to the description of a "chemical transection" of the axon (Bouldin and Cavanagh, 1979). Neuropathic OPs elicit a substantial reduction in the rate of retrograde axonal transport of exogenous radiolabelled tetanus toxin concomitant with, or slightly before, the first detectable ultrastructural changes in axons (Moretto et al., 1987).

The target protein for OP-induced neuropathy was detected by assaying brain homogenates, either for labelling with radiolabelled OP (diisopropyl-fluorophosphate; DFP), or for an esterase activity hydrolysing phenyl valerate. Neuropathy Target Esterase (NTE) was defined as the portion of DFP-labelled sites or of phenyl valerate hydrolysing activity which was sensitive to a neuropathic OP (mipafox) but resistant to a non-neuropathic OP (paraoxon)

(Johnson, 1969a,b). This esterase assay is now widely used to screen new commercial OPs for neuropathic potential.

Adult chickens have been the most frequently used experimental animal for studies on OP-induced neuropathy and NTE. The DFP-labelling and esterase assays have been applied to tissues from normal and OP-dosed chickens to characterise NTE and its role in OP-induced neuropathy. Some key results have been reviewed by Johnson (1992) and can be summarised thus:-

1. NTE is present throughout neural tissue (brain, spinal cord and peripheral nerve have relative specific activities of 25:5:1)

2 In brain, NTE is a quantitatively minor component (0.01% of total protein) and is associated with microsomal membranes.

3. Brain NTE has a subunit size of 155kDa on SDS-PAGE but migrates as a larger complex (>700kDa) on non-denaturing chromatographic systems.

4. NTE undergoes fast axonal transport.

5. Axonal NTE, rather than nerve-cell body NTE, appears to be the critical target for induction of OP-neuropathy.

6. Inhibition of NTE esterase activity alone is not sufficient to initiate neuropathy. After covalently binding to the active site serine of NTE, neuropathic OPs undergo a rapid chemical change which leaves a negatively charged moiety attached to the active site; this process has been described as "aging" of NTE.

Progress in immunolocalisation and molecular cloning of NTE

The above information on NTE was firmly established before the end of the 1980s. Further progress was impeded as, over a period of 10 years, three different laboratories tried unsuccessfully to isolate NTE in homogenous form using either DFP-labelling or phenyl valerate hydrolase activity to monitor the purification. Success was finally achieved when we synthesized a novel biotinylated OP (designated S9B) and used it as an affinity reagent to isolate NTE from brain microsomes in a simple two-step process (Glynn et al., 1994).

We have now raised rabbit antibodies reactive with chicken NTE and used these for immunohistochemical localisation studies on frozen, acetone-fixed sections. NTE is present at highest levels in neuronal cell bodies and in their proximal axon segments in brain, spinal cord and ganglia. Staining is not detected in glia, neural vasculature or muscle end plates, nor in normal sciatic nerve; however, immunoreactive material does accumulate within 3 to 12 h proximal to a ligature of the nerve, confirming the earlier biochemical demonstration that NTE undergoes fast axonal transport. The epitopes recognized by these antibodies do not survive the processes required to maintain morphology for ultrastructural studies, so immuno-electron microscopic localisation of NTE will have to await future antibodies. However, at the light microscope level, the NTE epitopes recognized by the present antibodies appear to be intracellular. Together with the earlier biochemical data, this suggests that, like liver carboxylesterases, NTE may be localized predominantly in the endoplasmic reticulum.

We have scaled-up our original isolation procedure by a factor of 100, and optimized conditions for proteolytic digestion of NTE to obtain sufficient material for N-terminal protein sequencing of resolved NTE fragments. From over 50 sequencing runs we have obtained potentially useful data for just 5 peptide sequences (between 6 and 24 residues). (Database searches with our NTE peptide sequences have revealed no obviously significant homologies). The peptide sequences are rich in amino acids encoded by maximally-redundant codons, and screens of a commercial brain cDNA library with oligonucleotides designed from these sequences have been unsuccessful. Some of the NTE proteolytic fragments contain the S9B affinity reagent (covalently bound to the active site serine), and from this information we have been able to infer the relative positions of two of the peptide sequences within intact NTE. PCR, using primers designed from these two sequences has generated a product of approximately the predicted size which encodes a potential short segment of protein sequence (about 50 amino acids) with discernable similarity to the N-terminal portion of a serine esterase domain. This PCR product is now being used to obtain longer NTE-encoding cDNA clones.

NTE and protein binding functions of serine esterases

The past few years have seen a radical shift not only in the techniques applied to study NTE, but also in our perspective on NTE – largely as a result of findings from various laboratories on other members of the serine esterase family (Table 1). Molecular cloning and partial characterization studies have revealed that certain members of this family lack some of the critical amino acid residues required for esteratic activity and contain an extra domain in addition to the esterase-like domain. Neuroligin, neurotactin and gliotactin are plasma membrane proteins with cytoplasmic non-esterase domains and extracellular catalytically-inactive esterase-like domains, which appear to mediate heterophilic protein binding interactions and thus facilitate cell-cell adhesion in the nervous system (Ichtchenko et al., 1995; Auld et al., 1995). The large subunit size of NTE (155kDa) implies possession of an extra domain in addition to the esterase domain but, unlike neuroligin, neurotactin and gliotactin, its esterase domain is both catalytically active and (from our immunohistochemical data) probably intracellular. However, two catalytically active serine esterases of liver endoplasmic reticulum can also mediate intracellular heterophilic protein binding via their esterase domains and, in one case, this binding has been shown to be OP sensitive (Medda et al., 1987; Macintyre et al., 1994).

Table 1. Some proteins with serine esterase-like domains

Name	Subunit (kDa)	Location	Catalytic	Function
AcChE	75	Synapse (PMand ECS)	+	Terminate transmission
Carboxyl-esterase	60	Liver (ER)	+	1. Metab/Detox 2. Regulate CRP secretion
Neuro-tactin	135	Cell-cell contacts in developing CNS (PM)	-	Cell adhesion?
Gliotactin	110	Peripheral nerve glia (PM)	-	Blood-Nerve barrier formation
Neuroligin	116	Synapse (PM)	-	Synaptogenesis
NTE	155	Neuronal (ER?)	+	?

(PM, plasma membrane; ECS, extracellular space; ER, endoplasmic reticulum; CRP, C-reactive protein).

Thus, the physiological roles of certain members of the serine esterase family may be attributable to their ability to bind other proteins, rather than to their esteratic activity. This is a particularly interesting concept for NTE, since it is clear (see above) that inhibition of its esterase activity alone is not a sufficient signal to initiate OP-induced neuropathy. In addition, the large apparent molecular size (>700kDa) of NTE under non-denaturing conditions (see above) could be due to its association, in vivo, with other cellular proteins.

From the foregoing, it seems reasonable to speculate that NTE's esterase domain can bind to other neuronal proteins and this binding is modified when NTE is "aged" by neuropathic OPs. Once the cloning of NTE cDNA is completed, recombinant NTE (which, unlike endogenous brain NTE, could be isolated under non-denaturing conditions) could be used as an affinity reagent to bind and identify putative cognate proteins in brain extracts. If these putative cognate proteins have a known function, then this would provide a valuable indicator of the process which is disturbed by neuropathic OPs.

References.

Auld VJ, Fetter RD, Broadie K, Goodman, CS (1995) Gliotactin, a novel transmembrane protein on peripheral glia, is required to form the blood nerve barrier in Drosophila. Cell 81:757–767

Bouldin TW, Cavanagh JB (1979) Organophosphorus Neuropathy. I. A teased-fiber study of the spatio-temporal spread of axonal degeneration. Amer J Pathol 94:241–252

Glynn P, Read DJ, Guo R, Wylie S, Johnson MK (1994) Synthesis and characterisation of a biotinylated organophosphorus ester for detection and affinity purification of a brain serine esterase: neuropathy target esterase. Biochem J 301:551–556

Ichtchenko K, Hata Y, Nguyen T, Ullrich B, Missler M, Moomaw C, Sudhof TC (1995) Neuroligin 1: a splice site-specific ligand for beta-neurexins. Cell 81:435–443

Johnson MK (1969a) A phosphorylation site in brain and the dedelayed neurotoxic effect of some organophosphorus compounds. Biochem J 111:487–495

Johnson MK (1969b) The delayed neurotoxic effect of some organophosphorus compounds: identification of the phosphorylation site as an esterase. Biochem J 114:711–717

Johnson MK (1992) Molecular events in delayed neuropathy: experimental aspects of neuropathy target esterase. In: Clinical and Experimental Toxicology of Anticholinesterases (Eds. D.B. Ballantyne and T.C. Marrs) pp90–113

Macintyre S, Samols D, Dailey P (1994) Two carboxylesterases bind C-reactive protein within the endoplasmic reticulum and regulate its secretion during the acute phase response. J Biol Chem 269:24496–24503

Medda S, Stevens AM, Swank, RT (1987) Involvement of the esterase active site of egasyn in compartmentalisation of beta-glucuronidase within the endoplasmic reticulum. Cell 50:301–310

Moretto A, Lotti M, Sabri MI, Spencer PS (1987) Progressive deficit of retrograde axonal transport is associated with pathogenesis of di-n-butyl dichlorvos axonopathy. J Neurochem 49:1515–1522

The Concept and Target of Promotion of Axonopathies

Marcello Lotti
Università degli Studi di Padova, Istituto di Medicina del Lavoro, Via Facciolati 71, I-35127 Padova, Italy

Introduction

A new phenomenon was recently observed whereby the treatment of experimental animals with certain esterase inhibitors elicits or intensifies the clinical expression of various neuropathic insults. This phenomenon has been described by two independent groups (Pope and Padilla, 1990; Lotti et al., 1991) while studying organophosphate-induced delayed polyneuropathy (OPIDP) and it was tentatively called promotion of OPIDP. Later on it was found to occur in other toxic as well as in traumatic axonopathies (Moretto et al., 1992a; Moretto et al., 1993).

A model for studying promotion: Organophosphate Induced Delayed Polyneuropathy (OPIDP)

OPIDP is a rare toxicity caused by certain organophosphorus esters (OP), characterized by the degeneration of some long axons in the central and peripheral nervous system (Johnson, 1982; Lotti 1992). The molecular target for OPIDP is considered to be an enzyme in the nervous tissues called neuropathy target esterase (NTE). In adult animals, OPIDP is initiated when, soon after dosing, at least 70% of NTE is inhibited by neurotoxic OPs.

It has long been known that other NTE inhibitors, such as certain carbamates, phosphinates, and sulfonyl halides (e.g., phenylmethanesulfonyl fluoride, PMSF), protect the animal from OPIDP when given before the neurotoxic OP (Johnson and Lauwerys, 1969; Johnson, 1974). This effect is related to the fact that NTE, inhibited by these compounds, does not undergo the aging reaction, i.e., the loss of a side chain leaving a charged unsubstituted residue attached to the protein. This intramolecular rearrangement of phosphorylated NTE was thought to be essential for OPIDP development (Johnson, 1982). However, it has recently been shown that this is not always the case (Johnson et al., 1991; Milatovic and Johnson, 1993) and that such protective compounds also cause mild neuropathy when they are given at extremely high doses which almost completely affect NTE (Lotti et al., 1993). It was concluded that these compounds might act as partial agonists on NTE, protecting from OPIDP with partial inhibition and causing OPIDP with near complete inhibition of NTE (Lotti et al., 1993).

The initial observation of promotion was made while studying chlorpyrifos (O,O-diethyl-O-3,5,6-trichloro-2-pyridyl-phosphorothioate) induced polyneuropathy (Lotti et al., 1991). Hens were treated with a single oral dose of chlorpyrifos and NTE inhibition was found to increase for several days after dosing, and to reach a peak (70-80%) 6 days thereafter. OPIDP developed 2–3 weeks later. The delayed biochemical response, due to the slow disposal of the compound, suggested a "therapeutic" use of PMSF to prevent the slow NTE phosphorylation by chlorpyrifos with the rapid sulfonylation by PMSF. Surprisingly, PMSF, given post-dosing, exacerbated the polyneuropathy induced by chlorpyrifos. This effect was also observed when a low dose of another known neurotoxic compound, di*iso*propylfluoro-phosphate (DFP), was followed by PMSF. The rapid clearance of DFP and PMSF, the experimental design (PMSF was given 24 hours after DFP) and time-course experiments ruled out pharmacokinetic interactions. Pope and Padilla (1990) made similar observations in hens treated with either mipafox or DFP followed by PMSF. Other non-neuropathy NTE inhibitors (such as phenyl-N-methyl-N-benzyl-carbamate and phenyl-*n*-pentyl-phosphinate) have also been found to promote OPIDP (Lotti et al., 1991; Johnson and Read, 1993). However, other inhibitors of esterases, but not of NTE (such as paraoxon and other sulfonyl fluorides) did not promote OPIDP when given at maximum tolerated doses (Lotti et al., 1991; Osman et al., 1996).

Dose-response relationship studies showed that the level of NTE inhibition by neuropathic OPs to initiate OPIDP was much lower than the typical 70–80% inhibition when PMSF was given afterwards (i.e. about 30/40%).

Retrograde axonal transport was selectively impaired during the development of OPIDP and its reduction preceded the onset of morphological and clinical signs (Moretto et al., 1987). Chicks were used to study the effect of promotion on retrograde axonal transport. These animals have been used because the mild OPIDP which is produced is reversible, as are the effects on retrograde axonal transport. It was shown that PMSF does not further impair retrograde axonal transport, when used to promote a dose of dibutyl-2,2-dichlorovinyl phosphate (DBDCVP) which caused moderate OPIDP in chicks. In fact, promoted and non-promoted chicks, while showing different severity of ataxia, had similar decrement and recovery of retrograde axonal transport (Moretto et al., 1992a). The conclusion was that promotion did not affect retrograde axonal transport and therefore it was likely to involve a mechanism separate from that of the progression of the initial lesion.

Neuropathology studies after OPIDP promotion have been performed in hens (Lotti et al., 1991) and, to a greater extent, in 35 to 70 day old chicks (Pope et al.,

1992; Pope et al., 1993). Promotion of DFP neuropathy by PMSF exacerbates existing damage since the types of lesion observed were not different from those observed in typical OPIDP. Moreover, brain regions (brainstem and cerebellum) were not affected in chicks treated with DFP followed by PMSF, indicating that degeneration does not affect areas which are not typically involved in OPIDP (Pope et al., 1992; Pope et al., 1993).

The search for the target of promotion

All compounds found to be promoters were also NTE inhibitors, and the question arose whether NTE was the target of promotion. However, since promotion was found to occur not only in OPIDP but also in other axonopathies, the hypothesis was made for a target for promotion other than NTE.

Chicks developed OPIDP when doses of neuropathic OPs were increased to obtain NTE inhibition higher than 90% (Pope et al., 1992; Peraica et al., 1993). Therefore in contrast to hens, larger doses of neuropathic OPs can be given to chicks without clinical OPIDP despite NTE inhibition higher than the threshold observed in hens. PMSF promoted OPIDP in 40-day old chicks when NTE was almost completely inhibited by DBDCVP (Peraica et al., 1993). Similar results have been obtained in rats, also known to be relatively resistant to OPIDP (Moretto et al., 1992b). The conclusion from these studies is that NTE is unlikely to be the target of promotion since this effect occurs when almost all NTE is inhibited. Pope et al. (1993) treated 35-day old chicks with DFP (2 mg/kg s.c.) causing almost complete inhibition of NTE. Four hours later, chicks were treated with either PMSF (90 mg/kg s.c.) or the same dose of DFP. Chicks treated with two doses of DFP showed very mild clinical signs, while those treated with DFP followed by PMSF were clearly ataxic. Therefore, all these data indirectly indicated that the target of promotion is not NTE. Finally, another compound was found, the phosphorothioic acid-O-(2-chloro-2,3,3-trifluorocyclobutyl)-O-ethyl S-propyl ester (Bayer KBR-2822), which caused promotion at doses which do not inhibit NTE (Moretto et al., 1994).

KBR-2822 also promoted OPIDP when given up to 6 days before the neuropathic OP (Moretto et al., 1994). Pretreatment with other promoters did not cause promotion of OPIDP because the doses used were always inhibitory to NTE. Therefore, they protected from OPIDP by occupying NTE and promotion did not occur because neuropathy could not be initiated (Lotti et al., 1991). Nevertheless, the promoter KBR-2822 exacerbates OPIDP when given at a dose just below that which would inhibit NTE. Moreover, both NTE (Lotti et al., 1987; Carrera et al., 1992) and the unknown promotion site (Peraica et al., 1995) must be affected in the peripheral nerve axons and not in the cell body to cause OPIDP and promotion, respectively. For these reasons it was suggested that the promotion site should be similar to and /or linked with NTE (Aldridge 1993).

Since NTE is a neural phenyl valerate (PV) esterase, it was first ascertained whether the target of promotion was also a PV esterase. Brain and sciatic nerve PV esterases of hens were incubated with neuropathic inhibitors, di*iso*propylphosphoro fluoridate (DFP, 5 μM) or N,N-di*iso*propylphosphoro-diamido-fluoridate (Mipafox, 50 μM) to inhibit NTE. Remaining activities, quantitatively similar after either inhibition, were incubated with PMSF (1 to 500 μM). Analysis of inhibition by PMSF suggested that one enzyme was sensitive to this compound. Mipafox-resistant PMSF sensitive activity was tested both *in vitro* and *in vivo* with several inhibitors. No correlation was found between inhibition of Mipafox (50 μM)-resistant PMSF, (500 μM)-sensitive activity and the capability of several inhibitors to promote OPIDP. It was concluded that the target of promotion was unlikely to be a PV esterase other than NTE (Milatovic et al., 1995).

NTE has only recently been characterized (Glynn et al., 1994) and it has so far been defined as a PV esterase resistant to paraoxon and sensitive to Mipafox (50 μM) (Johnson 1990). This definition of NTE was re-evaluated by titrating paraoxon-resistant PV esterases with mipafox. Another PV esterase sensitive to mipafox (I_{50} approx. 200 μM) was identified. This esterase was found to be affected by inhibitors when given to hens at doses causing promotion, but not by initiators of OPIDP (Moretto et al., 1996). This PV esterase resembles the soluble NTE (similar I_{50} for mipafox), which was identified in peripheral nerves (Vilanova et al., 1990), and it is possible that they are the same.

What is the mechanism of promotion?

The mechanism of promotion of axonopathies is unknown. However, the hypothesis was made that promotion might be the result of interference with compensation-repair mechanisms, based upon the following:

1. Promotion is unspecific because several axonopathies of different origin can be affected. Moreover, it occurs also when promoter is given before the neuropathic insult, suggesting that it affects mechanisms not initiated by the lesion itself.
2. Promotion both increases the severity of lesions, such as in OPIDP, and prolongs the recovery from the lesion such as in nerve crush. Moreover, inhibition of NTE less than 70% is apparently well compensated in experimental animals. However this threshold is lowered to 30-40% when promotion is given and OPIDP is elicited.
3. Promotion is less effective in conditions where compensation-repair mechanisms are thought to be more efficient, such as in chicks.

Based on the experiments in which the promoter was given before the neuropathic insult (either with KBR-2822 before neuropathic OP or with PMSF before nerve crush), the hypothesis was made that promotion might possibly affect functions involved in the physiological turnover of damage and compensation-repair of the axons. Thus it was ascertained whether repeated treatment with KBR-2822 causes neuropathy as result of unrepaired physiological damage. KBR-2822 was given by gavage fro 90 days at doses causing no or marginal NTE inhibition (0.2 or 0.4 mg/kg/day). No neuropathy was observed. These results however are not conclusive because they may be due to short duration of treatment, to low dose reaching the target or because physiological damage is minimal and not promotable (Jokanovic et al., 1996).

References

Aldridge WN (1993) Postscript to the symposium on organophosphorus compound induced delayed neuroppathy. Chem-Biol Interact 87:463–466

Carrera V, Barril J, Mauricia MC, Pellin MC, Vilanova E (1992) Local application of neuropathic organophosphorus compounds to hen sciatic nerve:inhibition of Neuropathy Target Esterase and peripheral neurological impairments.Toxicol Appl Pharmacol 117:218–225

Glynn PI, Read DJ, Guo R, Wylie S., Johnson MK (1994) Synthesis and characterisation of a biotinylated organophosphorus ester for detection and affinity purification of a brain serine esterase: neuropathy target esterase. Biochem J 301:551–556.

Johnson MK, Lauwerys R (1969) Protection by some carbamates against the delayed neurotoxic effects of di*iso*propyl phosphorofluoridate. Nature 222:1066–1067

Johnson MK (1974) The primary biochemical lesion leading to the delayed neurotoxic effects of some organophosphorus esters. J Neurochem 23:785–789

Johnson MK (1982) The target for initiation of delayed neurotoxicity by organophosphorus esters: Biochemical studies and toxicological applications. In: Hodgson E, Bend JR, Philipot RM (eds) Reviews in Biochemical Toxicology, Vol. 4, Elsevier, New York, pp 141–212

Johnson MK (1990) Organophosphates and delayed neuropathy - is NTE alive and well? Toxicol Appl Pharmacol 102:385–399

Johnson MK, Vilanova E, Read DJ (1991) Anomalous biochemical responses in test of delayed neuropathic potential of metamidophos (O,S-dimethyl phosphorothioamidate), its resolved isomers and of some higher O-alkyl homologues. Arch Toxicol 65:618–624

Johnson MK, Read DJ (1993) Prophylaxis against and promotion of DFP-induced delayed neuropathy by phenyl di-*n*-pentylphosphinate. Chem-Biol Interact 87:449–456

Jokanovic M, Moretto A, Lotti M (1996) Repeated dosing of hens with phosphorothioic acid O-(2-chloro-2,3,3-trifluorocyclobutyl)-O-ethyl-S-propyl ester (KBR 2822) causing marginal NTE inhibition does not induce clinical neuropathy. Toxicologist 30:302

Lotti M, Caroldi S, Moretto A, Johnson MK, Fish CJ, Gopinath C, Roberts NL (1987) Central-peripheral delayed neuropathy caused by diisopropyl phosphorofluoridate (DFP): segregation of peripheral nerve and spinal cord effects using biochemical, clinical and morphological criteria. Toxicol Appl Pharmacol 88:87–96

Lotti M, Caroldi S, Capodicasa E, Moretto A (1991) Promotion of organophosphate-induced delayed polyneuropathy by phenylmethanesulfonyl fluoride. Toxicol Appl Pharmacol 108:234–241

Lotti M (1992) The pathogenesis of organophosphate delayed polyneuropathy. Crit Rev Toxicol 21:465–487

Lotti M, Moretto A, Capodicasa E, Bertolazzi M, Peraica M, Scapellato ML (1993) Interactions between neuropathy target esterase and its inhibitors and the development of polyneuropathy. Toxicol Appl Pharmacol 122:165–171

Milatovic D, Johnson MK (1993) Reactivation of phosphorodiamidated acetylcholinesterase (AChE) and neuropathy target esterase (NTE) by treatment of inhibited enzyme with potassium fluoride (KF). Chem-Biol Interact 87:425–430

Milatovic D, Osman KM, Moretto A, Lotti M. (1995) Promotion of organophosphate induced delayed polyneuropathy (OPIDP): the target is not a phenyl valerate (PV) esterase. In: Proceedings of the International Congress of Toxicology - VII, Seattle, Washington (USA), p 22, P–34

Moretto A, Lotti M, Sabri MI, Spencer PS (1987) Progressive deficit of retrograde axonal transport is associated with the pathogenesis of di-n-butyl-dichlorvos axonopathy. J Neurochem, 49:1515–1522

Moretto A, Bertolazzi M, Capodicasa E, Peraica M, Richardson RJ, Scapellato ML, Lotti M (1992a) Phenylmethanesulfonyl fluoride elicits and intensifies the clinical expression of neuropathic insults. Arch Toxicol 66:67–72

Moretto A, Capodicasa E, Lotti M (1992b) Clinical expression of organophosphate-induced delayed polyneuropathy in rats. Toxicol Lett 63:97–102

Moretto A, Capodicasa E, Peraica M, Lotti M (1993) Phenylmethanesulfonyl fluoride delays recovery from crush of peripheral nerves in hens. Chem-Biol Interact 87:457–462

Moretto A, Bertolazzi M, Lotti M (1994) The phosphorothioic acid -O-(2-Chloro-2,3,3-Trifluorocyclobutyl)-O-ethyl S-propyl ester promotes organophosphate polyneuropathy without inhibition of neuropathy target esterase. Toxicol Appl Pharmacol 129:133–137

Moretto A, Jokanovic M, Lotti M (1996) Inhibition of phenyl valerate esterases (PVE) sensitive to mipafox and promotion of organophosphate induced polyneuropathy (OPIDP). Toxicologist 30:300

Osman KA, Moretto A, Lotti M (1996) Sulfonylfluorides and the promotion of diisopropylfluorophosphate neuropathy. Fund Appl Toxicol, in press

Peraica M, Capodicasa E, Moretto A, Lotti M (1993) Organophosphate polyneuropathy in chicks. Biochem Pharmacol 45:131–135

Peraica M, Moretto A, Lotti M (1995) Selective promotion by phenylmethanesulfonyl fluoride of peripheral and spinal cord neuropathies initiated by diisopropyl- phosphorofluoridate in the hen. Toxicol Lett 80:115–121

Pope CN, Padilla S (1990) Potentiation of organophosphorus-induced delayed neurotoxicity by phenylmethylsulfonyl fluoride. J Toxicol Environ Health 31:261–273

Pope C, Chapman ML, Tanaka D Jr, Padilla S (1992) Phenylmethylsulfonyl fluoride alters sensitivity to organophosphorus-induced delayed neurotoxicity in developing animals. Neurotoxicology 13:355–364

Pope C, Tanaka D, Padilla S (1993) The role of neurotoxic esterase (NTE) in the prevention and potentiation of organophosphorus induced delayed neurotoxicity (OPIDN). Chem-Biol Interact 87:395–406

Vilanova E, Barril J, Carrera V, Pellin MC (1990) Soluble and particulate forms of the organophosphorus neuropathy target esterase in hen sciatic nerve. J Neurochem 55:1258–1265

Mechanisms and Models of Neurotoxicity of n-Hexane and Related Solvents

Hanna Tähti[1], Maria Engelke[2] and Leila Vaalavirta[1]
[1] University of Tampere, Medical School, P.O. Box 607, SF-33101 Tampere
[2] University of Bremen, Department of Biophysics, P.O. Box 330440, D-28334 Bremen, Germany

Introduction

Organic solvents are widely used both in industry and in households. Misuse of organic solvents is common, and intoxications and sniffing episodes are not rare (Ikeda, 1992). Solvents belong to various chemical categories, including aliphatic, halogenated aliphatic, alicyclic and aromatic hydrocarbons, alcohols, ethers, ketones and other minor chemical groups. Adverse effects may be found at various sites in the central nervous system (CNS) and the peripheral nervous system (PNS). Prolonged exposure to aromatic hydrocarbons can cause permanent damage to the CNS (Hein et al., 1990), while long-term exposure to n-hexane is associated with development of peripheral neuropathy (Bachmann et al., 1993). It has been proposed that repeated acute exposures to low levels of organic solvents can cause permanent damage in the CNS. It is therefore important to clarify the mechanisms of solvent neurotoxicities, and to find appropriate sensitive methods for detecting even subtle signs of deleterious effects.

Mechanisms of PNS and CNS toxicity

The mechanism of PNS toxicity caused by n-hexane is based on the toxic biotransformation product 2,5-hexanedione, which causes peripheral neuropathy (Ruff et al., 1981; Abou-Donia et al., 1991). The neurotoxicity of methyl-n-butyl ketone (MBK) is based on the same toxic metobolite 2,5-hexanedione. MBK is roughly 12 times more potent than n-hexane (Bos et al., 1991). In MBK neuropathy, swellings have also been found in the cerebellar white matter in the CNS and in distal optic nerve, causing spastic gait, and impairments of vision and memory (Bos et al., 1991). The long ascending and descending tracts of the spinal cord are also particularly vulnerable. Carbon disulphide causes neurofilament-filled swellings of the distal axons similar to the metabolite of n-hexane, due to protein cross-linking which results in the accumulation of neurofilaments (Graham et al., 1995). The PNS seems to be the most sensitive target for n-hexane, while toluene mainly causes CNS toxicity. Long-term n-hexane exposure in rats caused significant changes in the PNS but failed to reveal any significant changes in the CNS in specific marker proteins (neuron-specific enolase, creatine kinase-B and beta-S100). In the CNS these marker proteins failed to reveal any significant changes after long-term n-hexane exposure, while chronic exposure of rats to toluene caused significant changes in these parameters in the CNS (Huang et al., 1992, 1993).

The mechanisms of the CNS toxicity of aromatic hydrocarbons is not known in detail. Generally it has not been possible to pinpoint specific sites of toxic effects in the CNS. Chronic heavy exposures to e.g. toluene cause multifocal disorders, cerebral, cerebellar and brain stem atrophy (Ikeda and Tsukagoshi, 1990). At low occupational exposure levels, organic solvents can cause many different symptoms of CNS origin: fatigue, anxiety, irritability, headache, weakness, depression, neurobehavioral changes and impairment of cognitive performance (Milanovic et al., 1990; Evans and Balster, 1991; Hakkola, 1994; Kishi et al., 1995). Also temporal epileptic seizures have been found due to occupational exposure to solvents (Jacobsen et al., 1994). The molecular mechanisms underlying these diverse actions are unknown. Repeated acute exposures to low levels of organic solvents can cause permanent changes in the CNS when they last for several years or for the whole working age. Occupational exposures to organic solvents for a period of 5 years or more have been shown to increase the risk of persistent memory difficulties and to decrease the concentration ability (Hein et al., 1990).

Aliphatic, alicyclic and aromatic hydrocarbons have in common the ability to interact with biological membranes (Engelke et al., 1996). This interaction has been thought to cause changes in neural membranes, which then cause the CNS-depressant effect (Seeman, 1972).

The membrane theory of the CNS-depressant effect caused by anaesthetics and organic solvents

According to the "lipid theory", the anaesthetic potency of inhalation anaesthetics and the CNS-depressant effects of organic solvents are related to their lipophilicity and to their effects on the lipid bilayer (Seeman 1972). Franks and Lieb (1978, 1981) showed that membrane integral proteins are the primary target sites of the anaesthetic action of lipid-soluble molecules.

The acute CNS-depressing effect of organic solvents depends on their anaesthetic potency. The mechanism of this anaesthetic effect is based on changes that solvents cause in neural cell membranes (Franks and Lieb, 1981, Korpela and Tähti, 1988). The general anaesthetic effect is produced by a wide range of compounds without common chemical structure or activity, such as noble gases, aliphatic, alicyclic, aromatic and halogenated hydrocarbons, alcohols, aldehydes, ketones, esters and ethers.

"The protein theory of anaesthetic effect" was confirmed by Franks and Lieb (1978), when they studied the firefly enzyme luciferase. The relative activity of this membrane-bound enzyme decreased according to the anaesthetic potency of various compounds. In their studies Franks and Lieb (1987, 1990) showed that in clinical concentrations of anaesthetics, the changes in the lipid bilayer structure and fluidity are too small to be significant factors causing an anaesthetic effect. A strong argument against lipid bilayer theories is the fact that changes in fluidity and other physical properties caused by anaesthetics at ED_{50} are equivalent to a temperature change of less than 1°C (Franks and Lieb, 1981). Since such a small change in temperature does not induce general anaesthesia, this data would seem to render the lipid theory of anaesthetic effects rather doubtful. The alternative view favours direct binding of anaesthetics to proteins at hydrophobic sites bound either by lipid or water. Franks and Lieb (1978) found that the anaesthetic binding site appeared to be a single amphiphilic pocket in the integral membrane protein containing both apolar and polar regions. They also found that primary target sites in general anaesthesia have some properties in common: they are both polar and apolar, and their polar regions are excellent H-bond acceptors but poor H-bond donors. They can interact with a wide variety of molecular sizes, shapes, and chemical groups, but they are unaffected by molecules that exceed a critical size (Franks and Lieb, 1978, 1990).

Our hypothesis has been that the CNS-depressing effect of organic solvents might depend on the same mechanisms as the effect of general anaesthetics. Many organic solvents have also been used as anaesthetics, and at high concentrations they can induce anaesthesia. Organic solvents readily pass the blood-brain barrier and distribute into the lipid membranes of the neural cells.

In our studies we have investigated the effects of organic solvents both on the lipid part of the neural membranes, by measuring the fluidity changes, and on the integral proteins acetylcholinesterase (AChE) and ATPases (Engelke et al., 1992, Naskali et al., 1993, Vaalavirta and Tähti, 1995, Engelke et al., 1996).

The effects of solvents on the fluidity of neural membranes

The effects of n-hexane and aromatic hydrocarbons on the lipid part in the neural membranes has been evaluated by measuring the relative membrane fluidity changes after solvent treatments of neural membrane isolations. In the fluidity studies, the lateral pyrene diffusion was determined based on the

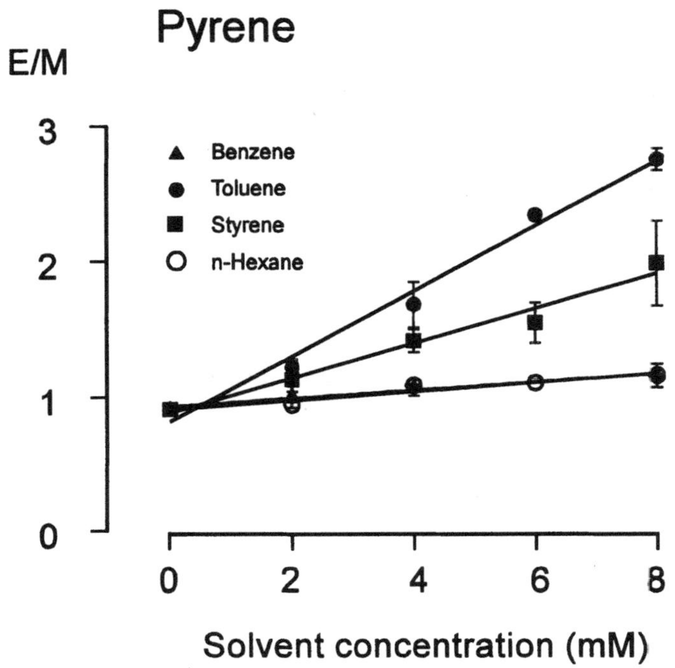

Fig. 1. Effects of aromatic hydrocarbons (benzene, toluene, styrene) and an aliphatic hydrocarbon (n-hexane) on the fluidity of rat synaptosomal membranes studied with the ratio of excimer/monomer (E/M) after pyrene incorporation and solvent exposure. The molecular weights (MW) and oil/water partition coefficients (p.c.) according to Leo et al. (1971) are: benzene MW 78, p.c. 177; toluene MW 92, p.c. 659; styrene MW 104, p.c. 1168; n-hexane MW 86, p.c. 8710.

formation of excimers (E) from pyrene monomers (M) (Galla and Hartman, 1980, Engelke et al., 1992). The E/M ratio was determined by fluorescence spectrometry. According to our results, solvents increase membrane fluidity. Toluene addition *in vitro* to the synaptosomal membranes led to a dose-correlated increase in the membrane fluidity (Fig. 1), but n-hexane, despite its high lipid solubility, increased the fluidity only slightly (Engelke et al., 1992; Naskali et al., 1993). This could be explained by the finding of King et al. (1985) and McIntosh et al. (1989) that n-hexane concentrates at the bilayer midplane in the cell membrane, without increasing the free volume of the membrane.

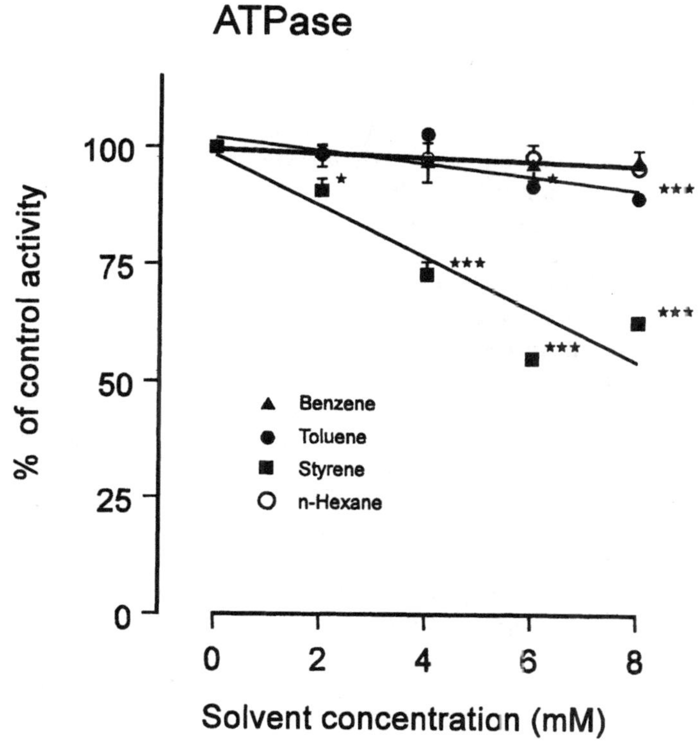

Fig 2. The effects of aromatic hydrocarbons (benzene, toluene, styrene), and an aliphatic hydrocarbon (n-hexane) on rat synaptosomal membrane integral enzyme ATPase activity.

The effects of solvents on neural membrane integral proteins

The effects of organic solvents on neural membrane structural proteins have been evaluated in recent studies by measuring the changes in synaptosomal membrane-bound enzymes (Korpela and Tähti, 1988; Edelfors and Ravn-Jonsen, 1992; Tähti et al., 1992; Naskali et al., 1993).

Aromatic hydrocarbons decreased the integral enzyme activities dose-dependently. The aliphatic hydrocarbon n-hexane did not cause any marked enzyme inhibition (Fig. 2). Solvents having an aromatic structure are supposed to be localised in hydrophobic pockets in the protein, or simply to be concentrated at the protein-lipid interface, whereas n-hexane molecules are localised at the bilayer midplane without causing any change in membrane fluidity or enzyme activity (Engelke et al., 1992).

The accumulation of organic solvents in neural cell membranes produces bilayer disorder (Engelke et al., 1992, 1996). The disorder in the phospholipid bilayer can indirectly affect the function of membrane integral enzymes (Edelfors and Ravn-Jonsen, 1992; Engelke et al., 1992).

Aromatic solvents decreased the ATPase activities in rat synaptosome membranes both *in vitro* and *in vivo* (Korpela and Tähti, 1988). They also showed that the inhibition is not only dependent on the lipid solubility of the solvents, but also on the molecular structure of the solvent. A direct relationship between the partition coefficient of various solvents and their effects on the structural integrity of the membrane has been found by Sikkema et al. (1994) and by Engelke et al. (1996). In isolated synaptosomal membranes (Naskali et al.1993) and in cells isolated from cerebral neuron cultures (Tähti and Naskali, 1992), aromatic hydrocarbons (benzene, toluene, styrene and xylene) caused ATPase inhibition. When the effects of the same solvents were tested on membranes isolated from astrocytic cultures (Naskali et al., 1994), the ATPase inhibition was even greater than in synaptosomes. Despite its high lipid solubility, n-hexane only had a slightly inhibiting effect on ATPases tested with different neural membrane models (Vaalavirta and Tähti, 1995). This may reduce the possibility of n-hexane interacting with membrane-associated proteins, e.g. ATPases.

The physical/chemical activity relationship has also been found important by Tanii et al. (1994) when studying the effects of organic solvents on synaptosomal membrane Na^+/K^+ -ATPase activity and membrane fluidity. According to them, partitioning into the lipid bilayer changes the membrane fluidity and leads to a direct inhibition of membrane-bound enzymes.

In vitro models in studies of solvent neurotoxic mechanisms

Due to the functional and structural complexity of the CNS, the detection of toxic effects on it can be more problematic than of those on many other target organs. All neurotoxic effects start in neural cells, and thus *in vitro* studies serve as good methods for studying the mechanisms of different neurotoxicities. They can also be recommended for screening studies of the neurotoxic potentials of industrial chemicals (Stoltenburg-Didinger, 1992; Tähti et al., 1995).

In vitro methods are valuable aids in studying different neural cell types and cells from different brain areas. Astrocytes have been shown to be vulnerable to solvent toxic effects (Ekblad-Sekund and Walum, 1991; Naskali et al., 1994). To the neuronal cultures, blood-brain functions and biotransformation reactions can be added via development of coculture methods (Walum et al., 1990; Tähti et al., 1995). For the moment, *in vitro* methods can be used in association with *in vivo* studies in improving overall risk evaluation, and in solving the neurotoxic mechanisms (Veronesi, 1992; Tähti et al., 1995).

Conclusions

Organic solvents can cause both PNS and CNS toxicity. The mechanisms of the PNS toxicity of n-hexane, MBK and carbon disulphide are well understood. However, the common CNS-depressing effect of most organic solvents is not clear. Occupational exposures to low levels of organic solvents for several years have caused various CNS effects, such as impairment of short-term memory and changes in personality. According to recent studies, these effects seem to be persistent.

We have investigated the mechanisms of this CNS-depressing effect with different *in vitro* membrane models. According to our studies, neural cell membrane integral proteins are targets for the actions of organic solvents. The integral proteins contain hydrophobic areas, where the lipid-soluble molecules can interact. Organic solvents increase fluidity, but the primary effect may be on the proteins.

The *in vitro* membrane models and neural cell cultures are good Lthods in studying the specific mechanisms of neurotoxic chemicals.

Acknowledgement

This study has been financially supported by the Academy of Finland.

References

Abou-Donia MB, Hu ZH, Lapadula DM and Gupta RP (1991) Mechanisms of joint neurotoxicity of n-hexane, methyl isobutylketone and O-ethyl-O-4-nitrophenyl-phenylphosphonothioate in hens. J Pharmacol Exp Ther 257:282-289

Bachmann MO, De Beer Z, and Myers JE (1993) n-Hexane neurotoxicity in metal can manufacturing workers. Occup Med Oxf 43:149-154

Bos PMJ, de Mik G and Bragt PC (1991) Critical review of the toxicity of methyl- n-butyl ketone: Risk from occupational exposure. Am J Ind Med 20:175-194

Edelfors S, Ravn-Jonsen A (1992) Effect of organic solvents on nervous cell membrane as measured by changes in the (Ca^{2+}/Mg^{2+}) ATPase activity and fluidity of synaptosomal membrane. Pharmacol Toxicol 70:181-187

Ekblad-Sekund G and Walum E (1991) Astrocytic regulatory functions: a possible target for CNS effects of organic solvents. Toxicol in Vitro 576:503-506

Engelke M, Diehl H and Tähti H (1992) Effects of toluene and n-hexane on rat synaptosomal membrane fluidity and integral enzyme activities. Pharmacol Toxicol 71:343-347

Engelke M, Tähti H and Vaalavirta L (1996) Perturbation of artificial and biological membranes by organic compounds of aliphatic, alicyclic and aromatic structure. Toxicol in Vitro 10:111-115

Evans EB and Balster RL (1991) CNS depressant effects of volatile organic solvents. Neurosci Biobehav Rev 15(2):233-241

Franks NP and Lieb WR (1978) Where do general anaesthetics act? Nature (London) 274:339-342

Franks NP and Lieb WR (1981) Is membrane expansion relevant to anaesthesia? Nature (London) 292:246-251

Franks NP and Lieb WR (1987) Anaesthetics on the mind. Nature (London) 328:113-114

Franks NP and Lieb WR (1990) Mechanism of general anaesthesia. Environ Health Perspect 87:199-205

Galla HJ and Hartman W (1980) Excimer-forming lipids in membrane research. Chem Phys Lipids 27:199-219

Graham DG, Amarnath V, Valentine WM, Pyle SJ and Douglas, CA (1995) Pathogenetic studies of hexane and carbon disulfide neurotoxicity. Crit Rev Toxicol 25:91-112

Hakkola M (1994) Neuropsychological symptoms among tanker drivers with exposure to solvents. Occup Med Oxf 44(5):243-246

Hein HO, Suadicani P and Gyntelberg F (1990) Mixed solvent exposure and cerebral symptoms among active and retired workers. An epidemiological investigation of 3387 men aged 53-75 years. Acta Neurol Scand 81(2):97-102

Huang J, Shibata E, Kato K, Asaeda N and Takeuchi Y (1992) Chronic exposure to n-hexane changes in nerve-specific marker proteins in the distal peripheral nerve of rat. Hum Exp Toxicol 11:323-327

Huang J, Kato K, Shibata E, Asaeda N and Takeuchi Y (1993) Nerve specific marker proteins as indicators of organic solvent neurotoxicity. Environ Res 63:82-87

Ikeda M and Tsukagoshi H (1990) Encephalopathy due to toluene sniffing. Eur Neurol 30:347-349

Ikeda M (1992) Public health problems of organic solvents. Toxicol Lett 64-65:191-201

Jacobsen M, Baelum J and Bonde JP (1994) Temporal epileptic seizures and occupational exposure to solvents. Occup Environ Med 51: 429-430

King GI, Jacobs RE and White SH (1985) Hexane dissolved in dioleoyllecithin bilayers has a partial molar volume of approximately zero. Biochemistry 24:4637-4645

Kishi R, Harabuchi I, Katakura Y, Ikeda T and Miyake H (1995) Neurobehavioral effects of chronic occupational exposure to organic solvents among Japanese industrial painters. Environ Res 62 (2):303-313

Korpela M and Tähti H (1988) Human and rat erythrocyte membranes as a model for studying the effects of organic solvents on membrane-bound acetylcholinesterase *in vitro*. Toxicol in Vitro 2: 135-139

Leo A, Hansch C and Elkins D (1971) Partition coefficients and their uses. Chemical Reviews 71:525-559

McIntosh TJ, Simon SA and MacDonald RC (1989) The organization of n-alkanes in lipid bilayers. Biochim Biophys Acta 597: 445-463

Milanovic L, Spilich G, Vucinic G, Knezevic S, Ribaric B and Mubrin Z (1990) Effects of occupational exposure to organic solvents upon cognitive performance. Neurotoxicol Teratol 12 (6):657-660

Naskali L, Engelke M, Tähti H and Diehl H (1993) The effects of selected organic solvents on rat synaptosomal membrane fluidity and integral enzyme activities. Neurosci Res Commun 13: 27-35

Naskali L, Oksanen H and Tähti H (1994) Astrocytes as targets for CNS effects of organic solvents in vitro. Neurotoxicology 15:609-612

Ruff RL, Petito CK and Acheson LS (1981) Neuropathy associated with chronic low level exposure to n-hexane. Clin Toxicol 18:515-519

Seeman P(1972) The membrane actions of anesthetics and tranquilizers. Pharmacol Rev 24:513-655

Sikkema J, de Bont JAM and Poolman B (1994) Interaction of cyclic hydrocarbons with biological membranes. J Biol Chem 269.8022-8028

Stoltenburg-Didinger G, Boegner F, Grüning W, Wagner M, Marx P and Altenkirch H (1992) Specific neurotoxic effects of different organic solvents on dissociated cultures of the nervous system. Neurotoxicology 13:161-164

Tähti H, Hyppönen S, Oksanen H and Korpela M (1992) Evaluation of the effects of organic solvents and solvent mixtures on cell membrane integral proteins *in vitro*. In Vitro Toxicol 5:1-6

Tähti H and Naskali L (1992) The effects of organic solvents on neural membrane integral protein tested in neural cell cultures. Neurosci Res Commun 10:71-77

Tähti H, Vaalavirta L and Toimela T (1995) *In vitro* methods as a tool in studies of neurotoxicity. ATLA-Altern Lab Anim 23:491-496

Tanii H, Huang J, Ohyashiki T and Hashimoto K (1994) Physical-chemical-activity relationship of organic solvent: effects on Na^+/K^+-ATPase activity and membrane fluidity in mouse synaptosomes. Neurotoxicol Teratol 16:575-582

Vaalavirta L and Tähti H (1995) Astrocyte membrane Na^+, K^+-ATPase and Mg^{2+} -ATPase as targets of organic solvent impact. Life Sci 57:2223-2230

Walum E, Hansson E and Harvey AL (1990) In vitro testing of neurotoxicity. ATLA-Altern Lab Anim 18: 153-179

Veronesi B (1992): In vitro screening batteries for neurotoxicants. Neurotoxicology 13:185-196

Bovine Chromaffin Cells as *in vitro* Model for the Study of non-Cholinergic Toxic Effects of Organophosphorus Compounds

M.A. Sogorb, S. Bas, L.M. Gutiérrez, E.Vilanova and S. Viniegra.
Department of Neurochemistry. Institute of Neuroscience. PO Box: 374. University of Alicante. E-03080 Alicante. Spain.

Introduction

Carboxylesterase activities (CbEs) are targets for the intoxication by organophosphorus compounds (OPs) (Aldridge and Reiner, 1972). They have been detected in homogenates of neural and non-neural tissues by employing a variety of substrates (Aldridge, 1993). OPs inhibit CbEs (Aldridge and Reiner, 1972) and are considered a powerful tool for the discrimination between different types of esterases. Most of these CbEs have not been well characterized; and, consequently, their biological functions are not well understood (Aldridge, 1993).

The mechanism by which certain OPs induce the neurodegenerative syndrome known as organophosphorus induced delayed polyneuropathy is related to the covalent modification of a CbE called neuropathy target esterase (NTE) (Clothier and Johnson, 1980). This protein is operatively defined as the phenyl valerate esterase activity (PVase) resistant to diethyl p-nitrophenyl phosphate (paraoxon) and sensitive to N, N'-diisopropyl diamidophosphorofluoridate (mipafox) (Vilanova et al., 1993). The study of OP delayed neurotoxicity has been mainly performed on the adult hen brain due to the relative abundance of NTE activity (Williams, 1983). In a recent study a partial characterization of NTE and related PVases in bovine adrenal medulla homogenates has been performed, demonstrating that this tissue posseses the greatest NTE activity reported to date (Sogorb at al., 1994).

Embryologically, chromaffin cells arise from the neural crest (Anderson, 1993), and they have been extensively used as *in vitro* models in the study of certain neuronal processes (Burgogne, 1991). In fact, chromaffin cells have been recently employed in studies of the toxic effects of the OPs using as endpoint alterations on catecholamine secretion and calcium uptake (Pei-Shan et al., 1994) and ultrastructural changes (Knoth-Anderson et al., 1991). However, NTE and related CbEs have not been characterized in these cultures.

The characterization and development of cellular models for studying the cellular functions of these enzymes would be a great advance. In this sense, the aim of this work was to establish the basis for the use of chromaffin cell cultures as an *in vitro* model for the study of the toxic effects of OPs and of the cellular functions of CbEs.

Materials and Methods

Cell isolation and culture: Chromaffin cells were prepared following Ballesta et al. (1989) and were employed between the third and sixth day after plating.

Enzymatic assays: The assay procedure used to establish the different activities was based in that described by Johnson (1977) for determining NTE activity in hen brain. The procedure started by three cell washings with 200 µl of modified Krebs/Hepes solution (K/H) (composition in mM: NaCl 134; KCl 4.7; $MgSo_4$ 1.2; $CaCl_2$ 2.5; HEPES 15; glucose 11; and 100 mg/l of ascorbic acid) pH 7.4. After that, cells were incubated for 30 min at 37 °C with 100 µl of K/H (total PVase), or with paraoxon (B-activity), or paraoxon plus mipafox (C-activity). Inhibitors were removed from the wells and 100 µl of 4.6 mM phenyl valerate were added, the cells being incubated with substrate at 37 °C. Reaction was stopped by addition of 100 µl of 2% SDS/0.25 mg/ml 4-aminoantipyrine. The mixture was left to rest for 30 min, after which 50 µl of 0.4% potassium ferrocyanide were added. After a further 60 min rest, absorbance was read at 510 nm. In all cases an enzymatic milliunit (mU) was defined as the amount of enzyme catalysing the formation of 1 nmol phenol in 1 min at 37 °C.

[^{3}H]noradrenaline release in intact cells: Cells were incubated with [^{3}H]noradrenaline (1mCi/ml) for 4 h and were then washed four times prior to the incubation (5 min) with the different OPs in K/H basal solution. Catecholamine secretion was induced for 5 min with basal solution or by plasma membrane depolarization with K/H high potassium (59 mM K^+ replacing isoosmotically NaCl by 59 mM KCl) or with 50 µM 1,1-dimethyl-4-phenyl-piperazinium (DMPP) in basal medium. Experiments with digitonin-permeabilized cells were performed as described by Gutiérrez et al. (1995).

Electrophysiological experiments: The whole-cell configuration of the patch-clamp was employed to study inward and outward currents as described by Gutiérrez et al. (1994).

Results

A total PVase of 12.7 ± 1.9 mU/10^6 cells (mean ± SD) was recorded from 10 different cultures. Because OPs are considered powerful tools for the study of CbEs (Aldridge, 1993), we tried to discriminate between them using the capability of OPs to inhibit these enzymes.

Inhibition of PVase by paraoxon: Cells were incubated at 37 °C during 30 minutes with different paraoxon concentrations, and with substrate under the same conditions. Activity was expressed as a percentage with respect to the control preincubated without paraoxon (Fig. 1A). Results showed that 40 µM paraoxon (which is habitually used in standard hen brain testing (Johnson, 1977)) is able to totally inhibit the sensitive component. We recorded a PVase activity resistant to 40 µM paraoxon (B-activity) of 10.5 ± 0.8 mU/10^6 cells (mean ± SD, n=8).

Inhibition of B-activity by mipafox: Standard assays of B-activity were made using different mipafox concentrations. The results were expressed as percentages with respect to the control (in absence of mipafox and with 40 µM paraoxon) (Fig. 1B). Results showed that 250 µM mipafox concentration inhibited 98% of total activity sensitive to 40 µM paraoxon. This is the concentration habitually employed in NTE determinations of bovine adrenal medulla (Sogorb et al, 1994) and hen sciatic nerve (Carrera et al., 1994). PVase resistant to 40 µM paraoxon plus 250 µM mipafox (C-activity) yielded 1.2 ± 0.5 mU/10^6 cells (mean ± SD, n=8).

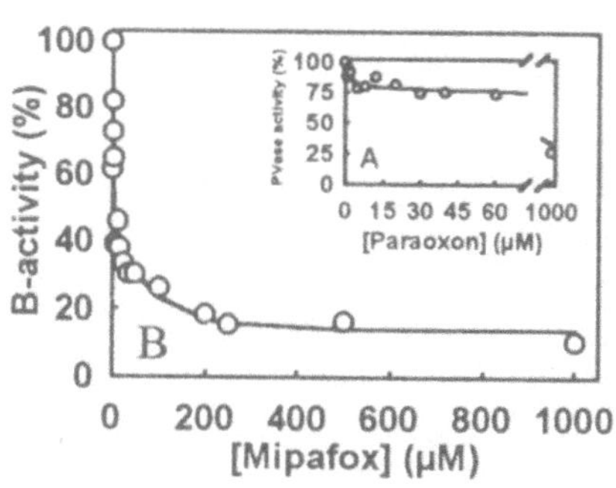

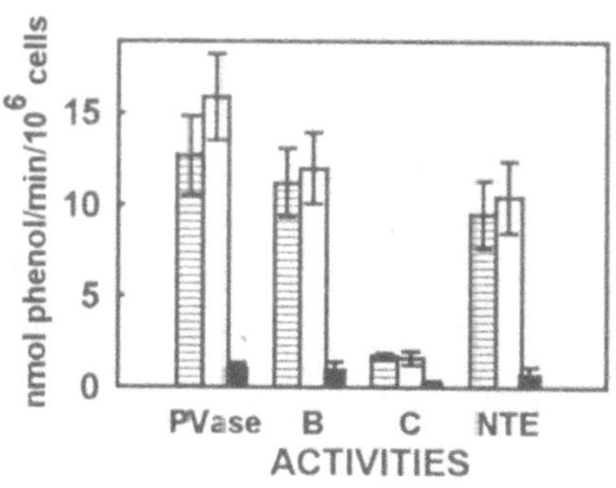

Fig. 1 (left panel): Inhibition of PVases by Paraoxon (A) and Mipafox (B). (A) Control without paraoxon, 12.9 mU/10^6 cells. (B) Control 40 mM paraoxon without mipafox, 10.3 mU/10^6 cells, respectively.

Fig. 2 (Right panel): Location of the Reaction Product. Control cells (▤). Extracellular medium (□). Intracellular medium (■).

NTE activity in bovine chromaffin cell cultures: With the above consideration about B and C activities, NTE was defined as the fraction of PVase resistant to 40 μM paraoxon and sensitive to 250 μM mipafox. With this criterion, we recorded a NTE activity (calculated as B-C) of 9.3 ± 0.8 mU/10^6 cells (mean ± SD, n=8).

Cell viability was evaluated by the trypan blue exclusion test after incubating the cells at 37 °C for 30 min with the following reagents: 40 μM paraoxon plus 250 μM mipafox, 4.6 mM phenyl valerate, 300 μM phenol and 0.9% (v/v) dimethylsulfoxide (substrate solvent), resulting in all cases in a total exclusion of the dye.

In order to establish whether the reaction product remains in the cell, we performed a standard activity assay for all activities. Before stopping the reaction, the substrate was removed together with the extracellular medium from a group of wells and 100 μl of modified K/H solution was added to them. Most of the phenol produced during the assays was detected in the extracellular medium, while the amount of phenol contained inside the cells was almost negligible (Figure 2).

Effect of mipafox on the catecholamine secretion of chromaffin cells: With the aim to determine the direct effect of NTE inhibition on catecholamine release we investigated the concentration-response relationship to mipafox. Catecholamine release can be triggered in chromaffin cells by direct opening of voltage-dependent Ca^{2+} channels (VDCC) using a high potassium solution (K/H 59 mM K^+) or, alternatively, by nicotinic receptor stimulation using 50 μM DMPP, a specific agonist which indirectly induces Ca^{2+} entry in response to a transient depolarization caused by the opening of the ionophore associated with nicotinic receptors. Both stimuli were used to test the effect of mipafox on the exocytotic process in chromaffin cells. Five hundred μM mipafox does not significantly affect basal secretion in non-depolarizing media during 5 min incubation period (data not showed), but caused a significant 28% inhibition ($p<0.05$) in the secretion induced by depolarization at concentrations over 100 μM (Fig 3A). The mipafox inhibition of nicotinic agonist induced secretion was greater, being 65% of the control release ($p<0.01$). Half maximal inhibition was reached at 117 and 140 μM mipafox respectively. These effects were reversible since washing the drug before secretion abolishes the inhibitory effect (data not shown).

It was postulated that OPs might act by altering Ca^{2+} influx (Liu et al., 1994). Experiments with digitonin-permeabilized chromaffin cells were performed to establish if the inhibition caused by mipafox could be explained in this way. In this system secretion can be triggered by controlling the free Ca^{2+} concentration in the external medium. Ca^{2+}-induced secretion in permeabilized chromaffin cells was unaffected by 500 μM mipafox (Fig 3B), thus the catecholamine secretion inhibition found in "intact" cells results from the alteration of the Ca^{2+} influx pathways.

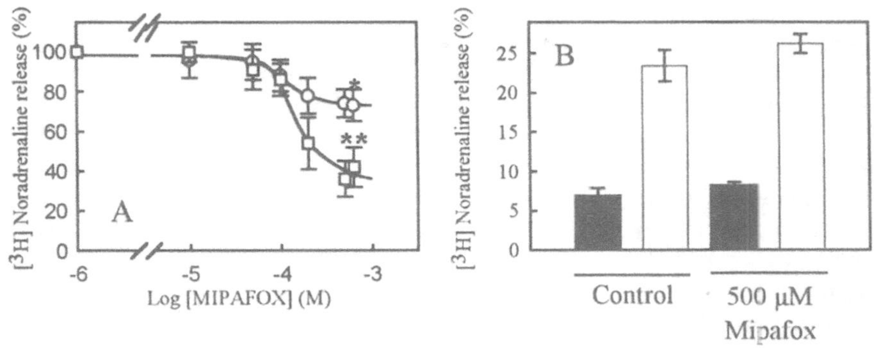

Fig. 3: Mipafox effects on catecholamine release from intact (A) and permeabilized (B) chromaffin cells. (A: ○ 59 mM K^+, □ 50 mM DMPP; B: ■ 5 mM EGTA, □ 10 mM Ca^{2+})

Effect of mipafox on the activity of voltage-dependent Ca^{2+} channels: The currents through these channels were elicited by using pulses from -80 mV to 0 mV and using Ba2+ as a charge carrier. Fig 4A shows that Ba2+ currents were partially blocked by short time incubations (from 30 to 120 sec) with 50 μM mipafox concentrations and over. This effect was totally reversible and control currents were recovered after washout of the compound. Half maximal effects were obtained with 50 μM mipafox, and a maximal effect of 35% inhibition compared with the control was obtained. Experiments were performed using 100 μM of methyl parathion, saligenin cyclic-o-phenylphosphate (PSP) and paraoxon, obtaining inhibitions of 26 ± 3, 31 ± 6 and 28 ± 4% respectively.

Effect of mipafox on the function of Na^+ and K^+ currents: The specificity of the inhibition was tested by studying mipafox effect on Na^+ and K^+ currents implicated in the action potential generation in chromaffin cells. Fig. 4B shows as the presence of 200 μM of mipafox did not significantly alter the Na^+ current. The K^+ currents observed showed a weak inhibition by 200 μM mipafox (Fig. 4C), this inhibition represented 12 ± 6 % of the control currents and was partially reversed by washing out the OP.

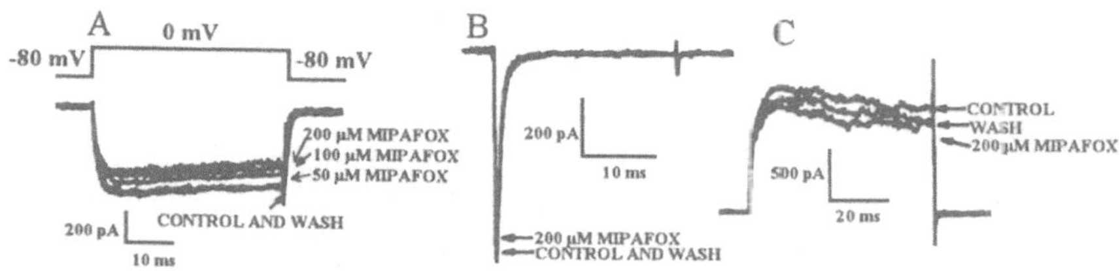

Fig. 4: Effect of mipafox on the Ba^{2+} (A), Na^+ (B) and K^+ (C) currents.

These results showed that mipafox has some specifity toward the blockade of VDCC and a minor effect on the K^+ current; as Ca^{2+} influx is coupled to secretion, blocking VDCC might be responsible for the inhibition of exocytosis observed in chromaffin cells which have been depolarized with high potassium. However, the higher level of inhibition obtained when the cells are stimulated using nicotinic agonists, suggests other possible mechanism of action of organophosphates; consequently we decided to study the effect of these compounds on the ionophore associated with nicotinic receptors.

Mipafox effect on the DMPP evoked currents: Experiments were conducted in patch-clamped cells by eliciting nicotinic currents using 50 µM DMPP. After 2 min incubation mipafox clearly inhibited the DMPP evoked current (Figure 5A). Maximal effects (50-60 % inhibition) were reached at 200 µM concentration, with this inhibition being reversed by mipafox washout. Figure 5B shows the dose-response relationship of mipafox effect on nicotinic currents: half maximal effects were reached with a 100 µM concentration. Other OP compounds like methyl parathion, PSP, and paraoxon at 200 µM have been tested. As inhibitions of 45 to 60 % were found, they seem to share similar characteristics with respect to the inhibition of the ionophore associated with nicotinic receptors.

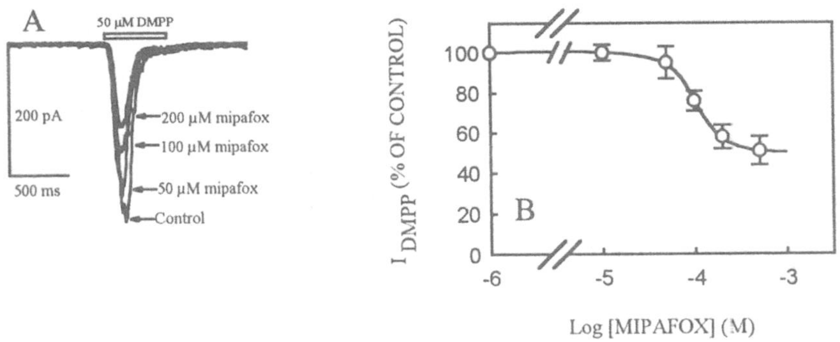

Fig 5. Mipafox effect on DMPP evoked currents (A) and dose-response relationship for the mipafox effect on the DMPP elicited currents (B)

Discussion

In the present study we have demonstrated that bovine chromaffin cells in culture express PVases, and that these activities are sensitive to OPs. This study establishes the basis for employing bovine chromaffin cells as an in vitro model for the study of the toxic effects of OPs and of the cellular functions of PVases.

The manifestation of PVases and their inhibition by paraoxon and mipafox demonstrates that both inhibitors and substrate are accessible to the active centres of these enzymes in cultured cells. Most of the phenol produced during the reaction is detected outside the cells, therefore, it is possible to measure PVases without destroying them. The cells excluded trypan blue dye from their interior and remained attached to the bottom of the well after treatment with all the chemical compounds that were present during the activity determination, suggesting that the cells are viable at least immediately after conclusion of the measurement process.

The neuropathic compound mipafox blocks catecholamine secretion in chromaffin cells through the inhibition of nicotinic currents and VDCC. The preferential block of the DMPP induced catecholamine secretion and the higher degree of inhibition of the DMPP stimulated ionic current demonstrates that nicotinic receptors are the main targets of OPs, causing the decrease in catecholamine release. Since the inhibition of nicotinic currents would affect cell depolarization, and in consequence the opening of VDCC, the cumulative effect resulting from a 50 % inhibition of nicotinic currents and a 30 % inhibition of VDCC accounts for an important inhibition of the secretion (estimated to almost 65-70 %), according to the values obtained. The present results are compatible with a previous study which showed that other OPs cause effects in catecholamine secretion and calcium influx (Liu et al., 1994). We have established by direct measurements the key role of VDCC and especially of nicotinic receptors in the molecular mechanism causing these effects. In addition, we showed that cholinergic (paraoxon), delayed neurophathy inducers (PSP) or both (mipafox) OPs caused inhibition of both nicotinic currents and VDCC.

The present work explains the effects of OPs on the secretion from bovine chromaffin cells as a consequence of a reversible alteration of the ionic currents contributing to secretion. The use of the patch-clamp technique was essential as a direct approach to evaluate the importance of the alteration of nicotinic currents and VDCC as new targets for the action of OPs. It also demonstrates that these effects are relatively specific, since only marginal decreases were found in the activity of Na^+ and K^+ currents. The reversibility suggests that the effects cannot be related to the well-known capability of the OPs for phosphorylating serine residues of the serine-esterases or NTE.

Conclusion

All the above results suggest that bovine chromaffin cells are an appropriate in vitro model for the study of the non-cholinergic toxic effects of organophosphorus compounds and for the study of the physiological role of NTE and related carboxylesterases.

Acknowledgements

We thank Mr. Miguel A. Company and Mr. Jose Luis Quintanar for expert cell culture, Dr. Francisco Sala for his help in electrophysiological data analysis. Work supported by FISSS 95/0053 and DGICYT PM 95-0110. MAS received a fellowship from the Spanish Ministry of Education.

References

Aldridge WN (1993) The esterases: perspectives and problems. Chem Biol Inter 87:5-13

Aldridge, W.N., and Reiner, E. (1972) Enzyme inhibitors as substrates. Interactions of esterases with esters of organophosphorus and carbarmic acids. (Tatum E.L. and Neuberger A, eds), pp 8-37. North-Holland Publishing Company. Amsterdam-London.

Anderson DJ (1993) Cell fate determination in the peripheral nervous system: The sympathoadrenal progenitor. J Neurobiol 24: 185-198

Ballesta JJ, Palmero M, Hidalgo MJ, Gutiérrez LM, Reig JA, Viniegra S, Garcia AG (1989) Separate binding and functional sites for ω-conotoxin and nitrendipine suggest two types of calcium channels in bovine chromaffin cells. J Neurochem 53:1050-1056

Burgogne RD (1991) Control of exocytosis in adrenal chromaffin cells. Biochim Biophys Acta 1071: 174-202

Carrera V, Díaz-Alejo N, Sogorb MA, Vicedo JL, Vilanova E (1994) In vivo inhibition by mipafox of soluble and particulate forms of organophosphorus neuropathy target esterase (NTE) in hen sciatic nerve. Toxicol Lett 71:47-51

Clothier B, Johnson MK (1980) Reactivation and aging of neurotoxic esterase inhibited by a variety of organophosphorus esters. Biochem J 185:739-747

Gutiérrez LM, Viniegra S, Quintanar, JL, Reig JA, Sala F (1994) Calyculin A blocks bovine chromaffin cell calcium channels independently of phosphatase inhibition. Neurosci Lett 178:55-58, 1994.

Gutiérrez LM, Quintanar, JL, Viniegra S, Salinas E, Moya F, Reig JA (1995) Anti-syntaxin antibodies inhibit calcium-dependent catecholamine secretion from permeabilized chromaffin cells. Biochem Biophys Res Commun 206(1):1-7

Johnson MK (1977) Improved assay of neurotoxic esterase for screening organophosphates for delayed neurotoxicity potential. Arch Toxicol 37:113-115

Knoth-Anderson J, Veronesi B, Jones K, Lapadula DM, Abou-Donia MB (1992) Triphenyl phosphite induced ultrastructural changes in bovine adrenomedullary chromaffin cells. Toxicol Appl Pharmacol 112: 110-119

Pei-Shan L, Lung-Sen K, Meng-Kai L (1994) Organophosphates inhibit catecholamine secretion and calcium influx in bovine adrenal chromaffin cells. Toxicology 90:81-91

Sogorb MA, Viniegra S, Reig JA, Vilanova E (1994) Partial characterization of neuropathy target esterase and related phenyl valerate esterases from bovine adrenal medulla. J Biochem Toxicol 9: 145-152

Vilanova E, Barril J, Carrera V (1993) Biochemical properties and possible toxicological meaning of various forms of NTE. Chem Biol Interact 87: 369-381

Williams DG (1983) Intramolecular group transfer is a characteristic of neurotoxic esterase and is independent of the tissue source of the enzyme. Biochem J 209: 817-829

Carcinogenesis Mechanisms in Transgenic Mouse Models

(Chair: E. Dybing, Norway)

Hepatic Tumor Induction in c-*Myc* mono-transgenic and TGF-α/c-*Myc* double-transgenic Mice.

Snorri S. Thorgeirsson, Eric Santoni-Rugiu, Cindy D. Davis, and Elizabeth G. Snyderwine
Laboratory of Experimental Carcinogenesis, Division of Basic Sciences, National Cancer Institute, National Institutes of Health, Bethesda, Maryland U.S.A.

Summary

Double transgenic mice bearing fusion genes consisting of mouse albumin enhancer/promoter-mouse c-*myc* cDNA and mouse metallothionein 1 promoter-human TGF-α cDNA were generated to investigate the interaction of these genes in hepatic oncogenesis and to provide a general paradigm for characterizing both the interaction of nuclear oncogenes and growth factors in tumorigenesis as well as to produce an experimental model to test how environmental chemicals might interact with these genes during the neoplastic process. Coexpression of c-*myc* and TGF-α as transgenes in the mouse liver resulted in a tremendous acceleration of neoplastic development in this organ as compared to expression of either of these transgenes alone. The two distinct cellular reactions that occurred in the liver of the double transgenic mice prior to the appearance of liver tumors were dysplastic and apoptotic changes in the existing hepatocytes followed by emergence of multiple focal lesions composed of both hyperplastic and dysplastic cell populations. These observations suggest that the interaction of c-*myc* and TGF-α, during development of hepatic neoplasia contributes to the selection and expansion of the preneoplastic cell populations which consequently increases the probability of malignant conversion. Treatment of the double transgenic mice with both genotoxic agents such as diethylnitrosamine and IQ as well as the tumor promoter phenobarbital greatly accelerated the neoplastic process. These results suggest that selective transgenic mouse models may provide important tools for testing both the carcinogenic potential of environmental chemicals and the interaction/cooperation of these compounds with specific genes during the neoplastic process.

Introduction

It is now well established that dietary factors play an important role in the etiology of human cancer (Doll and Peto, 1981). A number of rodent carcinogens and bacterial mutagens, formed during high-temperature cooking, have been isolated from cooked, protein-rich foods, and the most potent of these mutagens belong to the class of heterocyclic amines (HCA) (Sugimura and Sato, 1983). In order to exert their mutagenic and carcinogenic effects, the HCAs require metabolic activation (Kato, 1986). The first and most critical step in the activation process is N-hydroxylation of the exocyclic amino groups, a reaction catalyzed primarily by cytochrome P450IA2 (Snyderwine and Battula, 1989; Aoyama et al., 1990). The N-hydroxy derivatives of a number of the HCA have been shown to react directly with DNA *in vitro*, and specific HCA-DNA adducts have been isolated and quantitated in *Salmonella* (Asan et al., 1987; Schut et al., 1991), Syrian hamster embryo cells (Asan et al., 1987), Cynomolgus monkey (Snyderwine et al., 1988; Schut et al., 1991), CDF1 mice (Hall et al., 1990; Schut et al., 1991), and F-344 rats (Schut et al., 1988, 1991). Here we describe the development of a new transgenic mouse model that may be useful in testing the impact of HCA exposure as well as other suspected environmental carcinogens on the process of liver tumorigenesis.

The transgenic mouse system provides a unique model for assessing the effects of oncogenes acting either alone or in combination with other oncogenes within the organism (Compere et al., 1988; Hanahan, 1988). It therefore provides a research tool for exploring interactions of some of the genetic alterations observed in the fully-developed tumor phenotypes. This may be particularly important in the interaction of growth factors and nuclear oncogenes (Hunter, 1991). The *myc* gene family (c-*myc*, N-*myc* and L-*myc*) is an important member of the nuclear oncogenes. The *myc* oncoproteins act as sequence-specific transcription factors that regulate a variety of genes important in normal cellular growth and differentiation processes (DePinho et al., 1991; Luscher & Eisenman, 1990). Aberrant expression of all three members of the *myc* gene family has been implicated in the development of a wide variety of both experimentally-induced and naturally-occurring tumors (Moroy et al., 1991). Deregulation of c-*myc* expression is frequently observed in experimentally-induced hepatocellular carcinoma in rodents, as well as in primary human liver tumors (Chandar et al., 1989; Yaswen et al., 1985; Ren et al., 1986; Nagy et al, 1988). Similarly, expression of transforming growth factor-α (TGF-α), a potent hepatotrophic mitogen synthesized in hepatocytes during regeneration, is frequently observed in both human tumors and in transformed cultured cells (Derynck et al., 1987), including those of the liver (Liu et al., 1988; Raymond et al., 1989). TGF-α is structurally and functionally related to epidermal growth factor (EGF), and exerts it effects via the EGF receptor (Marquardt et al., 1983, 1984; Massague, 1983). Liver tumors occur in c-*myc* transgenic mice as a result of the selective expression of c-*myc* in the liver directed by the albumin enhancer/promoter or the α1-antitrypsin promoter (Sandgren et al., 1989; Perraud et al., 1991). However, the tumor incidence in these c-*myc* transgenic mice is relatively low with a long latency period (Sandgren et al., 1989; Perraud et al., 1991). Similarly, transgenic mice in which TGF-α is under the transcriptional control of the metallothionein promoter and is expressed at high levels in the liver and other tissues, develop liver tumors after a latency period of 10 to 15 months (Jhappan et al., 1990; Sandgren et al., 1990; Takagi et al., 1992).

Materials and Methods

Transgenic mice: The c-*myc* and c-*myc*/TGF-α transgenic mice were generated as reported before (Murakami et al., 1993, Santonii-Rugiu et al., 1996). Animal housing and animal care were in accordance with NIH guidelines.

Macroscopical and histological analyses: For morphological, immunohistochemical, and molecular analysis, parts of the livers were fixed in 10% phosphate buffered (pH 7.4) formalin, or promptly frozen in liquid nitrogen and stored at -80°C. All fixed tissues were routinely embedded in paraffin and sections stained with hematoxylin-eosin (H&E). Two sections from each lobe of nontumorous liver and all grossly visible tumors were examined. The histopathological diagnosis of HCA or HCC was independently done by three investigators on H&E-stained tissue sections obtained from all grossly visible tumors.

Tumor promotion by phenobarbital (PB). Mice were given PB 0.05% in their food pellets starting at 3 weeks of age. Ten to twelve transgenic mice and 5 wild type mice (wt) were sacrificed at each of the indicated time points and further analyzed as described for the untreated animals.

Treatment with chemical carcinogens: The double transgenic c-*myc*/TGF-α mice were treated with one dose (5 μg/g in saline, i.p.) of diethylnitrosamine (DEN) at 15 days of age, and 10 daily doses (100 μg/g in corn oil, i.g.) of 2-amino-3-methylimidazo[4,5-*f*]quinoline (IQ) beginning at 21 days of age.

Results

Development of hepatic neoplasia in c-*myc*/TGF-α transgenic mice. The appearance of both benign (HCAs) and malignant (HCCs) neoplastic lesions in c-*myc*/TGF-α transgenic mice started between 3 and 4 months and by the eigth month 100% of the males and 30% of the females had liver carcinomas in a background of hepatic dysplasia (Fig. 1). The HCCs more commonly displayed a trabecular histological pattern (57/82 cases, including the clear cell variant), but

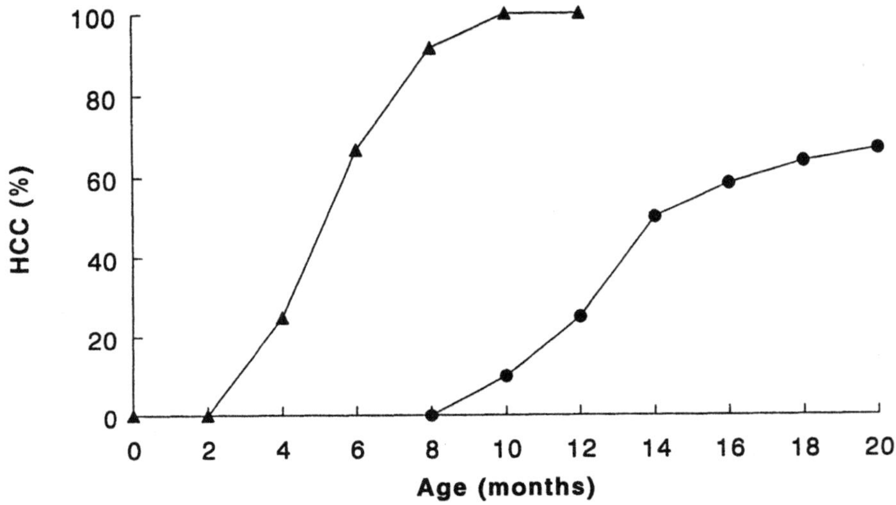

Fig. 1: Incidence and time course of HCC in c-*myc* (o) and c-*myc*/TGF-α ( )

the solid (19/82 cases) or pseudoglandular (adenocarcinoma, 6/82 cases) types were also detectable. Sometimes foci of trabecular carcinomas originated within HCAs producing the so-called "nodule in nodule" histotype. Each HCC histological pattern varied from well differentiated to poorly differentiated, the latter composed by anaplastic cells in intense mitotic activity, sinusoid dilatation, proliferation of neocapillaries, presence of unaccompanied arteries, and large areas of hemorragic necrosis. As the formation of neoplastic lesions progressed the peritumorous tissues presented some differences as compared to liver dysplasia present in mice not yet affected by tumors. In fact, the population of enlarged cells with large and convoluted nuclei showed a lower number of mitotic figures than in younger transgenic mice. In addition, the tissues adjacent to tumors also contained clusters of small dysplastic hepatocytes characterized by irregular nuclear contour, reduced cytoplasm, higher nuclear/cytoplasmic ratio, and consequentially nuclear crowding and increased cell density.

Tumor Promotion by Phenobarbital: Phenobarbital (PB) is considered a potent non-genotoxic liver tumor promoter in rodents treated with chemical carcinogens. Although several mechanisms by which PB promotes liver carcinogenesis have been proposed (Bursch et al., 1984; Jirtle & Meyer, 1991; Neveu et al., 1994; Jirtle et al., 1994), the notion that initiated hepatocytes have a distinct growth advantage over the mitoinhibited normal hepatocytes has recently attracted increased interest. PB has also been shown to promote growth of transplanted HCC (Seglen et al., 1990) and more recently to collaborate with

TGF-α in accelerating hepatocarcinogenesis of TGF-α transgenic mice (Takagi et al., 1993).

PB was administered in the diet to both transgenic lines and to wild type mice from 3 weeks to 10 months of age. The PB treatment resulted in an acceleration and a significant increase of the HCA and HCC incidence in c-*myc* transgenic mice as compared to the untreated c-*myc* mice (Table 1). The HCA and HCC appeared two months earlier, relative to untreated mice, and by 10 months of age 100% and 40% of PB-treated c-*myc* mice had HCA and HCC, respectively. These results suggest that hepatocytes overexpressing c-*myc* respond to PB promotion in a manner similar to hepatocytes initiated by chemical carcinogens.

Table 1. Tumor promotion by phenobarbital (PB)

Age	wt-PB	c-*myc*	c-*myc*-PB
		Incidence of HCA	
6 months	0/5 (0%)	0/10 (0%)	5/10 (50%)
8 months	0/5 (0%)	2/10 (20%)	8/10 (80%)
10 months	0/5 (0%)	4/12 (33%)	10/10 (100%)
		Incidence of HCC	
8 months	0/5 (0%)	0/10 (0%)	2/10 (20%)
10 months	0/5 (0%)	1/12 (8%)	4/10 (40%)

Effects of Chemical Carcinogens on Hepatocarcinogenesis in c-*myc*/TGF-α Transgenic Mice: Carcinogenesis is believed to be a multistep process that proceeds via discrete genetic events. We have recently established several transgenic mouse models aimed at exploring the interaction of nuclear oncogenes such as c-*myc* and growth factors. In one of these transgenic models, it was shown that coexpression of c-*myc* and TGF-α as transgenes in the mouse liver results in greatly accelerated neoplastic development in this organ as compared to either of these transgenes alone. Here we demonstrate the treatment of the c-*myc*/TGF-α transgenic mice with either DEN or IQ results in greatly enhanced carcinogenesis (Fig. 2). At 18 weeks of age, all untreated animals had areas of hyperplastic foci but none had macroscopically visible tumors. Treatment with IQ caused an increase of hyperplastic foci and macroscopically visible tumors. Microscopically. the tumors were well circumscribed and composed of clear, eosinophilic and basophilic cells. Animals treated with DEN had a dramatic increase in the number and the size of hyperplastic foci, dysplastic cells, and visible tumors. All male and most female

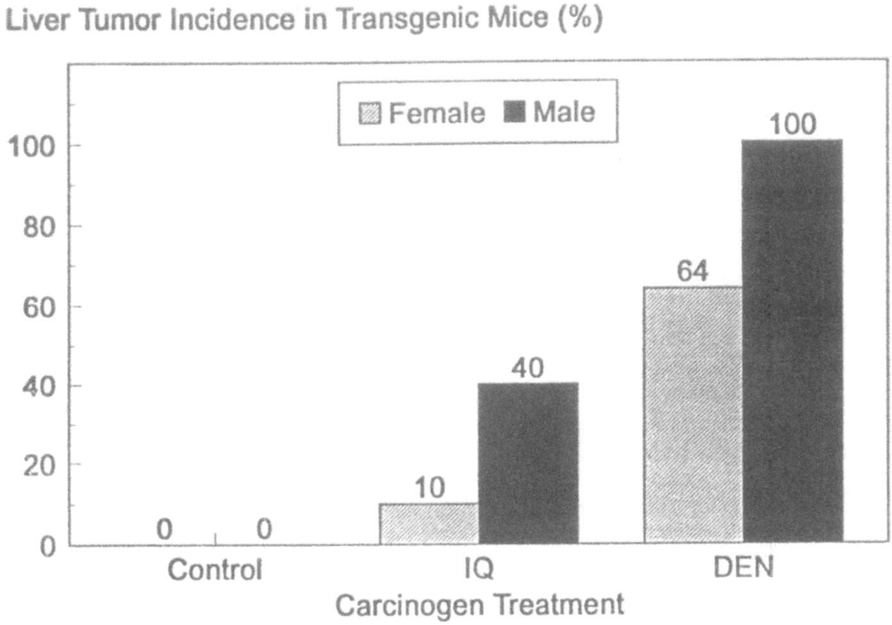

Fig. 2: Percentage of c-*myc*/TGF-α mice with visible tumors 4 months after treatment with IQ (100 μg/g, 10 doses), DEN (5 μg/g, single dose), or control vehicle (corn oil).

mice had gross tumors. Morphologically the tumors were well differentiated trabecular and glandular carcinomas. Therefore, treatment of the double transgenic mice with either IQ or DEN accelerates the tumorigenic process, and suggests that the c-*myc*/TGF-α transgenic mouse model may be useful for in vivo testing the impact of both suspected genotoxic and non-genotoxic agents.

References

Aoyama T, Gelboin HV, Gonzalez FJ (1990) Mutagenic activation of 2-amino-3-methylimidazo[4,5-f]quinoline by complementary DNA-expressed human liver P-450. Cancer Res 50: 2060-2063

Asan E, Fasshauer I, Wild D, Henschler D (1987) Heterocyclic aromatic amine-DNA-adducts in bacteria and mammalian cells detected by ^{32}P-postlabeling analysis. Carcinogenesis 8:1589-1593

Bursch W, Lauer B, Timmermann-Trosiener I, Barthel G, Schuppler J, Schulte-Hermann R (1984) Controlled death (apoptosis) of normal and putative preneoplastic cells in rat liver following withdrawal of tumor promoters. Carcinogenesis 5:453-458

Chandar N, Lombardi B, Locker J (1989) c-*Myc* gene amplification during hepatocarcinogenesis by a choline-devoid diet. Proc Natl Acad Sci USA 86:2703-2707

Compere SJ, Baldacci P, Jaenisch R (1988) Oncogenes in transgenic mice. Biochim Biophys Acta 948:129-149

DePinho RA, Schreiber-Agus N, Alt FW (1991) *Myc* family oncogenes in development of normal and neoplastic cells. Adv Cancer Res 57:1-46

Derynck R, Goeddel DV, Ullrich A, Gutterman JU, Williams RD, Bringman TS, Berger WH (1987) Synthesis of messenger RNAs for transforming growth factors α and β and the epidermal growth factor receptor by human tumors. Cancer Res 47: 707-712

Doll R, Peto D (1981) The causes of cancer: quantitative estimates of avoidable risks of cancer in the United States today. J Natl Cancer Inst 66:1191-1208

Hall M, She MH, Wild D, Fasshauer I, Hewer A, Phillips DH (1990) Tissue distribution of DNA adducts in CDF1 mice fed 2-amino-3-methylimidazo[4,5-*f*]quinoline (IQ) and 2-amino-3,4-dimethylimidazo[4,5-*f*]quinoline (MeIQ). Carcinogenesis 11:1005-1011

Hanahan D (1988) Dissecting multistep tumorigenesis in transgenic mice. Ann Rev Genet 22:479-519

Hunter T (1991) Cooperation between oncogenes. Cell 64:249-270

Jhappan C, Stahle C, Harkins RN, Fausto N, Smith GH, Merlino GT (1990) TGFα overexpression in transgenic mice induces liver neoplasia and abnormal development of the mammary gland and pancreas. Cell 61:1137-1146

Jirtle RL, Hankins GR, Reisenbichler H, Boyer IJ (1994) Regulation of mannose 6-phosphate/insulin-like growth factor-II receptors and transforming growth factor beta during liver tumor promotion with phenobarbital. Carcinogenesis 15:1473-1478

Jirtle RL, Meyer SA (1991) Liver tumor promotion: effect of phenobarbital on EGF and protein kinase C signal transduction and transforming growth factor-β1 expression. Dig Dis Sci 36:659-668

Kato R (1986) Metabolic activation of mutagenic heterocyclic aromatic amines from protein pyrolysates. CRC Crit Rev Toxicol 6:307-348

Liu C, Tsao MS, Grisham JW (1988) Transforming growth factors produced by normal and neoplastically transformed rat liver epithelial cells in culture. Cancer Res 48:850-855

Luscher B, Eisenman RN (1990) New light on *myc* and *myb*. Part. I. *Myc*. Genes Dev 4:2025-2035

Marquardt H, Hunkapillar MW, Hood LE, Twardzik DR, DeLarco JE, Stephenson JR, Todaro GJ (1983) Transforming growth factors produced by retrovirus-transformed rodent fibroblasts and human melanoma cells: amino acid sequence homology with epidermal growth factor. Proc Natl Acad Sci USA 80:4684-4688

Marquardt H, Hunkapillar MW, Hood LE, Todaro GJ (1984) Rat transforming growth factor type I: structure and relation to epidermal growth factor. Science 223:1079-1082

Massague J (1983) Epidermal growth factor-like transforming growth factor II. Interaction with epidermal growth factor receptors on human placenta membrane and A431 cells. J Biol Chem 258:13614-13620

Moroy T, Alt FW, DePinho RA (1991) Use of transgenic mice to study normal and abnormal mammalian development, in: S. Broder et al. (Eds.), Molecular Foundation of Oncology. Williams and Wilkins, Baltimore, MD, pp. 455-487.

Murakami H, Sanderson ND, Nagy P, Marino PA, Merlino G, Thorgeirsson SS (1993) Transgenic mouse model for synergistic effects of nuclear oncogenes and growth factors in tumorigenesis: interaction of c-*myc* and transforming growth factor-α in hepatic oncogenesis. Cancer Res 53:1719-1723

Nagy P, Evarts RP, Marsden E, Roach J, Thorgeirsson SS (1988) Cellular distribution of c-*myc* transcripts during chemical carcinogenesis in rats. Cancer Res 48:5522-5527

Neveu MJ, Babcock KL, Hertzberg EL, Paul DL, Nicholson BJ, Pitot HC (1994) Colocalized alterations in connexin 32 and cytochrome P450IIB1/2 by phenobarbital and related liver tumor promoters. Cancer Res 54:3145-3152

Perraud F, Dalemans W, Genrault JL, Dreyer D, Ali-Hadji D, Favre T, Pavirani A (1991) Characterization of trans-immortalized hepatic cell lines established from transgenic mice. Exp Cell Res 195:59-65

Raymond VW, Lee DC, Grisham JW, Earp HS (1989) Regulation of transforming growth factor a messenger RNA expression in a chemically transformed rat hepatic epithelial cell line by phorbol ester and hormones. Cancer Res 49:3608-3612

Ren GJ, Hu LF, Cheng YC, Wan DF (1986) Oncogenes in human primary hepatic cancer. J Cell Physiol Suppl 4:13-21

Sandgren EP, Luetteke NC, Palmiter RD, Brinster RL, Lee DC (1990) Overexpression of TGF-α in transgenic mice: induction of epithelial hyperplasia, pancreatic metaplasia, and carcinoma of the breast. Cell 61:1121-1135

Sandgren EP, Quaife CJ, Pinkert CA, Palmiter RD, Brinster RL (1989) Oncogene-induced liver neoplasia in transgenic mice. Oncogene 4:715-724

Santoni-Rugiu E, Nagy P, Jensen MR, Factor VM, Thorgeirsson SS (1996) Evolution of neoplastic development in the liver of transgenic mice co-expressing c-*myc* and transforming growth factor-α. Am J Pathol 149:407-428

Schut HAJ, Putman KL, Randerath K (1988) DNA adduct formation of the carcinogen 2-amino-3-methylimidazo[4,5-*f*]quinoline in target tissues of the F-344 rat. Cancer Lett 41:345-352

Schut HAJ, Snyderwine EG, Zu H-X, Thorgeirsson SS (1991) Similar pattern of DNA adduct formation of 2-amino-3-methylimidazo[4,5-*f*]quinoline in the Fischer-344 rat, CDF1 mouse, Cynomolgus monkey, and *Salmonella typhimurium*. Carcinogenesis 12: 931-934

Seglen PO, Saeterm G, Schwarze PE (1990) Liver tumor promoters stimulate growth of transplanted hepatocellular carcinomas. Hepatology 12: 295-300

Snyderwine EG, Battula N (1989) Selective mutagenic activation by cytochrome P_3-450 of carcinogenic arylamines found in foods. J Natl Cancer Inst 81:223-227

Snyderwine EG, Yamashita K, Adamson RH, Sato S, Nagao M, Sugimura, T , Thorgeirsson SS (1988) Use of the ^{32}P-postlabeling method to detect DNA adducts of 2-amino-3-methylimidazo[4,5-*f*]quinoline (IQ) in monkeys fed IQ: Identification of the N-(deoxyguanosin-8-yl)-IQ adduct. Carcinogenesis 9:1739-1743

Sugimura T, Sato S (1983) Mutagens/carcinogens in foods. Cancer Res (Suppl) 43:2415s-2421s

Takagi H, Jhappan C, Sharp R, Merlino G (1992) Hypertrophic gastropathy resembling Menetrier's disease in transgenic mice overexpressing transforming growth factor alpha in the stomach. J Clin Invest 90:1161-1167

Takagi H, Sharp R, Takayama H, Anver MR, Ward JM, Merlino G (1993) Collaboration between growth factors and diverse chemical carcinogens in hepatocarcinogenesis of transforming growth factor α transgenic mice. Cancer Res 53:4329-4336

Yaswen P, Goyette M, Shank PR, Fausto N (1985) Expression of *c-Ki-ras*, *c-Ha-ras* and c-*myc* in specific cell types during hepatocarcinogenesis. Mol Cell Biol 5:780-786

Interactions of TCDD with Signal Transduction and Neoplastic Development in c-myc Transgenic and TGF-alpha Transgenic Mice

D. Schrenk[1], M. Müller[1], G. Merlino[2], and S.S. Thorgeirsson[3]

[1] Institute of Toxicology, University of Tübingen, D-72074 Tübingen, Germany.

[2] Laboratory of Molecular Biology, National Cancer Institute, Bethesda, Maryland 20892, USA.

[3] Laboratory of Molecular Carcinogenesis, National Cancer Institute, Bethesda, Maryland, 20892, USA.

Introduction

The environmental pollutant 2,3,7,8-tetrachlordibenzo-p-dioxin (TCDD) is one of the most potent tumour promoters in rodent liver (Pitot et al. 1980). TCDD enhances the expansion of preneoplastic liver foci finally leading to the development of hepatocellular carcinoma. The exact mechanism of action of TCDD as a tumour promoter, however, in particular the role of dioxin receptor-mediated signalling pathways, is unknown. Recently, it has been shown, that TCDD inhibits apoptosis of preneoplastic hepatocytes rather than enhancing their proliferation (Stinchcombe et al. 1995).

Transforming growth factor (TGF) α is a member of the epidermal growth factor (EGF) family of peptides, and induces a variety of biological responses (Salomon et al. 1990). TGFα acts as a potent mitogen in many cell types including hepatocytes, and has been associated with malignant transformation (Derynck et al. 1987). Overexpression of a TGFα transgene induced the development of well differentiated hepatocellular carcinomas between 10 and 15 month of age in male mice (Jhappan et al. 1990).

c-Myc is a nuclear protooncogene that, when activated, is implicated in the development of neoplasia (Lüscher and Eisenman 1990). On the one hand, c-myc, in concert with its dimerization partner max, is thought to regulate genes mediating the mitogenic response. On the other hand, deregulated c-myc can induce apoptosis depending on the cell environment (Evan et al. 1992, Packham et al. 1996). c-Myc transgenic mice were shown to develop dysplastic liver lesions at the age of 2 - 3 months, and hepatocellular adenomas between 14 and 18 month of age (Murakami et al, 1993).

In the study presented here, we investigated the effect of TCDD treatment on the liver of TGFα or c-myc transgenic mice for two major reasons: i) TCDD is active as a comitogen enhancing the stimulating effect of EGF/TGFα on DNA synthesis in mouse hepatocytes in primary culture (Schrenk et al. 1994), and ii) a subtype of apoptosis elicited by DNA-damaging agents, is inhibited when rat hepatocytes are treated with TCDD (Wörner and Schrenk 1996).

Both findings led us to investigate if TCDD influences dysplasia and apoptosis in livers of TGFα or c-myc transgenic mice.

Materials and Methods

Metallothionein/TGFα homozygous mice (Jhappan et al. 1990) were bred with CD1 (wild-type) animals to expand the number of animals. Homozygous Alb/c-myc transgenic mice, obtained from a C57Bl/6JxCBA/J background (Murakami et al. 1993), were bred under standard conditions. All animals were kept on a 12-hours light-dark schedule, and had free access to standard diet and tap water. The transgenicity of all animals was checked by Southern analysis of tail tip DNA using cDNA probes for c-myc or TGFα. At the age of 4 - 6 weeks, male and female transgenic animals were divided into groups of seven animals, and received 1,400 ng TCDD/kg, dissolved in corn oil, every 2 weeks by i.p. injection, equivalent to an average dose of 100 ng TCDD/kg per day. Non-transgenic CD1 and C57Bl/6J mice of the same age were obtained from Charles River (Kisslegg, Germany), and were treated in parallel. After 6, 9, and 12 weeks of treatment, animals were sacrificed, livers were removed, frozen on dry ice, and microsections were prepared. The sections were stained with hematoxyline/eosin (HE) as described (Stinchcombe et al. 1995), and inspected microscopically.

The rate of giant nuclei (GN rate) was determined by counting the number of nuclei exceeding a diameter of 30 μm in a total of 1,000 nuclei. The apoptotic index was determined by counting the number of apoptotic bodies among 5,000 nuclei, using the fluoresence method of Stinchcombe et al. (1995).

CYP1A-catalyzed 7-ethoxyresorufin-O-deethylase (EROD) activity was measured in liver homogenates using the method of Burke and Mayer (1974), and expressed as pmol resorufin/mg protein per min.

Results

Dysplasia in transgenic mice - effects of TCDD: In male TGFα and c-myc transgenic mice, a marked dysplasia, characterized by large hepatocytes with abundant eosinophilic cytoplasm, and giant nuclei, preferably in the centrilobular area, was observed. Regular and irregular types of mitosis were also seen sporadically in HE-stained sections. The percentage of giant nuclei, exceeding a diameter of 30 μm (GN rate), was used as a first attempt to score the degree of dysplasia.

The GN rate was higher in male than in female TGFα transgenic mice (Fig. 1). TCDD treatment did not result in a significant change in the GN rate.

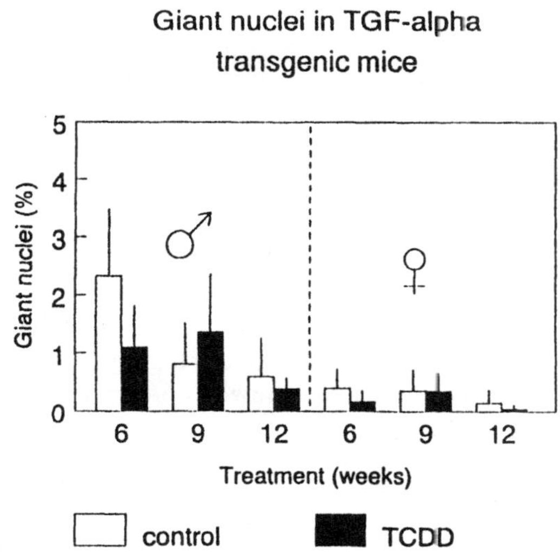

Figure 1. Percentage of giant nuclei in livers of male and female TGFα-transgenic mice treated with corn oil (open bars) or TCDD (closed bars). Means ± SD from seven animals.

A morphological inspection suggested, however, that the dysplasia was significantly aggravated in female animals (not shown).

In c-myc transgenic mice, the difference in GN rates between male and female animals was also highly significant (Fig. 2). Both the high rate in males, and the low rate in females were not affected by TCDD treatment. This was also true for the degree of dysplasia as concluded from morphological inspection of HE sections. Non-transgenic male CD1 mice showed a lower GN rate than male TGFα transgenic mice, while it was below 0.1 % in female CD1 mice or in C57Bl/6J mice of both sexes.

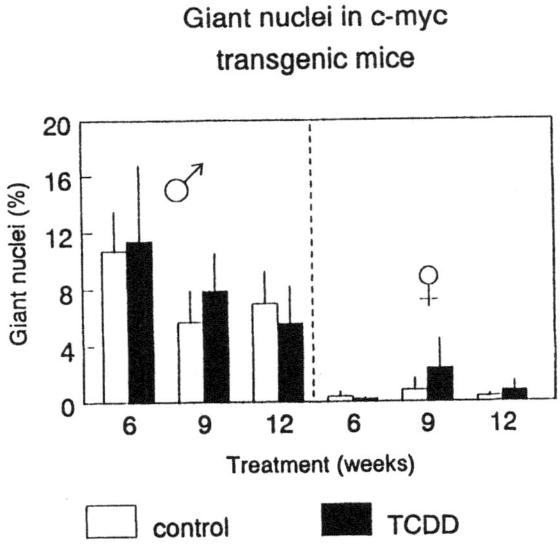

Figure 2. Percentage of giant nuclei in livers of male and female c-myc transgenic mice treated with corn oil (open bars) or TCDD (closed bars). Means ± SD from seven animals

Apoptosis in transgenic mice - effects of TCDD: The incidence of apoptotic bodies (AB) was much lower than the GN rate, ranging from 3 to 25 per 5,000 nuclei. However, c-myc transgenic mice showed a much higher AB index than TGFα transgenic animals. Furthermore, the AB index after 9 and 12 weeks was much higher in male than in female c-myc transgenics. Interestingly, the AB index was decreased by TCDD in female c-myc transgenic animals but not in males (Fig. 3).

In TGFα transgenic mice the low AB index was not changed significantly by TCDD treatment (Fig. 4). In non-transgenic animals, an AB index in the range of 0 - 10 per 5,000 nuclei was determined.

CYP1A activity - effects of TCDD: After six weeks of treatment, CYP1A-catalyzed EROD activity was determined in liver homogenates (Tab. 1). It was found that the transgenes had no significant influence on basal EROD activities. Basal values were slightly higher in CD1 compared to C57Bl/6J mice, and in female compared to male animals, except for TGFα transgenic mice, where they were almost equal in both sexes. A profound influence of the transgenes on EROD induction by TCDD was found, however. The c-myc transgene led to a significant attenuation of EROD induction in males, but to a pronounced increase in females. In contrast, the TGFα transgene resulted in a suppression of EROD induction in both sexes, which was dramatically reduced in females.

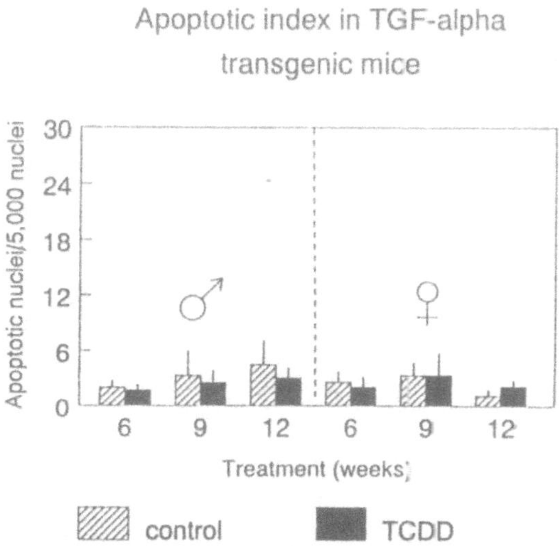

Figure 3. Apoptotic index in livers of male (A) and female (B) TGFα transgenic mice treated with corn oil (open bars) or TCDD (closed bars). Means ± SD from seven animals.

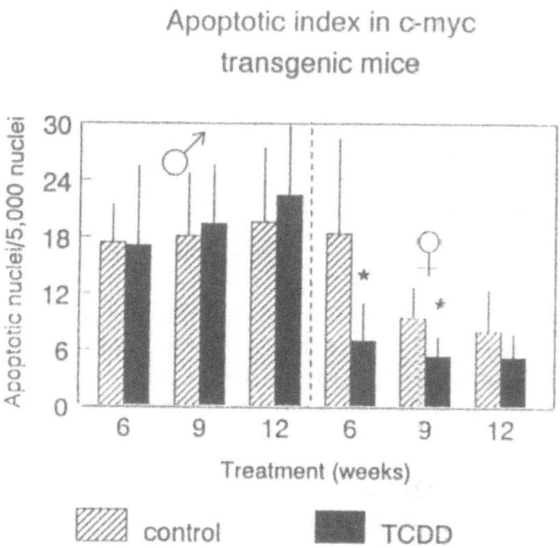

Figure 4. Apoptotic nuclei in livers of male (A) and female (B) c-myc transgenic mice treated with corn oil (open bars) or TCDD (closed bars). Means ± SD from seven animals.

Table 1. 7-Ethoxyresorufin O-deethylase (EROD) activity in liver homogenates of transgenic (TGFα and c-myc) and non-transgenic mice after 6 weeks of treatment with corn oil or 100 ng TCDD/kg per day.

Non-transgenic:	♂	♀	Transgenic: ♂	♀
	(pmol resorufin/min x mg protein)			
C57Bl/6J			c-myc	
corn oil	23.6 ± 12.0[a]	32.2 ± 18.8	16.8 ± 5.7	27.1 ± 14.1
TCDD	1628 ± 112	1235 ± 259	838 ± 211[b]	2208 ± 246[b]
CD1			TGF α	
corn oil	38.9 ± 13.8	46.4 ± 32.3	30.7 ± 2.3	29.5 ± 10.4
TCDD	1150 ± 412	468 ± 103	601± 96[b]	82 ± 78[b]

[a] Mean + SD of 3 - 5 animals.
[b] Significantly different from corresponding non-transgenic animals.

Discussion

TCDD is thought to influence growth and differentiation of epithelial-type cells leading to a number of adverse effects observable in laboratory animals and man. These effects, including hypertrophy/hyperplasia of certain target tissues such as the skin or liver, are most likely to play a key role in the tumour promoting potency of TCDD in these organs.

In particular in the liver, both enhancement of proliferation and suppression of apoptosis of preneoplastic hepatocytes have been discussed as possible mechanisms of action of TCDD. In rodent hepatocytes in primary culture, a synergistic effect of TCDD on EGF/TGFα-induced DNA synthesis (Schrenk et al. 1993), and a suppression of apoptosis after treatment with DNA-damaging agents (Wörner and Schrenk, 1996) was observed. However, the molecular basis, and the role of these mechanisms for liver tumour promotion are unknown. In particular, knowledge is lacking on the relevant molecular links between tumour promotion and the signalling pathways operated *via* the *Ah* (aryl hydrocarbon) or dioxin receptor. Transgenic mice provide the opportunity to study a possible interaction between TCDD and EGF/TGFα or c-myc, respectively.

In previous experiments, a high rate of liver tumours was found in male but not in female TGFα transgenic mice (Tagaki et al. 1992). Our findings demonstrate that dysplastic alterations of the liver including karyomegaly (giant nuclei) were abundant among male but not among female TGFα transgenic animals. The apoptotic indices were in the range of non-transgenic animals, suggesting that TGFa does not induce apoptosis or may even suppress it, as has been reported in other cell types (Amundadottir et al. 1996). TCDD had no significant influence on dysplasia or apoptosis in the livers of male TGFα transgenic animals but aggravated dysplasia in females. Since down-regulation of the EGF receptor is a characteristic effect of TCDD in the liver (Madhukar et al. 1984), our findings suggest that this effect may not prevent the development of dysplastic alterations when TGFα is permanently expressed. A comitogenic effect of TCDD in concert with ethinylestradiol on EGF/TGFα-triggered DNA synthesis, observed in rat hepatocytes (Schrenk et al. 1992), may explain the aggravation of dysplasia by TCDD observed in female TGFα transgenic animals.

TCDD exhibited a profound suppressive effect on apoptosis in female c-myc transgenic mice. This effect did not result in a significant increase in karyomegaly. Interestingly, the high rate of dysplasia/apoptosis in male c-myc transgenic mice was not affected by TCDD.

In accordance with the finding of a selective influence of TCDD on the liver of female c-myc transgenic mice are the findings on CYP1A-catalyzed EROD activity. The expression of CYP1A isozymes is regulated *via* the ligand-activated dioxin receptor which forms a heterodimer with the ARNT protein, thus acting as a nuclear transcription factor which enhances the transcription of CYP1A genes (Whitlock JP 1993). In female c-myc transgenic mice, TCDD-treatment resulted in a significantly higher EROD activity than in non-transgenic females. In contrast, induction was attenuated in male c-myc transgenic mice when compared to their non-trasnsgenic counterparts. Inducibility was suppressed throughout by the TGFα transgene, confirming findings on the inhibitory effect of TGFα on CYP1A induction in cultured mouse hepatocytes (Höhne et al. 1990).

In conclusion, our findings indicate that TCDD can suppress apoptosis in hepatocytes associated with the expression of a c-myc transgene in female mice. The interference of TCDD is also demonstrated by the marked enhancement of CYP1A induction in female c-myc transgenic mice. The findings suggest that the myc/max and dioxin receptor/ARNT signalling pathways are closely linked in an unknown manner, and that their interaction is under the influence of ovarian hormones.

Acknowledgements

This work was supported by the Deutsche Forschungsgemeinschaft (Forschergruppe 'Biochemische Toxikologie'), Bonn. The expert technical assistance of Mrs S. Vetter and Mrs E. Zabinski is gratefully acknowledged.

References

Amundadottir LT, Nass SJ, Berchem GJ, Johnson MD, Dickson RB (1996) Cooperation of TGFα and c-myc in mouse mammary tumorigenesis: coordinated stimulation of growth and suppression of apoptosis. Oncogene 13:757-765

Burke MD and Mayer RT (1974) Ethoxyresorufin: direct fluorimetric assay of a microsomal O-dealkylation which is preferentially inducible with 3-methylcholanthrene. Drug Metab Dispos 2:583-588

Derynck R, Goeddel DV, Ullrich A, Gutterman JU, Williams RD, Bringman TS, Berger WH (1987) Synthesis of mRNAs for transforming growth factors α and β and the epidermal growth factor receptor by human tumors. Cancer Res 47:707-712

Evan GI, Wyllie AH, Gilbert CS, Littlewood TD, Land H, Brooks M, Waters CM, Penn, LZ, Hancock DC (1992) Induction of apoptosis in fibroblasts by c-myc protein. Cell 69:119-128

Höhne M, Becker-Rabenstein V, Kahl GF, Taniguchi H (1990) Regulation of cytochrome P-450 CYPIA gene expression and proto-oncogene expression by growth factors in primary hepatocytes. FEBS Lett 273:219-222

Jhappan C, Stahle C, Harkins RN, Fausto N, Smith GH, Merlino GT (1990) TGF-α overexpression in transgenic mice induces liver neoplasia and abnormal development of the mammary gland and pancreas. Cell 61:1137-1146

Lüscher B and Eisenman RN (1990) New light on Myc and Mab. I. Myc. Genes Dev 4:2025-2035

Madhukar BV, Brewster DW, Matsumura F (1984) Effects of *in vivo* administered 2,3,7,8-tetrachlorodibenzo-p-dioxin on receptor binding of epidermal growth factor in the hepatic plasma membrane of rat, guinea pig, mouse, and hamster. Proc Natl Acad Sci USA 81:7404-7411

Murakami H, Sanderson ND, Nagy P, Marino PA, Merlino G, Thorgeirsson SS (1993) Transgenic mouse model for synergistic effcets of nuclear oncogenes and growth factors in tumorigenesis: interaction of c-myc and transforming growth factor α in hepatic oncogenesis. Cancer Res 53:1719-1723

Packham G, Porter CW, Cleveland JL (1996) c-Myc induces apoptosis and cell cycle progression by separable, yet overlapping, pathways. Oncogene 13:461-469

Pitot HC, Goldsworthy T, Campbell HA, Poland A (1980) Quantitative evaluation of the promotion by 2,3,7,8-tetrachlorodibenzo-p-dioxin of hepatocarcinogenesis from diethylnitrosamine. Carcinogenesis 8:1491-1499

Salomon D, Kim N, Saeki T, Ciardello F (1990) Transforming growth factor α: an oncodevelopmental growth factor.Cancer Cells 2:389-397

Schrenk D, Schäfer S, Bock KW (1994) 2,3,7,8-Tetrachlorodibenzo-p-dioxin as growth modulator in mouse hepatocytes with high and low affinity Ah receptor. Carcinogenesis 15:27-31

Stinchcombe S, Buchmann A, Bock KW, Schwarz M (1995) Inhibition of apoptosis during 2,3,7,8-tetrachlorodibenzo-p-dioxin-mediated tumour promotion in rat liver. Carcinogenesis 16:1271-1275

Whitlock JP (1993) Mechanistic aspects of dioxin action. Chem Res Toxicol 6:754-763

Wörner W and Schrenk D (1996) Influence of liver tumor promoters on apoptosis in rat hepatocytes induced by 2-acetylaminofluorene, ultraviolet light, or transforming growth factor β1. Cancer Res 56:1272-1278

Lymphoma Induction by Heterocyclic Amines in Eμ-*pim*-1 Transgenic Mice

Ilona Kryspin Sørensen[1,5], Eva Kristiansen[1], Alicja Mortensen[1], Coen van Kreijl[2], Richard H. Adamson[3,4], and Snorri S. Thorgeirsson[3]

[1] National Food Agency, Institute of Toxicology, Mørkhøj Bygade 19, DK-2860 Søborg Denmark,

[2] National Institute of Public Health and Environmental Protection, Bilthoven, The Netherlands,

[3] National Institutes of Health, National Cancer Institute, Bethesda, Maryland, USA,

[4] *present address:* National Soft Drink Association, Washington D.C.

[5] Corresponding author.

Abstract

The usefulness of transgenic Eμ-*pim*-1 mice bearing in their genome the *pim*-1 oncogene supplemented with an upstream immunoglobulin enhancer and a downstream murine leukaemia virus long terminal repeat, as sensitive test organisms was studied in two short-term carcinogenicity studies. The mice were fed standard diet Altromin 1314 supplemented either with 0.03% 2-amino-1-methyl-6-phenylimidazo[4,5-*b*]pyridine (PhIP) for 7 months or with 0.03% 2-amino-3-methylimidazo[4,5-*f*]quinoline (IQ) for 6 months. PhIP and IQ are heterocyclic amines formed during cooking of meat and fish and are mutagenic to bacteria and cultured mammalian cells. PhIP is a potent mouse lymphomagen, while IQ is a liver, lung and forestomach carcinogen in mice. We found that transgenic Eμ-*pim*-1 mice are highly susceptible to PhIP induced lymphomagenesis but do not respond to IQ treatment. PhIP feeding of Eμ-*pim*-1 mice not only increased the total number of T-cell lymphomas but also decreased the latency time compared to either transgenic or wild-type controls. The effect was most pronounced in the treated female Eμ-*pim*-1 mice, which showed a higher incidence of PhIP induced T-cell lymphomas than transgenic males and a strongly reduced latency period after PhIP treatment compared to non-transgenic mice. Our results suggest that the transgenic Eμ-*pim*-1 mouse may be a useful model for short-term carcinogenicity screening of potential genotoxic carcinogens having the lymphoid system as target tissue. Carcinogens that do not target this tissue, like IQ, however will not be recognised.

Introduction

Transgenic mice carrying specific oncogenes represent a novel approach for investigating neoplasia and are valuable models to assess the tissue specific action of oncogenes. These mice are usually predisposed to a specific type of neoplastic growth. We were interested whether the lymphomagenesis in Eμ-*pim*-1 transgenic mice could be accelerated by treating the mice with relatively low doses of the carcinogenic heterocyclic amines, PhIP (2-amino-1-methyl-6-phenylimidazo[4,5-*b*]pyridine) and IQ (2-amino-3-methylimidazo[4,5-*f*]-quinoline). Eμ-*pim*-1 mice, which bear the *pim*-1 gene supplemented with an upstream immunoglobulin enhancer and a downstream murine leukaemia virus (Mo-MuLV) long terminal repeat, overexpress *pim*-1 mRNA at high levels in both B and T cells (van Lohuizen et al. 1989). Only after a latency period which can last up to 8 months do 10% of the mice develop T cell lymphomas. Thus the spontaneous tumour incidence in the Eμ-*pim*-1 mice is very low. The usefulness of the transgenic Eμ-*pim*-1 mice in carcinogenicity testing has previously been investigated by Breuer et al. (1991), Storer et al. (1995) and Kroese et al. (submitted). They showed an increased lymphoma incidence of these mice in response to the action of certain chemical carcinogens and suggested the possibility that the transgenic Eμ-*pim*-1 mice might prove to be a valuable tool in carcinogenesis research.

In the present study we used the transgenic Eμ-*pim*-1 mice in a short-term carcinogenicity test with the carcinogens PhIP (2-amino-1-methyl-6-phenylimidazo[4,5-*b*]pyridine) and IQ (2-amino-3-methylimidazo[4,5-*f*]-quinoline). PhIP and IQ are formed in the crust of meat and fish during cooking. They have been shown to be potent carcinogens in rats and mice when added to the diet. PhiP induces tumours of colon and mamma in rats (Ito et al. 1991) and is a potent lymphomagen in mice (Esumi et al. 1989). IQ is a liver, lung, forestomach and hepatocarcinogen in mice (Takayama et al. 1984) and a colon, Zymbal gland and clitoral gland carcinogen in rats (Ohgaki et al. 1984). The use of these two carcinogens in a short-term carcinogenicity test in Eμ-*pim*-1 mice would enable us to determine whether the onset and the incidence of lymphoma in Em-*pim*-1 mice would be affected by the carcinogen PhiP which has the lymphatic system as a target versus IQ which does not target the lymphatic system.

Materials and methods

Test compounds and diets: Chemicals and diets were used as described previously (Sørensen et al. 1996). Briefly, synthetic 2-amino-1-methyl-6-phenylimidazo[4,5-*b*]pyridine (PhIP) and 2-amino-3-methylimidazo[4,5-*f*]-

quinoline (IQ), obtained from the National Cancer Institute (USA), were mixed into pelleted diets Altromin 1314 (Altromin GmbH u., Co KG, Lage, Germany) in concentrations of 0%, 0.03% PhIP and 0.03% IQ.

Animals, housing and clinical observations: In the study with PhIP, 59 Eμ-*pim*-1 heterozygous transgenic mice and 60 non-transgenic littermates, and in the study with IQ, 60 Eμ-*pim*-1 heterozygous transgenic mice and 60 non-transgenic littermates, of both sexes, age 9-12 weeks were used. All Eμ-*pim*-1 transgenic mice and the non-transgenic littermates were obtained from the National Institute of Public Health and Environmental Protection, Bilthoven, The Netherlands. All mice were housed 1 male/cage and 1 or 2 females/cage under controlled environmental conditions (Sørensen et al. 1996), and had free access to feed and tap water. Feed and water intake and body weight were recorded once weekly. All mice were observed at least twice a day for any abnormalities in clinical appearance.

Experimental design: The experimental design is seen from Table 1.

Table 1. Experimental design.

	Group	0.03% PhIP	0.03% IQ	Sex	No. of mice	Treatment (weeks)	Duration (weeks)
PhIP							
	I. WT	-	-	M	15	31	31
		-	-	F	15	31	31
	II. WT	+	-	M	15	31	31
		+	-	F	15	31	31
	III. TR	-	-	M	14	31	31
		-	-	F	15	31	31
	IV. TR	+	-	M	15	31	31
		+	-	F	15	31	31
IQ							
	I. WT	-	-	M	15	24	40
		-	-	F	15	24	40
	II. WT	-	+	M	15	24	40
		-	+	F	15	24	40
	III. TR	-	-	M	15	24	40
		-	-	F	15	24	40
	IV. TR	-	+	M	15	24	40
		-	+	F	15	24	40

WT: non-transgenic wild-type littermates TR: transgenic littermates

In the study with PhIP, wild-type mice received diet without PhIP (controls), and with 0.03% PhIP (experimental group) during 31 weeks, which was also the observation period. In the study with IQ the mice received diet without IQ (controls), and with 0.03% IQ (experimental group) during 24 weeks. The total observation period was 40 weeks. In both studies, all mice that became moribund or showed pronounced clinical manifestations were killed and autopsied. On day 218 in the PhIP study and on the day 280 in the IQ study all the remaining mice were killed and autopsied. Selected organs were fixed in 4% neutral buffered formaldehyde, and paraffin-embedded sections, 4-6 µm thick, were stained with haematoxylin and eosin for histopathological examination. In addition, lymphoid tissues were snap frozen for immunohistochemistry (Sørensen et al. 1996).

Results

PhIP study: The PhIP concentration in the diet was chosen to be 0.03% on the basis of a dose range finding study with 0.01%, 0.02%, 0.03%, and 0.04% of PhIP. The total number of mice that developed lymphomas is shown in Table 2.

Table 2. Total number of mice with lymphoma in non-transgenic wild-type and transgenic *pim*-1 littermates on a diet with 0% or 0.03% PhIP.

Group	0.03% PhIP	Sex	No. of mice	No of mice at termination[a]	
				With lymphoma	Lymphoma-free
I. WT	-	M	15	0 (0)[b]	15 (100)
	-	F	15	0 (0)	15 (100)
II. WT	+	M	15	0 (0)	15 (100)
	+	F	15	4 (27)	11 (73)
III. TR	-	M	14	1 (7)	13 (93)
	-	F	15	3 (20)	12 (80)
IV. TR	+	M	15	4 (27)	11 (73)
	+	F	15	12 (80)	3[cd] (20)

WT: non-transgenic wild-type littermates TR: transgenic littermates

[a]: day 218

[b]: values in parentheses are percentages

[c]: $p<0.01$ females from group IV compared to females from group III

[d]: $p<0.01$ females from group IV compared to females from group II

There was no statistically significant difference in the total number of spontaneous lymphomas in Eμ-*pim*-1 males and females (group III) compared with non-transgenic controls (group I). PhIP feeding increased the total number of Eμ-*pim*-1 mice (group IV) with lymphomas compared to the transgenic controls (group III). However, the increase in total number of mice with lymphomas was statistically significant for females only. Furthermore, transgenic Eμ-*pim*-1 females fed PhIP also had a statistically significantly higher incidence of lymphomas when compared to PhIP-fed non-transgenic littermates. Lymphoma onset and incidence are shown in Figs 1 and 2.

Fig 1.

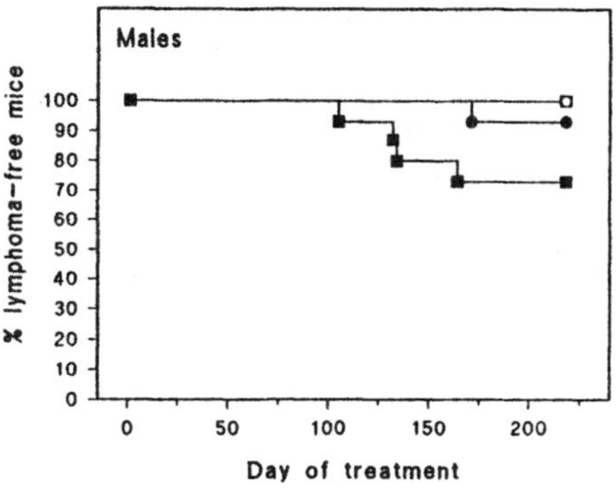

Fig 2.

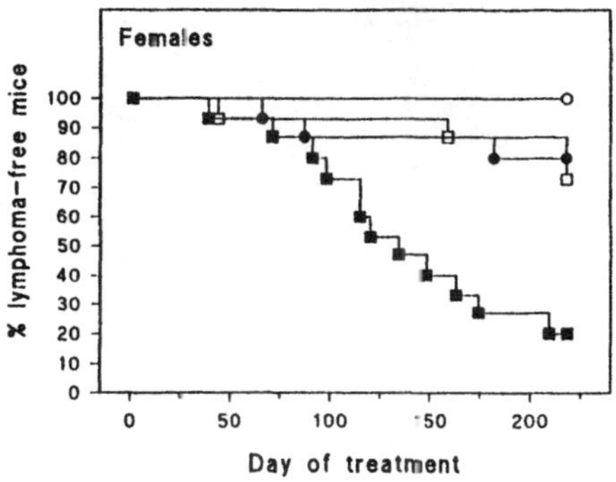

Fig. 1. Lymphoma onset and incidence in male non-transgenic wild-type (group I control: O and group II 0.03% PhIP: □) and transgenic Eμ-*pim*-1 littermates (group III control: ● and group IV 0.03% PhIP: ■).

Fig. 2. Lymphoma onset and incidence in female non-transgenic wild-type (group I control: O and group II 0.03% PhIP: □) and transgenic Eμ-*pim*-1 littermates (group III control: ● and group IV 0.03% PhIP: ■).

For males, the statistically significant difference in survival of lymphoma-free mice was seen only between groups II and IV ($p<0.05$). For females, a statistically significant difference in survival of lymphoma-free mice was seen between groups I and II ($p<0.05$), III and IV ($p<0.01$), and II and IV ($p<0.01$). However, no statistically significant difference in survival was seen between the groups I and III. At autopsy enlargement of thymus, spleen and lymph nodes were seen in affected animals (Fig 3).

Fig 3. Fig 4.

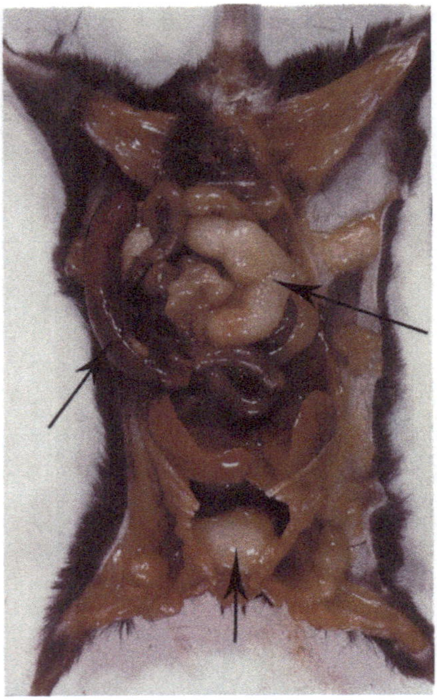

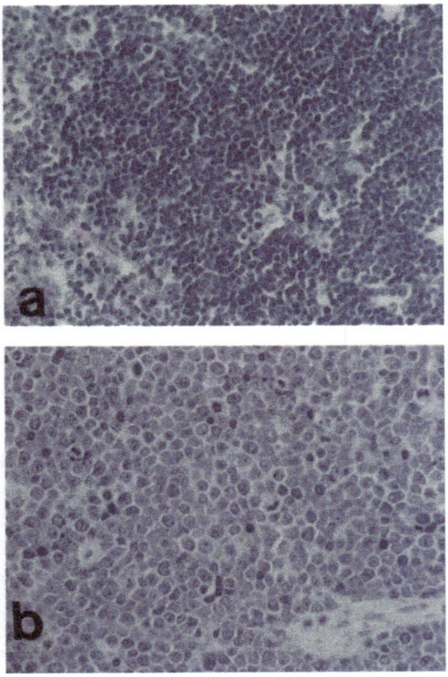

Fig. 3. Lymphoblastic lymphoma, transgenic Eμ-*pim*-1 mouse treated with 0,03% PhIP. Enlarged thymus occupying most of the thoracic cavity and enlarged spleen and mesenteric lymph node are indicated by arrows.

Fig. 4. Photomicrographs of control thymus and thymus with lymphoblastic lymphoma. H&E, X40 a: normal thymus. b: lymphoblastic lymphoma, transgenic Eμ-*pim*-1 mouse. Homogenous population of large size lymphoid cells.

Infiltrations of liver, kidneys and ovaries were sometimes also grossly apparent. Microscopically, the lymphoblastic lymphomas consisted of a homogeneous population of large size lymphoid cells with a moderate amount of cytoplasm (Fig 4). Mitotic figures were frequently seen. Both lymphatic and nonlymphatic organs were infiltrated by lymphoblasts. The type of lymphoblastic cells observed in lymphoid organs was confirmed by immunohistochemistry for T-cell markers.

IQ study: The IQ dose of 0.03% was chosen based on a previous report (Ohgaki et al. 1984). The total number of mice which have developed lymphomas is shown in Table 3.

Table 3. Total number of mice with lymphoma in non-transgenic wild-type and transgenic *pim*-1 littermates on a diet with 0% or 0.03% IQ.

Group	0.03% IQ	Sex	No. of mice	No of mice at termination[a] With lymphoma	Lymphoma-free
I. WT	-	M	15	0 (0)[b]	15 (100)
	-	F	15	0 (0)	15 (100)
II. WT	+	M	15	0 (0)	15 (100)
	+	F	15	1 (7)	14 (93)
III. TR	-	M	15	0 (0)	15 (100)
	-	F	15	2 (13)	13 (87)
IV. TR	+	M	15	1 (7)	14 (93)
	+	F	15	4 (27)	11 (73)

WT: non-transgenic wild-type littermates TR: transgenic littermates
[a]: day 280
[b]: values in parentheses are percentages

Fig 5.

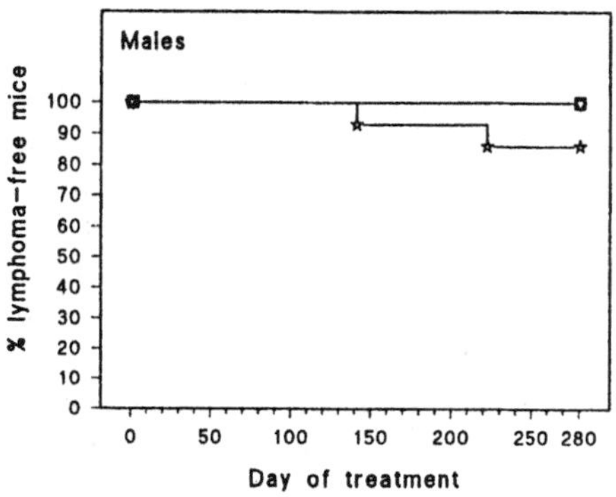

Fig 6.

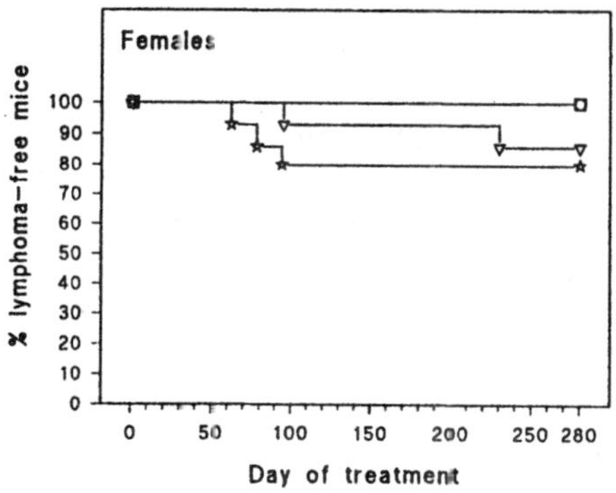

Fig. 5. Onset and incidence of lymphoma in male non-transgenic wild-type (group I control: O and group II 0.03% IQ: □) and transgenic Eμ-*pim*-1 littermates (group III control: ▽ and group IV 0.03% IQ: ★).

Fig. 6. Onset and incidence of lymphoma in female non-transgenic wild-type (group I control: O and group II 0.03% IQ: □) and transgenic Eμ-*pim*-1 littermates (group III control: ▽ and group IV 0.03% IQ: ★).

There was no statistically significant difference in the total number of spontaneous lymphomas between the transgenic Eμ-*pim*-1 and non-transgenic littermates of both sexes. IQ feeding did not statistically significantly increase lymphoma incidence in transgenic Eμ-*pim*-1 or non-transgenic littermates of both sexes. Lymphoma onset and incidence are shown in Figs 5 and 6.

No statistically significant difference in survival of lymphoma-free mice was seen between all experimental groups. In addition to the mice with lymphoblastic lymphomas (for description see PhIP results) the following tumours were observed among IQ dosed transgenic animals: one female had liver tumours (hepatocellular adenoma and carcinoma) and one male had a forestomach papilloma. In the IQ dosed non-transgenic animals two females had liver tumours (hepatocellular adenoma) and one female had lung tumours (bronchiolar/alveolar adenoma).

Discussion

So far, only a few studies have been conducted to assess the usefulness of the transgenic Eμ-*pim*-1 mice in carcinogenicity testing. In the study of Breuer et al. (1989) transgenic Eμ-*pim*-1 mice were exposed to a single low dose of the carcinogen N-ethyl-N-nitrosourea (ENU). Nearly 100% of the transgenics developed lymphoma versus 15% of ENU treated non-transgenics. In the study of Storer et al. (1995) four indirect acting genotoxic carcinogens were evaluated: 2-acetylaminofluorene (AAF), n-nitrosodiethylamine (NDEA), 1,2-dichloroethane (1,2-DCE), and benzene (BEN). These compounds require metabolic activation and with the exception of benzene are not mouse lymphomagens. Small but statistically significant increases in the incidence of malignant lymphoma were seen for three of the four carcinogens tested: in the high dose males treated with AAF, high and low dose females treated with NDEA, and high dose females treated with DCE. The results for benzene, the only mouse lymphomagen tested, did not show statistically significant increase in the incidence of malignant lymphomas in treated transgenic or non-transgenic mice compared to controls within the 40 weeks duration of the study. Their results demonstrated that overexpression of the *pim*-1 oncogene in lymphoid tissue does confer some susceptibility of this tissue to chemical carcinogenesis by indirect acting genotoxic carcinogens not targeting the lymphatic system. However, whereas these potent genotoxic carcinogens produced only slight increases in the incidence of lymphoma and since benzene, a mouse lymphomagen, was negative an obvious implication from these studies is that this system needs further characterization and validation in order to be useful for the screening of carcinogens in the short-term assays.

In the studies reported here we have tested whether the mouse lymphomagen PhIP and the non-lymphomagen IQ can increase the sensitivity towards lymphoma induction in the transgenic Eμ-*pim*-1 mice. We found that the transgenic Eμ-*pim*-1 mice were highly susceptible to PhIP induced lymphomagenesis, but did not respond to the IQ treatment. PhIP feeding of Eμ-*pim*-1 mice not only increased the total number of T-cell lymphomas but also decreased the latency time compared to either transgenic or wild-type controls. These data correlate well with the results of Breuer et al. (1991) who described the incidence of lymphoma in Eμ-*pim*-1 mice as a function of the dose of the carcinogen ENU (N-ethyl-N-nitrosourea). Either 200, 15, 4, 1 or 0.1 mg/kg ENU was injected into transgenic and control mice and the tumour incidence was monitored. These data show that the Eμ-*pim*-1 transgenic mice are about 25-fold more susceptible to ENU-induced lymphomagenesis than control mice. In another study Kroese et al. (submitted) administered the lymphomagen benzo[a]pyrene (B[a]P) by gavage for 4, 13 and 30 weeks at dose regimen of 13 mg/kg, three times weekly to the Eμ-*pim*-1 transgenic mice. This resulted in a dose-related rapid induction of lymphomas especially in male mice. It might be possible that the activity of metabolic enzymes involved are responsible for sex related differences in the response of Eμ-*pim*-1 transgenic mice to the PhIP and to the B(a)P treatment.

Our data indicate that the Eμ-*pim*-1 transgenic mice are a promising model in short-term carcinogenicity testing with genotoxic carcinogens which target the lymphatic system. However the present study with IQ and the study of Storer et al. (1995) show that the Eμ-*pim*-1 transgenic mice may lack sufficient sensitivity to established carcinogens which do not target the lymphatic system.

References

Breuer M, Slebos R, Verbeek S, van Lohuizen M, Wientjens E, Berns A (1989) Very high frequency of lymphoma induction by a chemical carcinogen in *pim*-1transgenic mice. Nature 340:61-63

Breuer M, Wientjens E, Verbeek S, Slebos R, Berns A (1991) Carcinogen-induced lymphomagenesis in *pim*-1 transgenic mice: Dose dependence and involvement of myc and ras. Cancer Res 51:958-963

Esumi H, Ohgaki H, Kohzen E, Takayama S, Sugimura T (1989) Induction of lymphoma in CDF_1 mice by the food mutagen, 2-amino-1-methyl-6-phenyl-imidazo[4,5-*b*]-pyridine. Jpn J Cancer Res 80:1176-1178

Ito N, Hasegawa R, Sano M, Tamano S, Esumi H, Takayama S, Sugimura T (1991) A new colon and mammary carcinogen in cooked food, 2-amino-1-methyl-6-phenyl-imidazo[4,5-*b*]pyridine (PhIP). Carcinogenesis 12:1503-1506

van Lohuizen M, Verbeek S, Krimpenfort P, Domen J, Saris C, Radaszkiewicz T, Berns A (1989) Predisposition to lymphomagenesis in *pim*-1 transgenic mice: cooperation with c-myc and N-myc in murine leukaemia virus-induced tumours. Cell 56:673-682

Ohgaki H, Kusama K, Matsukura N, Morino K, Hasegawa H (1984) Carcinogenicity in mice of a mutagenic compound 2-amino-3-methylimidazo[4,5-*f*]quinoline, from broiled sardine, cooked beef and beef extract. Carcinogenesis 5:921-924

Storer RD, Cartwright ME, Cook WO, Soper KA, Nichols WW (1995) Short-term carcinogenesis bioassay of genotoxic procarcinogens in *PIM*[R] transgenic mice. Carcinogenesis 16:285-293

Sørensen IK, Mortensen A, Kristiansen E, van Kreijl C, Adamson RH, Thorgeirsson SS (1996) Short-term carcinogenicity testing of 2-amino-1-methyl-6-phenylimidazo[4,5-*b*]pyridine (PhIP) and 2-amino-3-methylimidazo[4,5-*f*]quinoline (IQ) in Eμ-*pim*-1 transgenic mice. Carcinogenesis 17:101-107

Algal Toxins and Human Health

(Chair: G. Codd, United Kingdom)

Health Effects Associated With Algal Toxins From Seafood

Tore Aune
Norwegian College of Veterinary Medicine, Department of Farmacology, Microbiology and Food Hygiene, P.O.Box 8146 Dep., N-0033 Oslo, Norway.

Introduction

There are several different types of marine algal toxins which find their way into seafood. In Europe, consumption of mussels infected with toxin producing algae may lead to primarily two types of poisonings, the paralytic shellfish poisoning (PSP) and the diarrhetic shellfish poisoning (DSP). In many countries surveillance programs are in operation in order to reduce the number of intoxications as far as possible. Furthermore, quarantine levels are established to prevent marketing of mussels presenting a health threat to consumers.

Paralytic Shellfish Poisoning

Toxin Producing Organisms: Several species within the genera *Alexandrium*, *Pyrodinium* and *Gymnodinium* may produce PSP toxins. In addition, PSP toxins are produced by certain bacteria, some cyanobacteriae and some red algae. Nevertheless, the dinoflagellate *Alexandrium tamarense* is considered the dominating PSP producer in European waters.

The PSP Toxins: The PSP toxins are divided into three closely related groups; the carbamates, the sulphocarbamoyl compounds and the decarbamoyl compounds, each consisting of six toxins (Oshima et al., 1989). The best known PSP toxin is saxitoxin (STX) which belongs to the carbamates. The name is derived from the butterclam Saxidoma giganteus from which the toxin has been obtained (Kao, 1993). In recent years another three PSP toxins have been detected. In Fig. 1 the general structure of the PSP toxins is given.

The main difference between the three subgroups of the PSP toxins is substituent R4; in the carbamates it is a carbamoyl group (H_2N-CO-), while in the two others substituent R4 is a N-sulphocarbamoyl group (-O_3S-NH-CO-), and a decarbamoyl group (-H), referring to saxitoxin, respectively. In saxitoxin, substituents R1 to R3 are hydrogen atoms.

Concerning toxicity, the carbamates are the most potent, followed by the decarbamoyl compounds, while the sulphocarbamoyls are the least toxic. The sulphocarbamoyls are readily converted to the corresponding carbamates under acidic conditions (Kao, 1993). Usually, between 3 and 12 different PSP toxins are present in mussels.

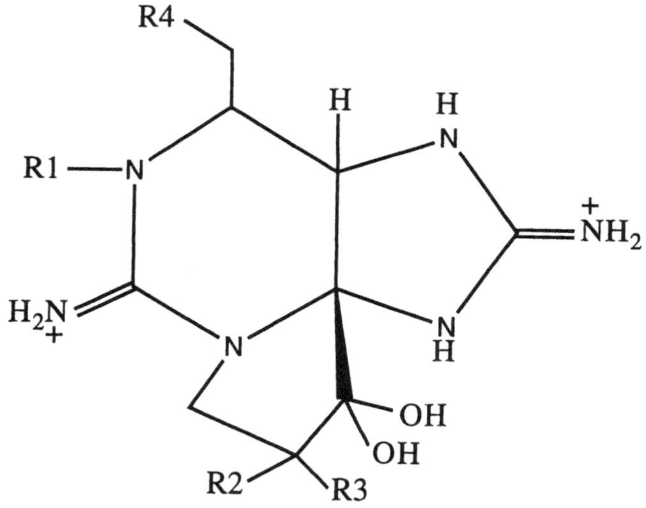

Fig.1: General structure of PSP toxins

Mechanisms of Action: Saxitoxin is the compound most thoroughly studied. Its toxicity is associated with high affinity binding to a certain location on the sodium channel in cell membranes. The voltage-gated sodium channel is a protein of molecular weight 250 kDa, which streches through the membrane of many excitable cells, for instance nerve cells, skeletal muscle fibres and most cardiac muscle fibers. STX blocks the flux of Na^+ ions into the cells, without affecting the current of K^+ ions out of the cells. The result is a suppression of the action potential of the cells, and paralysis to different degree depending upon dose. The binding of STX to the sodium channels is reversible. The PSP toxins have a biologic halflife of about 90 min (Kao, 1993).

Acute Toxicity: In animal experiments, the oral and intraperitoneal acute toxicity of saxitoxin has been studied. In mice, the oral LD_{50} is about 260 µg per kg body weight, compared with 10 µg per kg by the intraperitoneal route (Wiberg and Stephenson 1960). The oral acute toxicity is about the same in several other animal species (McFarren et al., 1960).

In humans, intoxications are reported upon intake of 144 - 1,660 µg STX per person (van Egmond et al., 1993). Fatal intoxications have been associated with intakes of saxitoxin estimated in the range from 500 - 1,000 µg (Tennant et al., 1955) up to 12,400 µg per person (Meyer, 1953). These wide variations can be due to a combination of differences in individual sensibility and to analytical errors, but they still indicate a level of potential danger.

Children are more at risk than adults: During an episode in Guatemala in 1987, lethality among children under six years of age was 50%, while the corresponding number for adults was 5% (Rosales-Loessener et al., 1989). The minimal lethal dose for children was estimated at 25 µg STX per kg body weight (van Egmond et al., 1993).

Clinical Symptoms: The PSP toxins are easily absorbed from the gastrointestinal tract, and symptoms of intoxication appear from about 30 min after ingestion of toxic mussels (Hughes and Merson, 1976). The first symptoms are paresthesia and numbness of the lips and mouth due to local absorption. This feeling spreads to the rest of the face and the neck. Next, those affected experience a prickling feeling in fingertips and toes. In cases of moderate to serious intoxication, paresthesia spreads to arms and legs. Those affected experience a feeling of lightness and floating, and their speach becomes incoherent. The motoric capacity is reduced, and, in serious cases, respiration becomes restricted and the patient may die from paralysis and respiratory arrest.

Until 1970 about 1600 cases of PSP poisoning were known (Prakash et al., 1971), while another 900 cases have become known in recent years (WHO, 1984). On average, lethality is estimated at 2-14% (Kao, 1993).

Treatment: There are no specific antidotes against PSP toxins. The clinical treatment consists of ventilatory support, and, if the patient comes early under treatment, of emptying of the stomach and/or supplementation with activated charcoal. Normally, the patients will fully recover with no late effects.

Limits of Acceptance: Many countries have established a quarantine level of PSP toxins in mussels corresponding to 80 µg STX per 100 g mussel meat (this corresponds to 400 „mouse units“ in the most widely used bioassay based on mouse lethality).

From a toxicological point of view, the difference between this acceptance level and the level potentially inducing toxic effects is relatively small, since intoxications are reported already from 144 µg STX, while lethal effects are recorded at levels above about 450 µg STX per person.

Diarrhetic Shellfish Poisoning

Toxin Producing Organisms: The DSP toxins are produced primarily by algae belonging to the genus *Dinophysis* (*D. acuta*, *D. norvegica*, *D. acuminata*, and *D. fortii*), but also by species within the genus *Prorocentrum* (*P. lima*, *P. minimum*).

The DSP Toxins: The so-called DSP toxin complex consists of three different chemical groups of toxins, namely okadaic acid (Fig.2) with the closely related dinophysistoxins (DTXs), of the pectenotoxins and of yessotoxin.

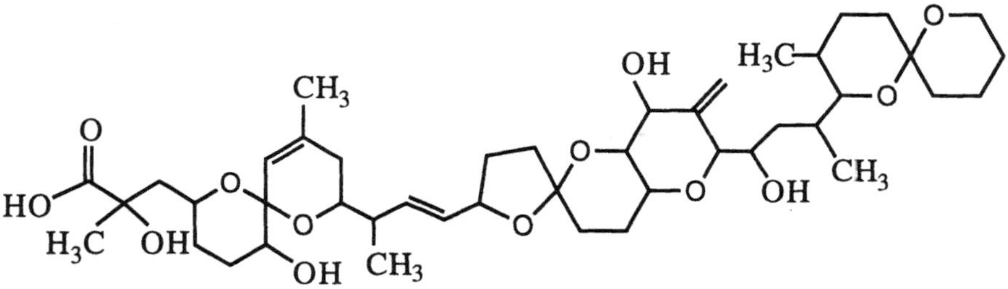

Fig. 2: Structure of the most important DSP toxin in Europe, okadaic acid.

Okadaic acid (OA) was first described by Tachibana et al. (1981) after its isolation from a sponge in the Pacific (*Halichondria okadai*). Later, OA and dinophysistoxins have been associated with DSP symptoms in people consuming mussels infected with toxin-producing algae. The structures of most DSP toxins were elucidated by professor Yasumoto and his coworkers (Yasumoto and Murata, 1993).

The DSP toxins are lipophilic and accumulate in the digestive gland of mussels exposed to toxic algae. In European mussels, okadaic acid is the dominating DSP toxin, but several DTXs, yessotoxin and pectenotoxins were also found in recent years.

Mechanism of Action: OA is a potent inhibitor of two protein phosphatases, protein phosphatase 1 (PP1) and protein phosphatase 2A (PP2A), which are two of the four major protein phosphatases in mammalian cells (Cohen, 1989; Cohen and Cohen, 1989; Haystead, 1989). Similar effects are also described for dinophysistoxin-1 (DTX1).

Phosphorylations and dephosphorylations are very important regulatory processes in eucaryotic cells, and consequently exposure to these DSP toxins might influence a whole range of cellular processes including metabolism, membrane transport, secretion, contractility and cell division (Cohen, 1989; Yamashita et al., 1990). OA inhibits PP2A about 200 times more strongly than PP1.

Induction of diarrhea by OA might be a result of the phosphorylation of proteins controlling the sodium secretion by intestinal cells (Cohen et al., 1990), or of enhanced phosphorylation of cytoskeletal elements which would influence the permeability of cell membranes (Dho et al., 1990).

In 1988 it was reported that OA and DTX1 are powerful cancer promoters in experimental systems (Suganuma et al., 1988). The mechanism of action is associated with their binding to PP1 and PP2A in cell membranes, which leads to increased phosphorylation of important cellular proteins, hyperphosphorylation of intermediate filaments and to changes in gene expression in DNA through hyperphosphorylation of suppressor genes. According to Fujiki and Suganuma (1993) this so-called okadaic acid pathway is a general mechanism of tumor promotion in various organs. Interestingly, they have found 30 compounds from four different chemical types which operate through binding to PP1 and PP2A, exemplified by okadaic acid, calyculin, microcystin and tautomycin. They also proposed a conformational change of the okadaic acid molecule, through hydrogen bond formation (between C-1 and C-24) which gives it a structure closely resembling that of microcystin. This might explain the binding of the different compounds to the same cellular receptors.

In an experiment with rats, the effects of OA on formation of neoplastic changes in the glandular stomach were studied (Suganuma et al., 1992). A combination of the cancer inducer N-methyl-N-nitro-N-nitrosoguanidine and OA at very high doses of OA (10 µg per rat per day for 46 weeks and 20 µg/day for another 16 weeks) increased the number of animals with neoplastic changes significantly, indicating a promoting activity of OA.

The genotoxic potential of okadaic acid has also been studied. OA is not mutagenic in the Ames mutagenicity assay, but was found to be mutagenic towards Chinese hamster lung cells in an assay where cytotoxicity also was recorded (Aonuma et al. 1991). Quite recently, professor Pfohl-Leszkowicz and coworkers reported formation of DNA adducts in human cells in culture upon incubation with OA (personal communication). More research is needed before the human health aspects from these results can be evaluated.

The pectenotoxins and yessotoxin are extracted from mussels with the same initial procedure as okadaic acid and the DTXs. Consequently they have been included in the DSP toxin complex. However, this is misleading from a toxicological point of view, since they are not, or only very weakly, diarrheagenic, in contrast to OA and the DTXs.

The pectenotoxins and yessotoxin are acutely toxic towards mice upon intraperitoneal injection. Pectenotoxins induce liver damage (Terao et al., 1986) after injection of 500 µg per kg, while yessotoxin at concentrations of 300 µg per kg and above induced intracytoplasmic edema in the cardiac muscle (Terao et al., 1990). Oral administration of yessotoxin at 500 µg per kg body weight to mice induced no toxic effects (Terao et al., 1990).

Clinical Symptoms: The dominating symptoms from ingestion of OA and dinophysistoxins are diarrhea, nausea, vomiting, and abdominal pain. The symptoms appear between 30 min and few hours after consumption of toxic mussels. Those affected feel very ill, but they recover fully after few days. Minimum dose of OA and DTX1 necessary to induce DSP symptoms are 40 μg and 36 μg, respectively (Hamano et al., 1985).

Treatment: Usually, people affected do not need medical attention. There is no known antidote against the diarrhea-inducing DSP toxins.

Acceptance Level: The maximum acceptance concentration of diarrheagenic DSP toxins, given as OA, varies slightly between countries. The main reason for this is the lack of precise analytical routine methods for their quantification. Due to lack of chemical standards for several DSP toxins, most countries use a mouse bioassay, where the amount of DSP toxins is determined from the death times. By means of titration of toxicity with OA, one mouse unit corresponds with 4-5 μg OA. Most countries with established regulations for DSP toxins apply 4-5 mouse units or about 20-25 μg equivalents of OA as acceptance limit. As soon as toxin standards will become available commercially, the mouse bioassay will be substituted with analytical methods. Before that, one has to agree upon which toxins should be included in the DSP complex, and at what levels.

Several authors discuss the possible human health risk associated with consumption of mussels containing toxins with cancer promoting properties. In this context it is important to consider the dose-respones relationships; in the promotor experiment with rats cited above (Suganuma et al., 1992), the animals received very high doses of OA (first 10 and then 20 μg per rat per day). This level is equivalent with daily doses of 1,500 and 3,000 μg for an adult human, a figure which has to be compared with the diarrhea-inducing level of OA in humans of about 40 μg. Furthermore, effects like tumor promotion display no-effect levels in contrast to genotoxic tumor inducers. Consequently, it is very important to find at what levels OA and DTX1 might represent a threat of tumor promoting activity towards humans. The rat experiment referred to above, indicates that relatively high dosages should be needed, possibly at levels far beyond those ingested by humans.

By the choice of the extraction procedure, one can include or exclude the pectenotoxins and yessotoxin from the DSP toxin complex, since they are extracted in the initial step with acetone, but not in the final step with ether (which is the most widespread method today). It is not reasonable to set quarantine levels for pectenotoxins and yessotoxin in seafood at the same level as that of the diarrheagenic DSP toxins, based on acute toxicity upon intraperitoneal injections in mice. The concentrations of mussel extracts used in the mouse bioassay (by the intraperitoneal route) represent the consumption of 75-100 kg mussel meat by an adult human. Even though these very high concentrations are of relevance for preventing humans from developing diarrhea, they are not relevant for estimating risks concerning organ toxicity from yessotoxin and the pectenotoxins.

One should also keep in mind the lower toxicity by the oral route, compared with intraperitoneal injections; in order to perform risk assessments, one needs oral toxicity data since in most cases oral toxicity is 1-2 orders of magnitude lower than intraperitoneal toxicity.

Other Marine Algal Toxins

Amnesic Shellfish Poisoning: In Canada, a new toxin was detected in mussels after an intoxication episode in 1987. About one hundred persons were taken ill after having eaten blue mussels from Prince Edward Island. The symptoms appearing within 24 hours consisted of nausea, vomiting, headache, abdominal cramps, and one or several neurological signs such as confusion, memory loss, and disorientation. The most serious cases ended in coma and death (Todd, 1993). The toxic agent was identified as an odd amino acid named domoic acid (DA). DA acts as an agonist to the glutamate receptor, and binds with high affinity (Takemoto, 1978). Ingestion of DA might lead to memory loss, thus the name of the intoxication. The causative organisms belong to the genus *Nitzschia* and closely related diatoms. Several of these algae appear regularly in European waters, but toxic levels of DA have not been detected in European mussels with one possible exception for low levels in Danish mussels on one occasion.

Unknown Toxins with Neurological Effects: From time to time, mice receiving DSP extracts display abnormal symptoms with cramps. In such cases, survival times often are much shorter than with the true diarrheagenic DSP toxins. The nature of the toxin(s) and the causative producers are under study in several countries.

Ichthyotoxins: Several algal genera produce compounds which are toxic towards fish and other marine organisms breathing via gills. The best known examples are *Prymnesium*, *Gyrodinium* and in recent years *Chrysochromulina*. When these algae appear in very high numbers (in the millions per litre sea water), marine organisms exposed are killed. The mechanism of action seems to be disruption of cell membranes in the gills. There are no indications that these ichthyotoxins represent a threat to humans. Furthermore, fish which is killed by environmental factors are not acceptable as food for humans. This general rule of hygiene prevents humans from being exposed to high levels of ichthyotoxins.

References

Aonuma S, Ushijima T, Nakayasu M, Shima H, Sugimura T, Nagao M (1991) Mutation induction by okadaic acid, a protein phosphatase inhibitor in CHL cells, but not in S. typhimurium. Mutat Res 250: 375-381

Cohen P (1989) The structure and regulation of protein phosphatases. Annu Rev Biochem 58: 453-508
Cohen P, Cohen PTW (1989) Protein phosphatases come of age. J Biol Chem 264: 21453-21438
Cohen P, Holmes CFB, Tsukitani Y (1990) Okadaic acid: a new probe for the study of cellular regulation. Trends Biochem Sci, March: 98-102
Dho S, Stewart K, Lu D, Grinstein S, Buchwald M, Auerbach W, Foskett JK (1990) Phosphatase inhibition increases tight junctional permeability in cultured human intestinal epithelial cells. J Cell Biol 111: 410A
Fujiki H, Suganuma M (1993) Tumor promotion by inhibitors of protein phosphatases 1 and 2A: The okadaic acid class of compounds. Adv Cancer Res 61: 143-194
Hamano Y, Kinoshita Y, Yasumoto T (1985) Suckling mice assay for diarrhetic shellfish toxins. In: Anderson DM, White AW, Baden DG (eds) Toxic Dinoflagellates. Elsevier, New York, pp 383-388
Haystead TAJ, Sim ATR, Carling D, Honnor RC, Tsukitani Y, Cohen P, Hardie DG (1989) Effects of the tumor promoter okadaic acid on intracellular protein phosphorylation and metabolism. Nature 337: 78-81
Hughes JM, Merson MH (1976) Fish and shellfish poisoning. New Engl J Med 295: 1117-1120
Kao CY (1993) Paralytic shellfish poisoning. In: Falconer IR (ed) Algal Toxins in Seafood and Drinking Water. Academic Press, London, pp 75-86
McFarren EF, Schafer ML, Campbell JE, Lewis KH, Jensen ET, Schantz EJ (1960) Public health significance of paralytic shellfish poison. Adv Food Sci 10: 135-179
Meyer KF (1953) Food poisoning. New England J Med 248: 843-852
Oshima Y, Sugino T, Yasumoto T (1989) Latest advances in HPLC analysis of paralytic shellfish toxins. In: Natori S, Hashimoto K, Ueno Y (eds) Mycotoxins and Phycotoxins, Elsevier, Amsterdam, pp 319-326
Prakash A, Medcof JC, Tennant AD (1971) Paralytic shellfish poisoning in Eastern Canada. Bull Fish Res Board Can 117: 1-88
Rosales-Loessener F, De Porras E, Dix MW (1989) Toxic shellfish poisoning in Guatemala. In: Okaichi T, Anderson DM, Nemoto T (eds) Red Tides: Biology, Environmental Science and Toxicology. Elsevier, New York, pp. 113-116
Suganuma M, Fujiki H, Suguri H, Yoshizawa S, Hirota M, Nakayasu M, Ojika M, Wakamatsu K, Yamada K, Sugimura T (1988) Okadaic acid: an additional non-phorbol-12-tetradecanoate-13-acetate tumor promoter. Proc Natl Acad Sci U.S.A. 85: 1768-1771
Suganuma M, Tatematsu M, Yatsunami J, Yoshizawa S, Okabe S, Uemura D, Fujiki H (1992) An alternative theory of tissue specificity by tumor promotion of okadaic acid in glandular stomach of SD rats. Carcinogenesis 13: 1841-1845
Tachibana K, Scheuer PJ, Tsukitani Y, Kikuchi H, van Engen D, Clardy J, Gopichand Y, Schmitz FJ (1981) Okadaic acid, a cytotoxic polyether from two marine sponges of the genus Halichondria. J Am Chem Soc 103: 2469-2471
Takemoto T (1978) Isolation and structural identification of naturally occurring exitatory amino acids. In: McGeer EG, Olney JW, McGeer PL (eds) Kainic acid as a tool in neurobiology. Raven Press, New York, pp 1-15
Tennant AD, Naubert J, Corbeil HE (1955) An out-break of paralytic shellfish poisoning. J Can Med Assoc 72: 436-439
Terao K, Ito E, Yanagi T, Yasumoto T (1986) Histopathological studies on experimental marine toxin poisoning. I. Ultrastructural changes in the small intestine and liver of suckling mice induced by dinophysistoxin-1 and pectenotoxin-1. Toxicon 24: 1141-1151

Terao K, Ito E, Oarada M, Murata M, Yasumoto T (1990) Histopathological studies on experimental marine toxin poisoning - 5. The effects in mice of yessotoxin isolated from Patinopecten yessoensis and of a desulfated derivative. Toxicon 28: 1095-1104

Todd ECD (1993) Domoic acid and amnesic shellfish poisoning - A review. J Food Protec 56: 69-83

Wiberg GS, Stephenson NR (1960) Toxicologic studies on paralytic shellfish poison. Toxicol Appl Pharmacol 2: 607-615

WHO (1984) Aquatic (Marine and Freshwater) Biotoxins. Environmental Health Criteria 37. International Programme on Chemical Safety, World Health Organization, Geneva.

Yamashita K, Yasuda H, Pines J, Yasumoto T, Nishitani H, Ohtsubo M, Hunter T, Sugimura T, Nishimoto T (1990) Okadaic acid, a potent inhibitor of type 1 and type 2A protein phosphatases, activates cdc 2/H1 kinase and transiently induces a premature mitosis-like state in BHK21 cells. EMBO J 9: 4331-4338

Yasumoto T, Murata M (1993) Marine toxins. Chem Rev 93: 1897-1909

Cyanobacterial Toxins: Occurrence, Modes of Action, Health Effects and Exposure Routes

G.A. Codd, C.J. Ward and S.G. Bell
Department of Biological Sciences, University of Dundee, Dundee DD1 4HN, UK

Summary

Cyanobacterial toxins are produced by terrestrial- fresh-, brackish- and sea-water cyanobacteria of cosmopolitan occurrence. These toxins present acute and chronic hazards to human and animal health and are responsible for isolated, sporadic animal fatalities (mammals, fish, birds) each year. Human health problems are associated with the ingestion of, and contact with cyanobacterial blooms and their toxins. Modes of action of cyanobacterial neurotoxins, hepatotoxins and skin irritants are considered. Recent indications of the accumulation of cyanobacterial toxins in fish, their effect on crop plants and their association with the deaths of human dialysis patients are discussed. These findings and events indicate an incomplete understanding of the exposure routes of these natural toxins and the need for greater awareness of their occurrence and properties among users of waterbodies which are prone to cyanobacterial bloom development.

Introduction

An awareness of hazards presented to human and animal health by blooms and scums of microalgae, including cyanobacteria (blue-green algae) has apparently existed in the folklore of hunting and nomadic indigenous people in Australia and N. America (Codd et al, 1994a). Over recent years, sporadic isolated, annual examples of animal deaths and human health problems associated with the ingestion of, or contact with cyanobacterial blooms and scums in waterbodies have exemplified the hazards presented by cyanobacterial toxins via two activities: drinking affected water and direct contact whether by washing or recreational activities (Carmichael and Falconer, 1993; Codd, 1994; Bell and Codd, 1994).

The formulation and application of policies for the protection of animal and human health from the actions of cyanobacterial toxins needs an understanding of several factors including: the occurrence and levels of the toxins; their acute and chronic toxicity to humans and animals; fate and behaviour of the toxins in aquatic environments, and exposure levels and routes encountered and present in man and animals. These requirements for hazard and risk assessment are gradually being met for some cyanobacterial toxins (Carmichael and Falconer, 1993; Ressom et al., 1994; Steffensen and Nicholson, 1994). Analytical capabilities and the availability of cyanobacterial toxin standards are obvious essential needs in understanding toxin occurrence, abundance, fate and behaviour. The classical mouse bioassay is being complemented by a range of invertebrate whole animal- in vitro cell-, immuno-, enzyme-, and physicochemical detection and analytical methods. Advances in these procedures are presented and assessed elsewhere (Codd et al., 1994b; Bell and Codd, 1996).

Here we consider advances in perception of the types and occurrence of cyanobacterial toxins, their toxicities, modes of action and effects on health. Evidence for the accumulation of cyanobacterial toxins in mussels and fish, and for effects on an agricultural crop plant at environmentally-relevant levels, indicate that human exposure routes may occur via foodstuffs, in addition to directly via water. We also present analytical data on materials from a recent human tragedy involving deaths at a kidney dialysis unit, in which exposure to cyanobacterial toxins is indicated (Dunn, 1996).

Occurrence of Toxic Cyanobacterial Blooms, Scums, Mats and Toxins

Cyanobacteria are common inhabitants of fresh-, brackish- and marine waters and of rock and soil surface layers (Fogg et al., 1973). Given sufficient mineral nutrients, and a neutral or alkaline pH, mass developments in aqueous suspensions (blooms), surface scums and mats of cyanobacteria develop annually in the summer and autumn in lakes, reservoirs, estuaries and slow-flowing rivers in temperate latitudes. Seasonal mats of cyanobacteria can also develop in polar environments. In sub-tropical and tropical latitudes, cyanobacterial blooms and scums may be longer-lived or year-round phenomena.

Animal-, bird- and fish-poisonings and human illnesses ascribed to cyanobacterial blooms and scums have been recorded in the scientific and popular press for over a century (Codd et al., 1994a). Many of the earlier reports could not, perforce, include cyanobacterial toxin identification. Nevertheless, based on more recent investigations and analyses, the cosmopolitan occurrence of toxic cyanobacterial blooms as scums can be assumed, with examples known from at least 42 countries and all continents (Carmichael, 1982; Codd, 1995). In a relatively small number of countries, more widespread assessments of cyanobacterial bloom and/or scum toxicity have been performed. A high incidence of toxicity by mouse bioassay was found (Table 1).

Table 1. National surveys of cyanobacterial blooms and scums for toxicity by intraperitoneal mouse bioassay

Country	Years	Number of sites sampled	Number of samples tested	% samples positive	Reference
UK	1981-89	24	23	75	See Codd et al. (1989). NRA (1990)
Norway, Sweden and Finland		51	30	59	
Sweden		27	15	56	
Finland		103	45	44	
UK		91	62	68	
Norway	1989-91	36	64	88	Skulberg et al. (1994)
Australia	1990-91	130+	231	42	Baker and Humpage (1994)
Belgium	pre-1993	NK	17	59	van Hoof et al. (1994)
UK	1989-93	48	51	51	Codd and Bell (1996)

The high probability of an individual sample being acutely toxic to mice supports the adoption of the following precautionary principle: a cyanobacterial bloom, scum or mat should be assumed to be toxic, unless tested at appropriate dose levels and found to be otherwise (Carmichael and Falconer, 1993; Bell and Codd, 1994). Since detection limits ranging from 10^{-2} to 10^{-5} of those achievable by mouse bioassay can be obtained by some of the chromatographic-, ELISA- and enzyme-based methods (Codd et al., 1994b) it is clear that mouse bioassays can underestimate the occurrence of cyanobacterial toxins. It follows that a cyanobacterial or other environmental sample should be defined as toxic or non-toxic in terms of the assay procedure and, in particular, of the organisms perceived to be at risk. Details of the toxigenic cyanobacteria in blooms and of axenic monocyanobacterial laboratory cultures, e.g. of *Microcystis, Anabaena, Aphanizomenon, Oscillatoria, Nodularia* and *Trichodesmium* are given elsewhere (Skulberg et al., 1994; Codd, 1995; Baker and Humpage, 1994).

Molecular and Toxicological Properties of Cyanobacterial Toxins

Over 60 cyanobacterial toxins have been described from fresh- and brackish water environments with a smaller number from marine sources (Carmichael and Falconer, 1993; Codd et al., 1994b; Rinehart et al., 1994. They include neurotoxins, hepatotoxins and endotoxins (Table 2).

Anatoxin-a and homoanatoxin-a are neuromuscular blocking agents and act as postsynaptic cholinergic, nicotine agonists. They cause staggering, gasping and muscle fasiculations in animals (with opisthotonos in birds) and death by respiratory arrest. The PSP toxins, also frequently produced by marine dinoflagellate "red tides", are sodium channel-blockers which inhibit nerve conduction. They bind reversibly to a high molecular weight protein (saxiphilin) spanning sodium channels. Anatoxin-a(s), a naturally-occurring organophosphorus compound, causes hypersalivation in animals with death occurring due to the inhibition of acetylcholinesterase (Carmichael, 1992).

Nodularins and microcystins cause hepatotoxicosis in animals with initial external signs including weakness, pallor, cold extremities, heavy breathing, recumbency, vomiting and diarrhoea. Massive disruption of liver structure and function occurs with haemorrage into the liver and death by respiratory arrest. At the molecular level these toxins bind irreversibly (covalently) to protein phosphatases (PPases) (Carmichael, 1992; Codd, 1994; Bell and Codd, 1994). The implications of the inhibition of PPases are far-reaching in terms of the significance of nodularins and microcystins to human and animal health and their present and potential impact on other organisms in aquatic and terrestrial environments (see later). The effects of cylindrospermopsin at the whole animal level are widespread with necrotic tissue injury to the liver, kidneys, adrenals, lungs and intestine. Cylindrospermopsin is a protein synthesis inhibitor. Cyanobacterial LPS endotoxins are associated with gastroenteritis and skin irritations. Little toxicological information is available on purified cyanobacterial LPS, the principal study indicating *Microcystis* LPS to be only 10% toxic to rodents as *Salmonella* LPS (Raziuddin et al., 1983). Since cyanobacteria are Gram negative prokaryotes and LPS is a characteristic component of Gram negative cell outer layers, it is likely that cyanobacterial LPS levels in blooms and scums can be high, resulting in high exposure levels to, albeit, relatively low toxicity endotoxins.

Table 2. Properties and sources of cyanobacterial toxins

Toxin	Sources	Molecular structure	Number of variants
NEUROTOXINS			
Anatoxin-a	*Anabaena, Microcystis, Oscillatoria, Phormidium, Aphanizomenon*	Secondary amine alkaloid	1
Homoantotoxin-a	*Phormidium*	As above	1
PSP toxins (saxitoxin, neo-saxitoxin, GTX and CTX toxins)	*Aphanizomenon, Anabaena*	Alkaloids	at least 8
Anatoxin-a(s)	*Anabaena*	Guanidinium methyl phosphate ester	1
HEPATOTOXINS			
Nodularin	*Nodularia*	Cyclic pentapeptide	6
Microcystin	*Microcystis, Anabaena, Nostoc, Oscillatoria*	Cyclic heptapeptide	> 50
Cylindrospermopsin	*Cylindrospermopsis, Umezakia*	Cyclic guanidine alkaloid	1
ENDOTOXINS			
LPS	many genera	Lipopolyaccharide	> 3

Summarised from Codd et al, 1989; Carmichael and Falconer, 1993; Rinehart et al., 1994; Codd, 1995).

Accumulation and Actions of Cyanobacterial Toxins in Animals and Plants

The biological significance of cyanobacterial toxins is likely to extend beyond the deterrent effects of some of the toxins on aquatic phytoplankton grazers (Codd, 1995). As summarised in Table 3, a history of animal poisonings is available, involving wild and domestic mammals, birds, fish and invertebrates. The most common activities leading to mammal and bird deaths have been drinking from the margins and surfaces of waterbodies when scums of planktonic cyanobacteria have accumulated. In these circumstances, an acute oral dose of the toxins can be presented in less than the daily drinking water needs of the animal. Dogs have also died due to nodularin and anatoxin-a, after their habits of grooming adhered scum from their coats following swimming, and of scavenging from shoreline deposits of neurotoxic benthic cyanobacterial mats (Codd et al., 1992). Evidence also exists for the deaths of fish-eating birds, and mussel-eating muskrats, due to microcystins which have been accumulated via food chains from a hepatotoxic *Oscillatoria* bloom (Eriksson et al., 1986). We have exposed the water flea *Daphnia magna* to purified radiolabelled microcystin-LR in water at subacute levels and found a linear accumulation of radiolabel in *D. magna* tissue

Table 3. Animal deaths attributed to cyanobacteria and/or their toxins (ca. 1851 to 1996)

	Hepatotoxic blooms and hepatotoxicoses	Neurotoxic blooms and neurotoxicoses
1. Domestic animals	Cattle, sheep, horses, pigs, dogs	Cattle, sheep, horses, pigs, dogs, cats
	Turkeys, geese, ducks, chickens	Turkeys, ducks, chickens
2. Wild animals	Rhinoceros, giraffe, zebra, deer, monkeys, bats, frogs	Muskrats, squirrels, bats
	Bird species	Swans, coots, mallards, gulls, terns, pheasants
	Bees	
	Snakes, salamanders	
	Brown trout, rainbow trout	Perch

Summarised from Carmichael (1982, 1992), Codd et al. (1989, 1994a), Yoo et al (1995).

Subsequent consumption of the radiolabelled *D. magna* by zooplanktivorous fish (roach, *Rutilus rutilus*) resulted in accumulation of the radiolabel in fish tissues, further suggesting the accumulation of cyanobacterial toxins and/or their transformation products via aquatic food chains (Codd and Bell, 1996). Indeed, a situation analogous to the accumulation of algal toxins from marine dinoflagellates and diatoms by marine mussels, occurs between cyanobacterial toxins and mussels: blue mussels (*Mytilus edulis*) collected from estuarine waters containing *Nodularia spumigena* were hepatotoxic and contained nodularin

(Falconer et al., 1992). PSP toxins have been extracted from freshwater mussels (*Alathyria condola*) and from *Anabaena circinalis* cells which were accumulated by the mussels in filter-feeding (Negri and Jones, 1995).

Estimation of the toxicity of cyanobacterial toxins to aquatic organisms is of particular importance since they, in contrast to laboratory animals, are exposed to the toxins in their own habitats. Such estimations are needed in order to formulate appropriate policies for aquatic wildlife (including biodiversity) conservation, aquaculture and recreational fishing in waters affected by cyanobacterial blooms. Little information exists on the toxicity of cyanobacterial toxins to aquatic wildlife and farmed organisms, with most data having been obtained with laboratory mammals. We have determined the toxicity of purified microcystin-LR and anatoxin-a by intraperitoneal bioassay using laboratory mice and rainbow trout (Table 4).

Table 4. Comparative toxicities of purified microcystin-LR and anatoxin-a by intraperitoneal administration to mice and rainbow trout (*Salmo gairdneri* Richardson)

		Acute toxicity (µg per kg body weight)	
		Microcystin-LR	Anatoxin-a
Mouse:	LD_{50}	50	ND
	MLD	100	186
Rainbow trout:	LD_{50}	840	1400
	MLD	10	12

ND, not determined.

These findings illustrate the need for inclusion of at least a 10-fold uncertainty factor for interspecies variability in cyanobacterial toxin risk assessments for aquatic biota. Table 4 also clearly shows that the neurotoxin was more toxic to the fish than the hepatotoxin, whereas the reverse applies in the mouse, by the intraperitoneal route in acute tests.

The effects of cyanobacterial toxins on plants have been little studied. The only obvious circumstances where contact between cyanobacterial toxins and plants are likely to occur is in waterbodies which support cyanobacterial blooms and macrophytes. However, some agricultural and horticultural practices may also result in exposure of plants to the toxins: hydroponic growth of plants and crop spray irrigation may expose plants to the toxins (NRA, 1990). The latter possibility is exemplified by recent findings from this laboratory (Table 5).

These whole plant effects are likely to be due to the inhibition of plant PPases by microcystin-LR in vivo. The inhibition of these regulatory enzymes of plants was previously found in vitro (MacKintosh et al., 1990). The whole plant findings imply that the toxin can be taken up following topical exposure of the leaves to the toxin in water. These circumstances would occur if crop plants were sprayed with water containing microcystin-producing cyanobacterial blooms and if the toxins were subsequently released into the water e.g. following cyanobacterial cell lysis (NRA, 1990; Abe et al., 1996).

Table 5. Summary of actions of purified microcystin-LR on the primary leaves of French bean (*Phaesolus vulgaris*) via topical application to whole plants

Treatment	Response
Single high level exposure	Leaf necrosis Irreversible in vivo inhibition of net photosynthetic rate Reduction of CO_2 -saturated and light-saturated rates of net photosynthesis in vivo
Multiple low level exposures	Inhibition of net photosynthetic rate in vivo; reversible, becoming irreversible

Abe et al. (1986). Primary leaves exposed topically in vivo to microcystin-LR in aqueous solution. Low level exposures carried 20 µg microcystin-LR per litre.

Human Health Effects and Exposure Routes of Cyanobacterial Toxins

A series of recent reviews discusses human health incidents attributed to cyanobacterial toxins (Carmichael and Falconer, 1993; Falconer, 1994; Codd, 1994; Bell and Codd, 1994; Ressom et al., 1994; Yoo et al., 1995). These include gastroenteritis outbreaks, pulmonary consolidation, skin irritations, allergenic responses and liver damage following contact with and ingestion of water containing cyanobacterial blooms or scums during water-based recreation and drinking. As pointed out by Hunter (1994), incomplete post-event investigations into human health incidents, suspected to be due to these toxins, do not satisfy all of the epidemiological criteria of Bradford-Hill. However, the balance of evidence from the investigations of human illnesses, taken with the effects of the toxins in animals, and at the molecular level, shows the hazardous nature of cyanobacterial toxins to human health. For example, the implications of the inhibition of PPases by microcystins and nodularins include tumour promotion, which has been shown in laboratory rodents (Carmichael and Falconer, 1993; Falconer, 1994). Epidemiological investigations in China in which human primary liver cancer is thought to be associated with drinking water containing microcystins (Yu, 1989) are in progress.

This year, a human tragedy occurred at a haemodialysis clinic in the town of Caruaru in north-east Brazil (Dunn, 1996). More than 50 people have died, with liver damage being consistent throughout. Microcystins are suspected to have been present in the water used for haemodialysis. We have received filter materials used for water treatment prior to closure of the clinic and are examining these for microcystins. Extraction from the granular activated carbon (GAC) and analysis for microcystins yields the following: a peak in high-performance liquid chromatography (HPLC) which coelutes with authentic microcystin-LR and has a similar UV-absorption spectrum. Quantification of the putative microcystin-LR by HPLC yields a higher estimate of toxin per g wet weight of GAC than analysis of filter eluates by immunoassay and PPase inhibition assay (C.J. Ward et al., in preparation). Mass spectrometry to resolve these discrepancies and analysis of patient serum samples is in progress. Aside from the eventual outcome of these investigations, there is earlier evidence to include haemodialysis water as a cyanobacterial toxin exposure route (Hindman et al., 1975)

Table 6. Some possible cyanobacterial toxin exposure routes in human water and consumers

1. Skin contact	: toxic scum or mat material : raw water containing toxic blooms or free toxins : treated water containing toxic blooms or free toxins
2. Drinking water	: accidental ingestion of toxic scum : raw water containing toxic blooms or free toxins : treated water containing toxic blooms or free toxins
3. Inhalation	: toxins during water-sports, showering or work practices
4. Food consumption	: shellfish or finfish if containing toxins : plant products if containing toxins
5. Haemodialysis	: using water containing free toxins

Investigations into human exposure routes for cyanobacterial toxins have centred in recent exercises into hazard identification and risk assessment on drinking water and recreational contact. Table 6 lists these and other perceived exposure routes. The toxicity of microcystin-LR and anatoxin-a to mice via intranasal aerosol exposure is as high as via intraperitoneal administration (Fitzgeorge et al., 1994), justifying the inclusion of inhalation as a likely exposure route in several activities (Table 6). The addition of food consumption is prudent in the case of cyanobacterial accumulation in shellfish and of the possibility of toxin accumulation in finfish and plants (section 4).

Finally, the inclusion of haemodialysis represents perhaps one of the most direct routes of exposure of these low molecular weight toxins. The need to increase awareness of the health hazards presented by cyanobacterial toxins among water-users and environmental agencies, conservation and wildlife organizations, water industry- veterinary- and public health professionals, is being acknowledged and addressed in some countries (e.g. NRA, 1990; Steffensen and Nicholson, 1994; Codd, 1994; Yoo et al., 1995). This need is, however, closely dependent on primary research and a fuller understanding of cyanobacterial toxin exposure routes is needed to formulate effective preventative and control measures.

References

Abe T, Lawson T, Weyers JDB, Codd GA (1996) Microcystin-LR inhibits photosynthesis of *Phaseolus vulgaris* primary leaves: implication for current spray irrigation practice. New Phytologist 133: 651-658

Baker PD, Humpage AR (1994) Toxicity associated with commonly occurring cyanobacteria in surface waters of the Murray-Darling Basin, Australia. Australian Journal of Marine and Freshwater Research 45: 773-786

Bell SG, Codd GA (1994) Cyanobacterial toxins and human health. Reviews in Medical Microbiology 5: 256-264

Bell SG, Codd GA (1996) Detection, analysis and risk assessment of cyanobacterial toxins. In: Hester RE, Harrison RM (eds) Agricultural Chemicals and the Environment. The Royal Society of Chemistry, Cambridge, pp 109-122

Carmichael WW (1982) Chemical and toxicological studies of the toxic freshwater cyanobacteria *Microcystis aeruginosa, Anabaena flos-aquae* and *Aphanizomenon flos-aquae.* South African Journal of Science 78: 367-372

Carmichael WW (1992) Cyanobacteria secondary metabolites - the cyanotoxins. Journal of Applied Bacteriology 72: 445-459

Carmichael WW, Falconer IR (1993) Diseases related to freshwater blue-green algal toxins and control measures. In: Falconer IR (ed) Algal Toxins in Seafood and Drinking Water, Academic Press, London, pp 187-209

Codd GA (1994) Blue-green algal toxins: water-borne hazards to health. In: Golding AMB, Noah N, Stamwell-Smith, London R (eds) Water and Public Health, Smith-Gordon, London, pp 271-278

Codd GA (1995) Cyanobacterial toxins: occurrence, properties and biological significance. Water Science and Technology 32: 149-55

Codd GA, Bell SG (1996) The Occurrence and Fate of Blue-Green Algal Toxins in Freshwater. NRA R & D Report 29, HMSO, London, pp 30

Codd GA, Bell SG, Brooks WP (1989) Cyanobacterial toxins in water. Water Science and Technology 21: 1-13

Codd GA, Edwards C, Beattie KA, Barr WM, Gunn GJ (1992) Fatal attraction to cyanobacteria? Nature 359: 110-111

Codd GA, Jefferies TM, Keevil CW, Potter E (1994b) (eds) Detection Methods for Cyanobacterial Toxins. The Royal Society of Chemistry, Cambridge, pp 191

Codd GA, Steffensen DA, Burch MD, Baker PD (1994a) Toxic blooms of cyanobacterial in Lake Alexandrina, South Australian - learning from history. Australian Journal of Marine and Freshwater Research 45: 731-736

Dunn J (1996) Algae kills dialysis patients in Brazil. British Medical Journal 312: 1183-1184

Eriksson JE, Meriluoto JAP, Lindholm T (1986) Can cyanobacterial peptide toxins accumulate in aquatic food chains? In: Megusar F, Gantar M (eds) Proceedings of the IV International Symposium on Microbial Ecology. Slavonic Society for Microbiology, Ljubljana, pp 655-658

Falconer IR (1994) Health problems from exposure to cyanobacteria and proposed safety guidelines for drinking and recreational water. In: (see Codd et al., 1994b) pp 3-10

Falconer IR, Choice A, Hosja W (1992) Toxicity of edible mussels (*Mytilus edulis*) growing naturally in an estuary during a water bloom of the blue-green alga *Nodularia spumigena*. Environmental Toxicology and Water Quality 7: 119-124

Fitzgeorge RB, Clark SA, Keevil CW (1994) Routes of intoxication In: (see Codd et al., 1994b) pp 69-74

Fogg GE, Stewart WDP, Fay P, Walsby AE (1993) The Blue-Green Algae. Academic Press, New York and London, pp 459

Hindman SH, Favero M, Carson LA, Peterson NJ, Schonberger LB, Solans JT (1975) Pyrogenic reaction during hemodialysis caused by extramural endotoxin. The Lancet 2: 731-732

Hunter PR (1994) An epidemiological critique of reports of human illness associated with cyanobacteria. In: (see Codd et al., 1994b) pp 11-18

MacKintosh C, Beattie KA, Klumpp S, Cohen P, Codd GA (1990) Cyanobacterial microcystin-LR is a potent and specific inhibitor of protein phosphatases 1 and 2A from both mammals and higher plants. FEBS Letters 264: 187-192

Negri AP, Jones GJ (1995) Bioaccumulation of paralytic shellfish poisoning (PSP) toxins from the cyanobacterium *Anabaena circinalis* by the freshwater mussel *Alathyria condola*. Toxicon 33: 667-678

NRA (1990) Toxic Blue-Green Algae. Water Quality Series No. 2. National Rivers Authority, London, pp 125

Raziuddin S, Siegelman HW, Tornabene TG (1983) Lipopolysaccharides of the cyanobacterium *Microcystis aeruginosa*. European Journal of Biochemistry 137: 333-336

Ressom R, Soong FS, Fitzgerald J, Turczynowicz L, El Saadi O, Roder D, Maynard T, Falconer IR (1994) Health Effects of Cyanobacteria (Blue-Green Algae). National Health and Medical Research Council, Canberra, pp 108

Rinehart KL, Namikoshi M, Choi BW (1994) Structure and biosynthesis of toxins from blue-green algae (cyanobacteria). Journal of Applied Phycology 6: 159-176

Skulberg OM, Carmichael WW, Codd GA, Skulberg R (1993) Taxonomy of toxic cyanophyceae (cyanobacteria). In: Falconer IR (ed) Algal Toxins in Seafood and Drinking Water, Academic Press, London

Steffensen DS, Nicholson BC (1994) (eds) Toxic Cyanobacteria: Current Status of Research and Management. Australian Centre for Water Quality Research, Adelaide, pp 172

van Hoof F, van Es T, D'hout D, de Pauw N (1994) Screening of cyanobacterial toxins in *Microcystis aeruginosa* collected from blooms and cultures. In (see Codd et al., 1994b), pp 136-138

Yoo RS, Carmichael WW, Hoehn RC, Hrudey SE (1995) Cyanobacterial (Blue-Green Algal) Toxins: A Resource Guide. American Water Works Association Research Foundation, Denver, pp 229

Yu S-Z (1989) Drinking water and primary liver cancer. In: Tang ZY, Wu MC, Xia SS (eds) Primary Liver Cancer, China Academic Publishers, Beijing, pp 30-37

The Young Scientist Poster Award

The Young Scientist Poster Award is given to the first author of the best poster, who must be younger than 35 years.

The winner is announced at the EUROTOX Congress Dinner, and the abstract is published in the Congress Proceedings.

The winner 1996 was M. Lush, Leicester.

Molecular Cloning of Neuropathy Target Esterase

Michael Lush, David Read and Paul Glynn
MRC Toxicology Unit, Hodgkin Building, University of Leicester, P.O. Box 138, Lancaster Road, Leicester, LE1 9HN, UK

Organophosphate Induced Delayed Polyneuropathy (OPIDP) is a condition, characterised by a paralysing neuropathy with axonal degeneration, developing after a latent period of approximately one to three weeks. The primary target of these neuropathic OPs has been shown to be a 155 kDa neural protein with serine esterase activity designated Neuropathy Target Esterase (NTE). The OPs toxic effects are apparently due to the covalent inhibition and subsequent secondary modification of this protein. Recently, we have purified NTE to apparent homogeneity (Glynn et al. (1994) Biochem J 301: 551-556) using a novel biotinylated OP, designated S9B (1-[Saligenin-cyclic-phosphoryl]-9-biotinyl-diaminoane), and have produced sufficient pure protein for Edman degradation sequencing. Seven pig NTE sequences from V8 protease digest fragments have been produced. Of these, two have been positioned relative to the active site serine and a further three have been putatively positioned on a serine esterase multiple alignment. The two catalytic serine related peptides have been successfully used to design a pair of degenerate PCR primers which generate a short (170 bp) product designated ORF1 from pig brain cDNA. ORF1 contains an open reading frame with significant homology to the same region of other serine esterases. ORF1 is currently being used in hybridisation screening of a cDNA library, 3' RACE and PCR with degenerate primers from the other peptides to produce further nucleotide sequences.

Subject Index

Springer and the environment

FSC
www.fsc.org
MIX
Papier | Fördert
gute Waldnutzung
FSC® C083411

Kolibri 360 GmbH
Ferdinand-Jühlke-Straße 7
99095 Erfurt, Deutschland
produktsicherheit@kolibri360.de